A Guide to the

Wireless Engineering

Body of Knowledge

(WEBOK)

A Guide to the

Wireless Engineering Body of Knowledge

(WEBOK)

Second Edition

Edited by
Andrzej Jajszczyk

IEEE COMMUNICATIONS SOCIETY

IEEE Press

A JOHN WILEY & SONS, INC., PUBLICATION

Library of Congress Cataloging-in-Publication Data:

A guide to the wireless engineering body of knowledge (WEBOK) / Andrzej Jajszczyk, Editor. -- Second edition.
    pages cm
    ISBN 978-1-118-34357-9 (pbk.)
1. Wireless communication systems. I. Jajszczyk, Andrzej, editor of compilation.
TK5103.2.G83 2012
621.384--dc23
                            2012015862

10 9 8 7 6 5 4 3 2

Table of Contents

Contributing Editors and Authors

The following volunteers contributed to the writing, updating, editing, and reviewing of this Second Edition of *A Guide to the Wireless Engineering Body of Knowledge (WEBOK)*. This edition would not have been possible without their efforts and dedication. It builds on the work of the editors and authors of the 2009 edition, several of whom participated in this revision. IEEE and the IEEE Communications Society gratefully acknowledge their contributions.

We especially acknowledge the efforts of **Gustavo Giannattasio,** the Editor in Chief of the 2009 Edition of the WEBOK, who provided many constructive comments as a reviewer for this Second Edition.

WEBOK 2.0

Editor in Chief	Andrzej Jajszczyk, AGH University of Science and Technology
Editorial Reviewer	Rolf Frantz, Independent Consultant

Chapter 1 WIRELESS ACCESS TECHNOLOGIES

Editor and Lead Author	Javan Erfanian, Bell Mobility Canada
Contributing Authors	Haseeb Akhtar, Ericsson
	Anne Daviaud, France Telecom R&D
	Paul Eichorn, Bell Mobility
	Derek McAvoy, Bell Mobility
	Mona Mustapha, Vodafone
	Rémi Thomas, Orange France Telecom
	Jin Yang, Verizon Wireless

Chapter 2 NETWORK AND SERVICE ARCHITECTURE

Editor and Lead Author	K. Daniel Wong, Daniel Wireless LLC
Contributing Authors	Dharma Agrawal, University of Cincinnati
	Javan Erfanian, Bell Mobility Canada
	Thomas Magadanz, Fraunhofer
	Julius Mueller, Fraunhofer
	Roberto Sabella, Ericsson
	Vijay Varma, Applied Communication Sciences
	Hung-Yu Wei, National Taiwan University
	Qinquing Zhang, Johns Hopkins University

Chapter 3 NETWORK MANAGEMENT AND SECURITY

Editor and Lead Author	Peter Wills, NBN Co., Ltd.
Contributing Authors	Bernard Colbert, Deakin University
	Paul Kubik, Telstra
	Santiago Paz, Ort University

Chapter 4 **RADIO ENGINEERING AND ANTENNAS**
        Editor and Lead Author   Javan Erfanian, Bell Mobility Canada
        Contributing Authors   Hung Nguyen, The Aerospace Corporation
                                          Eva Rajo Iglesias, Universidad Carlos III de Madrid
                                          Matilde Sánchez Fernández, Universidad Carlos III de Madrid
                                          Mojca Volk, University of Ljubljana Telecommunications Laboratories

Chapter 5 **FACILITIES INFRASTRUCTURE**
        Editor and Lead Author   Mehmet Ulema, Manhattan College
        Contributing Authors   Richard Chadwick, Joslyn Electronics Corporation

Filomena Citarella, Independent Consultant
Thomas Croda, CSI Telecommunications
Rolf Frantz, Independent Consultant
Z.A. Hartono, Lightning Research Pte.
Barcin Kozbe, Ericsson
Krishnamurthy Raghunandan, New York City Transit/MTA
K. Daniel Wong, Daniel Wireless LLC

Chapter 6 **AGREEMENTS, STANDARDS, POLICIES, AND REGULATIONS**
        Editor and Lead Author   K.C. Chen, National Taiwan University
        Contributing Authors   Karl Rauscher, EastWest Institute

Irene S. Wu, Georgetown University

Chapter 7 **FUNDAMENTAL KNOWLEDGE**
        Editor and Lead Author   Xavier Fernando, Ryerson University
        Editor and Lead Author   Niovi Pavlidou, Aristotle University
        Contributing Authors   Anurag Bhargava, Ericsson

Joseph Bocuzzi, Broadcom Corporation
Naveen Chilamkurti, La Trobe University
Ali Grami, University of Ontario Institute of Technology
Stylianos Karapantazis, Aristotle University
George Koltsidas, Aristotle University
Wookwon Lee, Gannon University
Evangelos Papapetrou, University of Ioannina
Xianbin Wang, University of Western Ontario
Traianos V. Yioultsis, Aristotle University

Introduction

The enormous success of wireless technology has changed peoples' lives forever. We are surrounded by an impressive number of mobile and fixed wireless devices enabling us to communicate with the entire world using a variety of media: voice, text, and video. Behind a plethora of services offered by these devices there is sophisticated technology as well as a complex fabric of standards and regulations. Understanding and applying the relevant knowledge in practice by wireless engineering professionals requires years of study and practical experience. Moreover, continuous progress in this area forces engineers to update their knowledge to meet the current needs of their employers.

This Second Edition of *A Guide to the Wireless Engineering Body of Knowledge (WEBOK)* provides a broad overview of the technical and other areas with which wireless practitioners should be familiar. Along with the tutorial material, the book lists additional topics and contains suggestions for further study. Unlike most regular academic textbooks, the WEBOK is focused on practical engineering issues, although it does not neglect their theoretical backgrounds. The text was written by a large group of international experts, mostly practitioners having extensive experience at either network operators, equipment manufacturers and vendors, or research and development organizations. Their backgrounds played a key role in defining the scope of this book by selecting topics that have proven their practical importance.

One of the original aims of the WEBOK was to facilitate the development of a common technical understanding, language, set of tools, and approach among wireless professionals educated and working in different parts of the world. This aim is strengthened in the current edition. Despite considerable differences in the teaching curricula used in educating wireless technology engineers in various countries, as well as in their professional paths, this book can help them communicate and work together.

The first edition of the WEBOK was put together by nine chapter editors led by chief editor Gustavo Giannattasio, an engineer, manager, and professor having impressive global experience in wireless and related technologies. The market success of the book well reflects both the competencies and hard work of the entire editorial team. The current, second edition takes into account the observed progress in wireless technologies and standardization as well as filling some gaps, removing unwanted repetitions, improving readability, and adding valuable new references that can be used for individual study. The size limits did not allow the inclusion of all possible subject areas within wireless technologies, but the editorial team has made a considerable effort to select those that are of primary importance for the telecommunications industry.

Although the WEBOK should not be viewed as a study guide for any wireless certification exam, it outlines the major areas that should be known and understood by a wireless engineering practitioner, employed in the industry, as well as suggesting sources for further information and study. It should be noted that this book does not cover all the topics that may be covered by certification exams.

The WEBOK is organized as follows:

Chapter 1: Wireless Access Technologies

The spectacular success of wireless technologies in the access area is due to the fact that they allow user mobility, although they also play an important role in fixed access environments. The chapter begins with fundamental access layer considerations, giving a general picture of the topic. Then mobile cellular architectures are discussed, including capacity and coverage issues as well as mobility management. Various mobile cellular technologies are covered, reflecting the evolution from GSM to LTE. *Long Term Evolution* is a considerably extended topic compared with the first edition of WEBOK, while the discussion of femto cells is a considerably expanded entry in this second edition. Separate sections describe wireless local and personal area networks, WiMAX, and RFID. Some other initiatives are also discussed.

Chapter 2: Network and Service Architecture

This chapter focuses on the core network. The authors cover voice-centric solutions from the traditional telephony world as well as the IP-centric network architectures that play a more and more important role today. The chapter presents basic operation procedures, signaling flows, and message formats. Some important protocols are discussed, including TCP/IP, SIP, RTP, and RoHC. The concepts of mesh and mobile ad hoc networks are explained. The coverage of the IP Multimedia Subsystem (IMS) is extended in this WEBOK edition along with some important issues, such as traffic engineering and the role of the control plane. New sections on wireless service technologies have been added.

Chapter 3: Network Management and Security

Keeping a wireless network in the operational state and making it secure are two of the key challenges facing network operators. The chapter begins with some newly added sections introducing the reader to network management, overviewing the fundamental concepts. A separate section is devoted to operations process models developed by the TeleManagement Forum. Then the Simple Network Management Protocol (SNMP) is presented. The second part of the chapter covers various security issues related to wireless networking. It starts with security basics and continues with network access control, wireless LAN security, Robust Security Networks (RSNs), and 3G security.

Chapter 4: Radio Engineering and Antennas

The chapter, considerably re-written in the second edition of WEBOK, begins with radio propagation issues, characterizing different properties of frequency ranges and describing such phenomena as free-space and atmospheric loss, reflection, diffraction, refraction, and scattering. Then the most important antenna parameters are defined and explained. Antenna types are reviewed, including phased arrays and smart antennas. Basics of antenna design and measurement are given. A separate section deals with radio engineering and wireless link design, presenting approaches to link budget analysis and radio frequency engineering. Radio system considerations concentrate on practical receiver issues.

Chapter 5: Facilities Infrastructure

This chapter outlines the information needed to specify, design, and implement wireless facilities and sites. Its length in the second edition of WEBOK is doubled with respect to the first edition. After presenting AC and DC power system issues, the authors discuss various aspects of lightning protection as well as heating, ventilation, and air conditioning. Separate sections are devoted to mounting various types of equipment, waveguides and transmission lines, and tower specifications and standards. Then physical security, alarms, and surveillance systems are discussed. The chapter concludes with an overview of industry standard specifications and national and international standards.

Chapter 6:   Agreements, Standards, Policies, and Regulations

The role of agreements, standards, policies, and regulations can hardly be overestimated in the wireless engineering environment. The authors define and discuss their meaning, scope, and importance. Several practical examples of agreements, standards, policies, and regulations are given.

Chapter 7:   Fundamental Knowledge

The role of this chapter is to overview the fundamental knowledge that, in the opinion of the editors and authors, is necessary for a practicing wireless communications engineer. Since it does not make any sense to repeat the vast contents of many available textbooks in this area, the chapter lists major topics, issues, and definitions, rather than presenting them in detail.

Appendices

The WEBOK contains four appendices: a Glossary of acronyms, a Summary of Knowledge Areas addressed in the book, a short description of what was involved in Creating the WEBOK, and brief information about the producer of the book, i.e., the IEEE Communications Society, which is a community comprised of a diverse group of industry professionals with a common interest in advancing all communications technologies.

Chapter 1

Wireless Access Technologies

1.1 Introduction

Wireless links are broadly utilized for point-to-point, point-to-multi-point, and mesh applications, in fixed or mobile, satellite or terrestrial applications, as backhaul or as user access networks, and from scan-zone to ad-hoc, relay, and wide-area applications. A great asset of wireless access is that it enables user mobility in its broad sense, whether nomadic or at high speeds. Phased evolution of true user mobility is enabled through seamless connectivity at multiple levels. Geographically, the user may be connected through one or more personal area (PAN), home or office (e.g., Femto cell), local area (LAN), metropolitan area (MAN – campus, hot-zone, municipality, mesh), or wide area (WAN) network(s). There is further granularity in wireless access, increasingly enabled through sensing, mobile tags, and near-field communication (NFC). Sensors are expected to enable communication between connected "things" (sensors, grids, tags), people, and machines, reaching tens of billions within five-plus years, and providing a broad range of (new) applications such as health and bio-engineering, environment and geo-engineering, robotics, and many others. A user's mobility is also maintained both through inter-technology and intra-technology handoff within an increasingly multi-cell and heteregenous communication space. The former may occur when moving from one cell to another in a given cellular network, and the latter may occur when the user's session and application is maintained while the access moves from one technology (e.g., Wireless LAN) to another (e.g., 3G or 4G). Although the user may have some level of awareness with respect to the access or connectivity mechanism, the user's communication space is ultimately (and increasingly) virtual, with the user aware of his or her intention, application, preferences, interaction, and experience, but typically not the access mechanism or network technology. This goal of creating an increasingly natural communication, which is more user-centric and less technology-centric, makes the enabling role of technologies and technologists more significant, more exciting, and perhaps more complex (certainly not less). In addition, a so-called natural communication may be enhanced by the application enabler, or the user terminal, as it discovers and utilizes the smart user space (proximity) capabilities.

A wireless access network must obviously allow the end user(s) to access the network. This includes the signaling, transmission, and communication aspects over wireless links, and provides coverage, capacity, and a user experience with such attributes as data rate, latency, and quality of service, among others. A group of users share the resources and are awarded access, governed by a certain discipline. At the heart of this discipline lies a multiple access mechanism. How can an increasingly large number of users access the same network, and even the same channel, at the same time? Multiple access mechanisms (e.g., FDMA, TDMA, CDMA, or OFDMA, all of which are defined later in this chapter) have evolved through generations of wireless systems to enable this, while enhancing such capabilities as data speed, capacity, flexibility, and cost efficiency.

Coverage is particularly significant in wireless network design, in reach, indoor penetration, and continuity. Capacity is another design fundamental. This is an end-to-end attribute but greatly impacted by the wireless access component. There is a need for small cell sites and more transmission carriers (and use of bandwidth) to provide sufficient capacity when there is a greater number of users, or more accurately, greater simultaneous traffic. Generally, wide-area cellular network design optimization is limited (or dictated) by coverage in low traffic areas, and by capacity in high traffic areas. With the explosive growth in wireless traffic, addressing capacity constraints requires multiple strategies above and beyond the traditional methods. A significant emerging paradigm is dynamic configuration, optimization, and management enabled by Self-Organizing Network (SON) mechanisms.

This brings us to the important concept of the frequency spectrum. A radio tone has a frequency, and a radio signal carrying information has a range of frequency content. Modulation at the transmitter, based on and coupled with a given multiple access mechanism, allows wireless communication over particular frequency bands. These bands are designated by local regulatory authorities, and generally coordinated by regional and global (e.g., the International Telecommunication Union [ITU]) bodies. They are typically licensed (e.g., bands used by service providers in mobile cellular technologies), but also unlicensed (e.g., typical bands used by WLAN and Bluetooth, among others).

Mobile communications systems were traditionally designed and optimized for voice communication. Although voice continues to be the dominant application, data applications have grown dramatically over the years, from basic messaging, downloads, browsing, and positioning applications (enabled by second generation (2G) systems and their enhanced versions) to an incredible growth of multimedia and content-based applications (particularly enabled by third generation − 3G − systems and beyond). The core network (discussed in the next chapter) is evolving to provide a ubiquitous application environment in conjunction with a common service architecture, and seamless access, interfacing with one or more access techniques. This is a phased evolution to an all-IP or packet (heterogeneous) network, already introduced, with a flat (or flatter) architecture which allows seamless mobility across different access technologies. A new era is already here, with content-centric and IP-centric wireless networks.

The ITU has defined the family of 3G systems (International Mobile Telecommunications for the Year 2000, IMT-2000) and has set out the goals and attributes of the next-generation IMT-Advanced systems as shown in Figure 1-1 [ITU03]. Industry standards bodies (e.g., the 3rd Generation Partnership Project, 3GPP, and 3GPP2; also the IEEE 802.x committees) have developed definitions for generations of mobile and nomadic communication access (and core) technologies (e.g., LTE, Long-Term Evolution), working with other standards groups such as the Internet Engineering Task Force (IETF) (to leverage universal protocols and elements) and the Open Mobile Alliance (to align on and seek service enabler definition and interfaces).

This chapter starts with access network concepts and moves on to introduce access technologies and standards. As this is truly a broad topic, this chapter highlights key concepts and technologies but does not claim to be inclusive of all forms of access layer concepts, technologies, or implementations. Furthermore, it does not intend to promote or validate any particular technology. While significant technology attributes and design goals are highlighted, it must be noted that the products, innovations, implementation environment and wisdom, customer solutions, operational ingenuity, and inter-carrier initiatives can provide a variety of prospects and capabilities, above and beyond fundamental concepts, standards and reference models, technology goals, and common core practices.

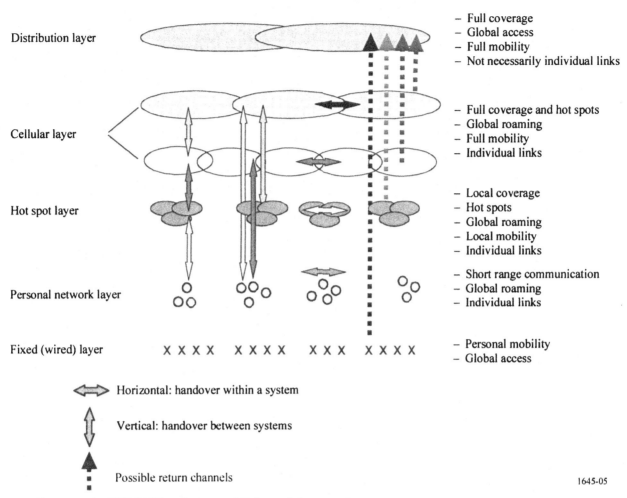

Distribution layer — Full coverage, Global access, Full mobility, Not necessarily individual links

Cellular layer — Full coverage and hot spots, Global roaming, Full mobility, Individual links

Hot spot layer — Local coverage, Hot spots, Global roaming, Local mobility, Individual links

Personal network layer — Short range communication, Global roaming, Individual links

Fixed (wired) layer — Personal mobility, Global access

Horizontal: handover within a system

Vertical: handover between systems

Possible return channels

1645-05

Figure 1-1: ITU IMT-Advanced Vision of Complementary Interconnected Access Systems

1.2 Wireless Access

1.2.1 Fundamental Access Layer Considerations

Wireless access technologies are concerned with the lower layers, particularly transmission and link layer functions. These functions are not fixed, and through generations of technologies, particularly mobile cellular, some of the core network functions (e.g., in mobility management) have moved to the edges and been taken up by the so-called access technologies. Figure 1-2 identifies some of the key functions.

Key attributes of access technologies include:

- Modulation and Coding
- Multiple Access Mechanism
- Spectral Efficiency
- Frame Structure and Radio Signal Generation
- Bandwidth Support
- Antenna Structure/Function Support (e.g., MIMO, Multiple Input Multiple Output)
- Resourcing and Channel Mapping

- Mobility Management Support and Handovers
- Short Transmission Time Interval
- Access Layer Security and QoS
- Detection and Equalization
- Re-Transmission Support
- Frequency Re-use Support (e.g., Fractional)
- Other Access Layer Innovations, such as
  - Link Adaptation
  - Interference Cancellation
  - Fast Scheduling
  - Bandwidth Aggregation

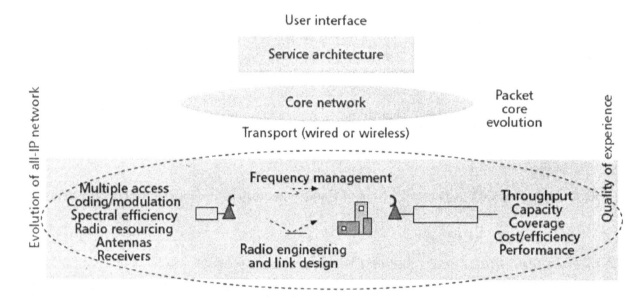

Figure 1-2: Wireless Access Functions Foundational to an End-to-End Communication System

Research, standardization, and a great deal of innovation have led to the realization of wireless access technologies. To make use of these technologies, access networks are designed, configured, put in operation, and managed. This includes such important considerations as:

- Wireless Link, Cell, Antenna, and Architecture Design
- Coverage
- Capacity
- Frequency Management
- Performance Requirements
- Interactions with Core, Servicees, Network Management, and Terminal Equipment

- End-to-End Design, Configuration, and Operation

A number of key wireless access technology attributes and network design considerations are discussed in detail in the following sections and broadly addressed in the context of wireless technology standards evolution. Furthermore, Chapter 4 of this book is dedicated to Radio Engineering, and Antennas. Finally, Chapter 7 on Fundamental Knowledge gives reference to access layer and transmission concepts.

1.2.2 Spectrum Considerations

Wireless communication is obviously based on the transmission of (radio) signals with a certain frequency content as part of the electromagnetic spectrum. The frequency spectrum is apportioned among users based on the variety of different applications and its suitability for each, competition, regional and global (International Telecommunication Union – Radiocommunication Sector, ITU-R; World Radio-communication Conference, WRC) alignment, historical reasons, and other regulatory considerations.

Some significant spectrum considerations, in relation to access technologies are briefly highlighted here:

- Frequencies below ~3 GHz have been suitable for non-line-of-sight (NLOS) applications such as mobile cellular. Higher frequencies, however, have also been used extensively (e.g., for satellite communication).

- Higher frequencies have higher (free-space) power loss and shorter reach, for the same transmit power. For example, frequencies such as those in the 1.7–2.7 GHz range may (particularly) be used if or where capacity is a concern and lower ones such as 0.7–1 GHz may (particularly) be used if coverage is a consideration. In practice, a combination of considerations, however, may result in the use of one or more bands.

- To meet growing capacity requirements (e.g., for mobile applications), sufficient frequency spectrum is needed. This is to meet broadband speed and high-traffic volume (in addition to technology) requirements. ITU-R leads in the identification of global spectrum needs for future systems.

- In a communication session, the two-way communication paths (duplex) need to be divided, either in frequency (Frequency Division Duplex, FDD) or in time (Time Division Duplex, TDD). The frequency channel plan is obviously different in each case. Specifically, uplink and downlink frequencies are distinct and separated in the FDD case, as in many mobile cellular systems today. There is a growing trend to support a dual FDD/TDD mode seamlessly, to leverage spectrum resources and to roam between networks of the same or different service providers.

- Wireless systems are designed with ingenuity, particularly to avoid interference. Interference can potentially come from a variety of sources including the spectrum band structure. Interference cancellation and coordination schemes (e.g., in a multi-cell environment where low-power small cells are inserted within a macro network) are among the new and emerging standards, innovations, and best deployment practices.

- Significant considerations are regional and global frequency band alignment, along with multi-band multi-mode user device support. The latter is enabled by technology feasibility evolution (e.g., receiver front-end innovations) and also new breakthrough paradigms such as carrier-aggregation (3GPP standards in Release 10 [3GP11a, 3GP11b] and beyond). These trends enable user roaming, a strong technology ecosystem, and cost-effective (user-terminal) product availability and innovations.

As examples, mobile cellular operation in the Americas includes some or all of the frequency bands at 700 MHz, 850 MHz, 1900 MHz, 1700 / 2100 MHz (Advanced Wireless Services, AWS), and 2.5+ GHz. Similarly, such frequency bands as 800/850 MHz, 900 MHz, 1800 MHz, and 2.1 GHz, among others, are used in much of the rest of the world. In addition, there are frequency bands used for satellite communications, Wireless LANs (2.4 GHz and 5 GHz), point-to-point, distribution, microwave, radar, and other systems.

1.2.3 Generic Picture

Wireless access technologies allow connectivity and communication over wireless link(s). They are based on principles of radio engineering, with such attributes as propagation, power, antenna technology, and link analysis and design. Furthermore, a wireless access network sets up intelligent wireless connectivity, with increasing sophistication in speed, performance, flexibility, and efficiency, to enable user access, networking ,and applications.

Figure 1-3(a) shows a generic wireless transmission system with common functions of transmission, propagation, and reception. Figure 1-3(b) shows an example of a wireless (mobile) access system architecture as a key component of the end-to-end communication network.

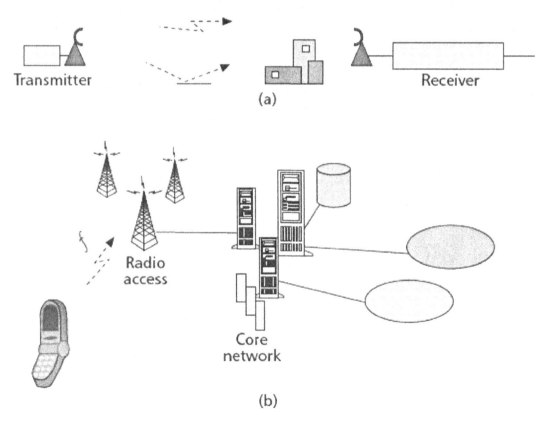

Figure 1-3: (a) Simplified View of a Generic Wireless Transmission System;
(b) a (Cellular) Network Example

Although overly simplified, this figure helps to illustrate how all wireless systems, both in their variety and in their evolution, deal with such concepts as transmission/reception (e.g., coding and modulation), antennas, link and propagation attributes, spectrum, multiple access etiquette (for shared systems), and others. For example, moving from third generation (3G) mobile access technologies to beyond 3G is enabled by technologies that leverage advanced forms of coding and modulation, multiple access, and antenna technologies, among others.

All wireless communication systems, satellite or terrestrial, fixed or mobile, personal, local, or wide-area, dedicated or shared, transport (backhaul) or access, regardless of frequency bands or topologies (point-to-point, point-to-multipoint, or mesh), have a similar fundamental anatomy. Furthermore, they have such similar concerns and considerations as coverage, capacity, transmitting power, interference, received signal power, infrastructure, and of course, performance and efficiency. The detailed attributes, however, vary depending on design and application goals, access technology, links, mobility, or frequency spectrum. Some details are provided for the case of mobile cellular systems in section 1.3.

1.2.4 Multiple Access Mechanisms

1.2.4.1 FDMA

Frequency Division Multiple Access or FDMA is an access technology that is used by radio systems to share the radio spectrum. In an FDMA scheme, the given Radio Frequency (RF) bandwidth is divided into adjacent frequency segments. Each segment is provided with bandwidth to enable an associated communications signal to pass through a transmission environment with an acceptable level of interference from communications signals in adjacent frequency segments.

FDMA also supports demand assignment (e.g., in satellite communications) in addition to fixed assignment. Demand assignment allows all users apparently continuous access to the transponder bandwidth by assigning carrier frequencies on a temporary basis using a statistical assignment process.

FDMA has been the multiple access mechanism for analog systems. It is not an efficient system on its own, but can be used in conjunction with other (digital) multiple access schemes. In this hybrid format, FDMA may be viewed as a higher-level frequency band plan, facilitating the splitting of channel bandwidths, upon which sophisticated digital multiple access schemes can be applied, allowing the system to be shared by an increasing number of users.

1.2.4.2 TDMA

Time division multiple access (TDMA) is a channel access method for shared-medium (radio) networks. It allows several users to share the same frequency channel by dividing the signal into different timeslots. The users' information is transmitted in rapid succession, each individual using its own timeslot, one after the other. This allows multiple stations to share the same transmission medium (e.g., radio frequency channel) while using only the part of its bandwidth that is required. TDMA is used in second-generation (2G) digital cellular systems, and is part of their evolution. Examples include the Global System for Mobile Communications (GSM), Interim Standard IS-136, Personal Digital Cellular (PDC), Integrated Digital Enhanced Network (IDEN), General Packet Radio Service (GPRS), and Enhanced Data Rates for GSM Evolution (EDGE). It is also used in the Digital Enhanced Cordless Telecommunications (DECT) standard for portable phones and in some satellite systems. Figure 1-4 shows the TDMA frame structure.

TDMA features (and concerns) include simpler handoff and less stringent power control, while potentially more complexity in cell breathing (borrowing resources from adjacent cells), synchronization overhead, and frequency/slot allocation (in comparison to CDMA). On its own, TDMA is limited by the number of timeslots and the fast transition between them. However, in addition to being used in many existing systems, it has the potential to be further leveraged in future hybrid multiple-access systems.

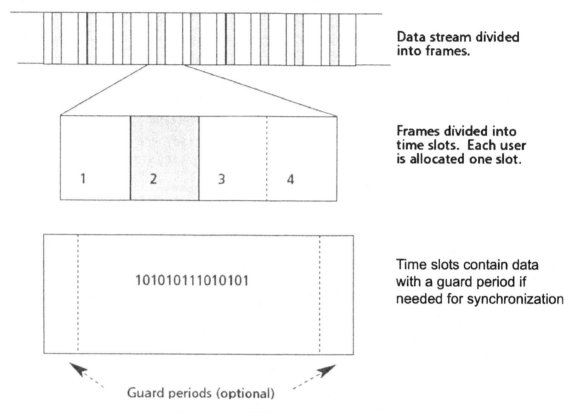

Data stream divided into frames.

Frames divided into time slots. Each user is allocated one slot.

Time slots contain data with a guard period if needed for synchronization

Guard periods (optional)

Figure 1-4: TDMA Frame Structure

1.2.4.3 CDMA

Code division multiple access (CDMA) is a channel access method utilized by various radio communication technologies. It employs a form of spread spectrum and a special coding scheme (where each transmitter is assigned a code). The spreading ensures that the modulated coded signal has a much higher bandwidth than the user data being communicated. This in turn provides dynamic (trunking) efficiency, allowing capacity versus signal-to-noise ratio tradeoffs.

The multiple user signals share the same time, set of frequencies, and even space, but remain distinct as each is modulated (or correlated) with a distinct code. The codes are (quasi-) orthogonal such that a cross-correlation of a received signal with the "wrong" codes results in a spread (and hence suppressed) "noise," while the auto-correlation with the "right" code results in the (de-spread) desired output. The signal-to-noise power ratio decreases with increasing number of users (or load on the system). This implies that with lower load, higher quality is achievable, while conversely, if some degradation is tolerable, the system allows higher capacity.

CDMA has been used in many communication and navigation terrestrial and satellite systems. Most notably, it has been used in third-generation (3G) mobile cellular systems (Universal Mobile

Telecommunications System, UMTS; cdma2000; Time-Division Synchronous CDMA, TD-SCDMA) for its strong features such as capacity/throughput, spectral efficiency, and security, among others.

1.2.4.4 TD-CDMA and TD-SCDMA

Time-Division CDMA and Time-Division-Synchronous CDMA use CDMA channels (5 and 1.6 MHz, respectively) and apply TDMA by slicing in time. As such, TDD transmission is used and the same frequencies can be used for uplink and downlink transmission. This is a key feature that the scheme exploits to improve capacity.

TD-CDMA and TD-SCDMA are 3G technologies standardized by 3GPP with different chip-rate options: UMTS Terrestrial Radio Access (UTRA) TDD-HCR (High Chip Rate) and UTRA TDD-LCR (Low Chip Rate), respectively.

TD-SCDMA has been introduced in China. For more information, the reader is directed to www.tdscdma-forum.org and www.tdscdma-alliance.org.

1.2.4.5 OFDM and OFDMA

Orthogonal Frequency Division Multiplexing (OFDM) is a multiplexing technique that subdivides the available bandwidth into multiple frequency sub-carriers as shown in Figure 1-5. In an OFDM system, the input data stream is divided into several parallel sub-streams of reduced data rate (and thus increased symbol duration), and each sub-stream is modulated and transmitted on a separate orthogonal sub-carrier. The increased symbol duration improves the robustness of OFDM to delay spread. Furthermore, the introduction of the CP (Cyclic Prefix) can completely eliminate ISI (Inter-Symbol Interference) as long as the CP duration is longer than the channel delay spread. The CP is typically a repetition of the last samples of data portion of the block that is appended to the beginning of the data payload.

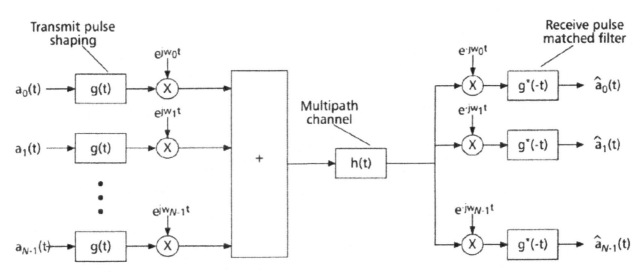

Figure 1-5: Basic Architecture of an OFDM System

OFDM exploits the frequency diversity of the multi-path channel by coding and interleaving the information across the sub-carriers prior to transmission. OFDM modulation can be realized with efficient IFFT (Inverse Fast Fourier Transform), which enables a large number of sub-carriers (up to 2048) with low complexity. In an OFDM system, resources are analyzed in the time domain by means of OFDM symbols and in the frequency domain by means of sub-carriers. The time and frequency resources can be organized into sub-channels for allocation to individual users.

OFDMA (Orthogonal Frequency Division Multiple Access) is a multiple-access/multiplexing scheme that provides for the multiplexing of data streams from multiple users onto the downlink sub-channels, and uplink multiple accesses by means of uplink sub-channels. This allows simultaneous low data rate transmission from several users. Based on feedback information about the channel conditions, adaptive user-to-subcarrier assignment can be achieved. If the assignment is done sufficiently fast, this further improves the OFDM robustness to fast fading and narrow-band co-channel interference, and makes it possible to achieve even better system spectral efficiency.

The OFDMA symbol structure consists of three types of sub-carriers as shown in Figure 1-6:

- Data sub-carriers for data transmission;

- Pilot sub-carriers for estimation and synchronization purposes; and

- Null sub-carriers for no transmission; used for guard bands and DC carriers.

Active (data and pilot) sub-carriers are grouped into subsets of sub-carriers called sub-channels.

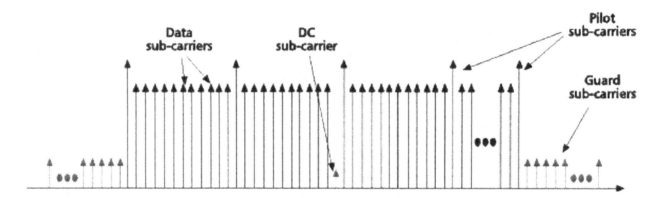

Figure 1-6: OFDMA Sub-Carrier Structure

OFDMA has certain elements of resemblance to CDMA, and even a combination of other schemes (considering how the resources are partitioned in the time-frequency space).

Put simply, OFDMA enhances the capacity of the system significantly and yet efficiently. Advanced OFDMA systems address such concerns as required flexibility in wide-area mobility, complexity in adaptive sub-carrier assignment, co-channel interference mitigation, and power consumption.

Advanced technologies such as systems beyond third-generation mobile are defined to leverage OFDMA's great potential for significant capacity and efficiency improvement, together with other innovations (e.g., in coding and modulation and in antenna technologies).

1.3   Mobile Cellular Architecture and Design Fundamentals

Mobile cellular networks have grown rapidly since the inception of commercial services in 1983. The network has evolved from pure circuit voice communications to high-quality voice and multimedia support and high-speed connectivity (access). The evolution of wireless mobile networks has been driven by the need to support mobile services, high efficiency, and the evolving user experience.

A simplified wireless network architecture is illustrated in Figure 1-7. The user terminal is wirelessly connected to and thus supported by a Base Transceiver Station (BTS). This base station and a number of others are connected to a Base Station Controller (BSC). Traditional circuit voice is supported through a Mobile Switching Center (MSC) both directly (not shown) and in connection to a Public Switched Telephone Network (PSTN). The BSC can also be connected to an IP Gateway to support various packet data services.

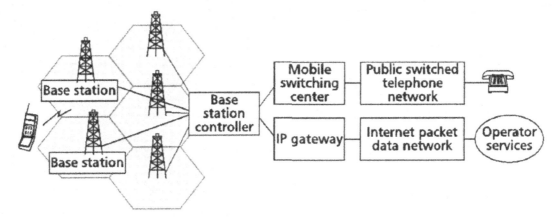

Figure 1-7: A Basic Wireless Cellular System

The quality of the wireless access connectivity is measured by the call drop rate, access failure rate, blocking probability, packet loss rate, and network reliability.

1.3.1 Capacity and Coverage

Capacity and coverage engineering are needed to achieve the desired access connectivity. Coverage is defined as the geographical area that can support continuous wireless access connectivity with the desired reliability and minimum guaranteed quality of service. It is heavily impacted by the terrain, RF environment, applications, and interference. Capacity is defined as the maximum number of users a network can serve with given resources and reliable access connectivity. It relies on traffic loading, traffic pattern, cell site equipment capability, and hardware dimensioning.

Some access mechanisms have a theoretically deterministic capacity based on their channel structure (e.g., FDMA and TDMA systems), while others may have dynamic allocation and allow some degradation (say in voice quality) in exchange for capacity (e.g., CDMA systems).

For example, a FDMA-based analog system (Advanced Mobile Phone System, AMPS) has a channel bandwidth of 30 kHz. Therefore, a 10-MHz cellular band can support 333 FDMA channels. With a frequency reuse of seven, this is equivalent to a radio channel capacity of 15 channels per sector for a site with three sectorized cells.

A TDMA system can further divide the time slot (typically) among three users and thus increase the capacity to 45 channels per sector.

CDMA capacity is a function of the required signal bit-energy-to-noise-density ratio (E_b/N_o), spreading factor (chip rate B_{ss} divided by data rate R), channel activity factor (D), sectorization gain (G_s), and frequency reuse factor (K). The maximal number of users a CDMA sector can support, or the reverse link (uplink) pole capacity, is

$$N = 1 + \frac{B_{ss}}{R} \cdot \frac{1}{E_b / N_o} \cdot \frac{1}{D} \cdot G_s \cdot K \tag{1}$$

For example, assuming 1.2288 Mb/s chip rate, 9.6 kb/s channel data rate, a frequency reuse factor of 0.66, and a channel activity factor 0.4, the uplink (or reverse) channel capacity is around 36 traffic channels in a three-sector cell for cdmaOne with a 7 dB requirement for E_b/N_o. This number increases to 72 for cdma2000-1x with the required E_b/N_o reduced to 4 dB. Typically, the cdma2000 operational capacity is around 50% of those maximal pole capacity numbers due to forward link interference limitations. This results in a commercial operational capacity of 36 users in 1.25 MHz, or around 288 users over 10 MHz.

The coverage area is determined by the operating frequency, radio receiver sensitivity, and required signal-to-noise ratio that an access technology can support. Typically, cellular network coverage is determined by the reverse link due to limited mobile station transmit power.

In a CDMA based system, the capacity and coverage enhancement are achieved by optimization of various power management components. This includes sector level and link level power management. Therefore, a CDMA system must be optimized from a system point of view, so that the system can tolerate a maximal interference level. Figure 1-8 implies that when the system loading is above 75% of the reverse link pole capacity, as specified in (1), the coverage will shrink dramatically.

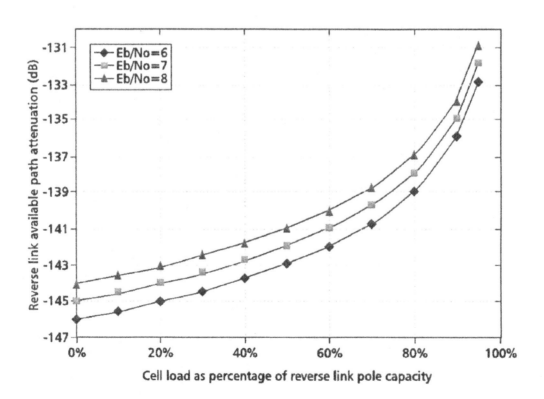

Figure 1-8: Capacity and Coverage of a CDMA Radio Network

This capacity and coverage tradeoff becomes even more important in support of IP multimedia services, where both impact the overall quality of service. Commercial cellular networks deployed worldwide have continuously grown through cell splitting and sectorization, in addition to technology advancements, to optimize capacity, coverage, quality, and cost considerations and trade-offs.

It is important to note that as in any other engineering practice, innovation, implementation, and the desired service and experience should be targeted while maximizing the cost efficiency. As indicated earlier, generally, the design of a low-traffic area (cell) is governed by coverage, and that of a high-traffic area by capacity, given performance requirements.

1.3.1.1 Wireless traffic load analysis
Distribution of traffic
The role of a wireless network is to supply voice and data over a sufficiently wide geographic area. On the demand side of the equation, there are assumptions regarding the number of subscribers who will use the network and the amount of traffic each subscriber will generate. It is not sufficient for the network to be able to supply traffic in aggregate to meet the aggregate demand, since demand varies with time and location. The goal of network design is for it to be highly probable that there is a sufficient supply of locally available network resources to satisfy the demand at any given point in time. This section will examine the distribution of traffic in both geography and time.

In the design of a network consisting of distinct cell sites and sectors, it is common practice to deploy higher cell site densities in densely populated areas since there is good correlation between population density and the level of traffic. However, there will always be sectors that carry much higher traffic levels than surrounding sectors despite attempts to equalize traffic. Consideration must be given to ensure that subscribers in busy sectors see acceptable performance, balanced with the service provider's desire to maximize utilization of network resources.

Traffic will also not be evenly distributed in time. Voice traffic exhibits a daily pattern which is observable at both the network level and sector level. Data traffic at the sector level appears to change randomly from one hour to the next making it much more unpredictable.

Spatial distribution of traffic
Once commercial service has begun, it is common practice to analyze sector traffic loading on a weekly basis. There are some differences between how voice traffic and data traffic are measured. It will be explained in the next section that voice traffic for a sector follows a fairly regular daily pattern while the sector data traffic level appears much more random due to its volatility. Because voice traffic appears more regular, a weekly traffic report will typically indicate how much voice traffic each sector in the network carries during a typical weekday busy hour. Busy hour voice traffic is measured in erlangs or equivalently minutes of use: 1 erlang is equivalent to 60 minutes of voice traffic. Since sector data traffic experiences much more volatility, there is no standard method for the weekly reporting of data traffic. For the exercise of examining the distribution of sector data traffic across a network, it will be assumed that sector data traffic is measured in bytes transmitted during the busy weekday hours (7AM – 10PM). Table 1-1 is an example of a weekly traffic report.

A large network is typically subdivided into a number of clusters. The average cluster in an urban area will consist of several dozen cell sites, with each cell site typically supporting three sectors. Rural clusters tend to have many fewer cell sites, typically 10 to 20 at most. When analyzing sector traffic distribution, it is advisable to restrict the analysis to one cluster at a time. Cell sites in the same cluster share the same morphology (i.e., urban, suburban, rural). The variation in sector traffic within a cluster is a function of some less observable random factor, whereas variation in sector traffic across different morphologies can be explained by differences in population densities. The sector traffic in Table 1-1 represents only a portion of the cell sites in a newly deployed urban cluster. The average busy hour sector voice traffic for the three dozen sectors shown in the table is 14 erlangs, while the average daily sector data traffic transmitted during the busy weekday hours is approximately 2.5 Gbytes. As can be seen, there are large variations in sector voice traffic and even larger variations in sector data traffic.

Sector ID	Voice Sector Traffic (Erlangs)	Data Sector Traffic (Bytes)	Sector ID	Voice Sector Traffic (Erlangs)	Data Sector Traffic (Bytes)
137_2	28.284	4,365,099,849	502_1	7.1	698,434,405
137_3	11.7	2,127,861,090	502_2	0.567	21,995,600
137_4	20.649	4,168,707,627	505_1	6.4	1,413,580,646
169_1	20.55	3,207,151,391	507_1	9.85	1,062,948,327
169_2	11.266	1,818,055,685	519_3	3.55	1,671,208,893
169_3	22.483	5,526,755,843	522_1	15.983	1,790,521,069
170_1	0.718	40,416,440	588_1	14.183	1,632,050,063
170_2	2.133	125,678,960	588_2	14.684	3,538,614,073
170_3	2.083	170,029,760	588_3	13.05	3,089,046,865
201_1	27.75	5,360,934,127	601_1	6.183	715,005,462
201_3	14.2	4,294,135,440	604_2	2.767	102,456,262
373_1	5.717	1,380,927,814	606_1	21.334	3,358,938,686
402_1	12.451	1,731,847,955	606_2	13.2	3,066,103,932
406_1	2.7	243,596,411	606_3	9.7	2,358,217,166
406_3	0.667	39,325,800	867_1	9.55	2,570,042,812
407_1	13.817	1,516,597,744	867_2	20.5	3,981,103,772
407_2	21.183	2,621,158,140	867_3	17.434	3,618,978,927
407_3	38.084	8,923,106,692	881_1	22.817	3,311,845,732

Table 1-1: Sample Weekly Traffic Report

Although obtaining the average and standard deviation of the sector traffic values for a cluster is useful, more insight can be gained by comparing the observed sector traffic values with different probability density functions. Below are histograms for voice (Figure 1-9) and data (Figure 1-10) traffic for two typical clusters alongside some possible probability density functions.

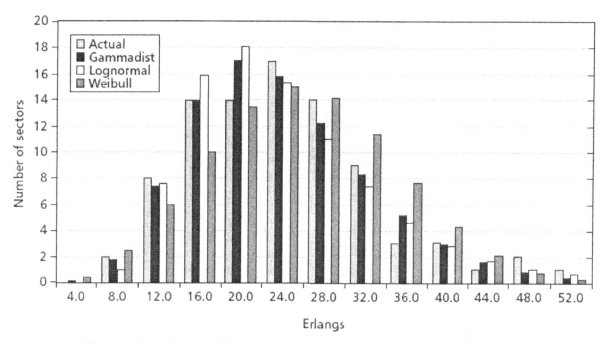

Figure 1-9: Voice Traffic Histogram for an Urban Cluster of Cell Sites

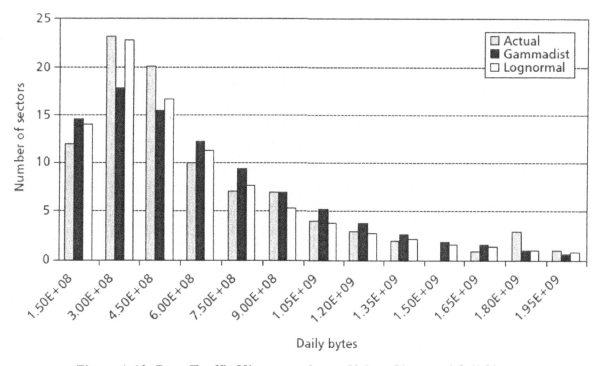

Figure 1-10: Data Traffic Histogram for an Urban Cluster of Cell Sites

Looking at the charts, it would appear that both the gamma and lognormal distributions match the voice and data traffic histograms quite nicely. The chi-square test is a tool for testing goodness-of-fit, and both the gamma and lognormal distribution pass the test for most clusters. It would be desirable if some explanatory model for traffic distribution were developed. However, there are insights that can still be obtained by assuming certain probability distributions. For example, it was mentioned above that sector data traffic exhibits greater variability than sector voice traffic. Assume that a sector is considered busy if its traffic is in 95th percentile. Although only 5% of the cell sites are considered busy, those cell sites carry a disproportionate amount of the traffic. Assuming a certain probability distribution for data traffic, the busiest 5% of the sectors can carry 20% of the traffic in the cluster. Since voice traffic is more evenly distributed, the busiest 5% of the sectors will only carry 10% of the cluster traffic. Compared to voice traffic, it is twice as likely for data traffic in a sector to reside in the 95th percentile.

The distribution of traffic for a cluster is not stationary over long periods of time. When a network is first deployed, there will be large variations in sector traffic for both voice and data. Due in part to the network planner's effort to equalize traffic among the sectors, traffic tends to become more evenly distributed with cell splitting over time.

Time distribution of traffic

There is much literature on the modeling of voice and data traffic. Much of it is based on the study of stochastic processes and in particular Poisson models and queuing theory for the study of connection-oriented circuit switched voice traffic. Data traffic has not been so easily modeled since the Internet is used for a wider variety of applications. There has been evidence that shows that data traffic is statistically self-similar and that Poisson models used to analyze voice traffic do not capture the fractal behavior exhibited by data traffic. In any case, it is not the intention of this section to analyze the short term characteristics of voice and data traffic. Rather, the analysis here focuses on the daily patterns of voice and data traffic at the sector level. An analogy can be made between the characterization of RF fading and what is being proposed here in the analysis of the time distribution of traffic. RF signal strength fading is divided into two components: short term and long term fading. Short term fading is generally recognized as being Rayleigh distributed, while it is generally recognized that long term fading is log-normal and independent of the nature of the short term fading component. This section assumes that data traffic can similarly be divided into two components. Only the longer term components (hourly variations) will be examined while the short term statistics (traffic variations within the hour) will be ignored, or in the case of circuit switched voice traffic it can be assumed that the Poisson model can be applied. It should be mentioned that adherents of the fractal model of data traffic will argue that longer term traffic patterns should be analyzed simultaneously with the short term traffic patterns. To make the analysis simpler for those who do not have a strong background in probability, the short term and long term traffic patterns are analyzed separately with a focus on the longer term component.

Hourly sector voice traffic

Voice traffic has an easily recognizable daily pattern as can be seen in Figure 1-11, which is a five-day snapshot of the hourly traffic for three sectors of a particular cell site.

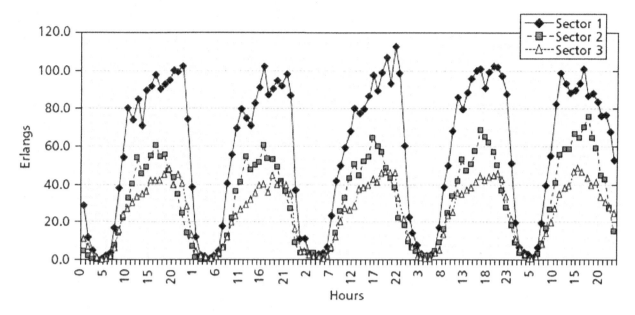

Figure 1-11: Hourly Voice Traffic for Three Sectors

The hourly traffic has a strong cyclical component with a less dominant random component superimposed. It is relatively easy to plan sufficient network resources to accommodate the voice traffic demand, due to the lack of day-to-day volatility in voice traffic. It is also relatively easy to calculate the relationship between monthly voice traffic and busy hour traffic for a particular sector. There are a couple of definitions for the term busy hour traffic, but it suffices to say it is a measure of the amount of traffic in the busiest hour of the day. On average, the monthly voice traffic for a sector is 300 times the busy hour traffic. Hence, if one is given the monthly voice traffic for a sector, the busy hour traffic can be calculated by dividing the monthly traffic by 300. If voice traffic were evenly distributed, the monthly voice traffic would be 720 times the busy hour traffic assuming there are 30 days in the month (744 times for a 31 day month). Ideally it would be beneficial to have the ratio of monthly traffic to busy hour traffic be as big as possible, since the network must be designed to carry the busy hour traffic.

Note the imbalance of sector traffic loading for the cell site shown in the above figure. This was examined in the section describing the spatial distribution of traffic. Depending on the technology deployed, it is possible for the busier sector to expropriate resources from less busy sectors to somewhat mitigate the effects of traffic imbalance.

Hourly sector data traffic

The hourly pattern for sector data traffic is much more volatile than for voice as can be seen in the traces shown in Figure 1-12, representing hourly data traffic for three different sectors over a one-week span.

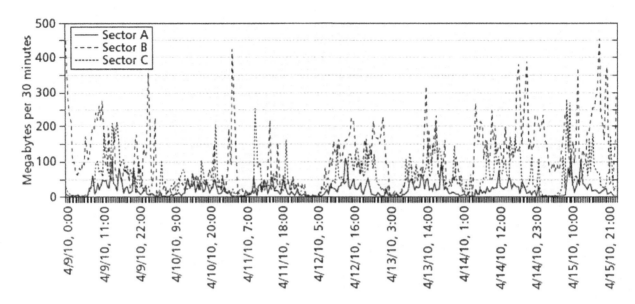

Figure 1-12: Half-Hourly Data Traffic for Three Sectors

As with voice traffic, there is a cyclical component since usage is definitely lower in the early hours of the morning. However, the random component of the data traffic is much stronger. If one takes the logarithm of the hourly sector data traffic, it appears that the resulting signal looks like white noise as can be seen in Figure 1-13, which is a 15 day trace for one sector.

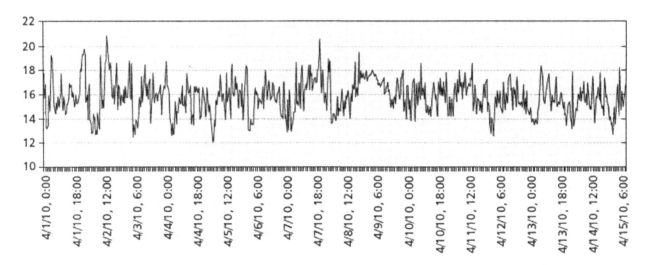

Figure 1-13: Logarithm of Half-Hourly Data Traffic for a Particular Sector

This suggests that the hourly traffic during the busy hours of the day should conform to a lognormal distribution. The histogram in Figure 1-14 compares the hourly sector data traffic with the lognormal probability distribution for a particular sector (early morning data traffic, from 1 AM to 8 AM is excluded).

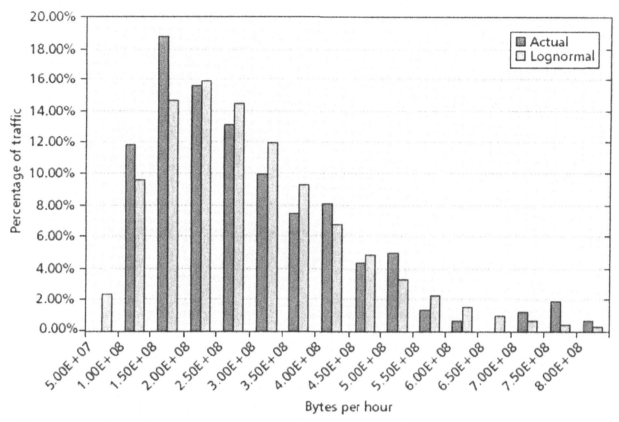

Figure 1-14: Histogram Comparing the Hourly Traffic with the Lognormal Distribution

The data in above graph pass the chi-square goodness-of-fit test for the lognormal distribution if the data traffic in the early morning hours is excluded.

Due to the volatility of data traffic, the concept of busy hour traffic is a bit more difficult to apply. If one were to use the same definition for busy hour traffic as used for circuit switched voice traffic planning, the monthly traffic would be 200 times the busy hour traffic. This means that data traffic is less evenly distributed in time than voice traffic. However, unlike voice, there could be certain hours where data traffic can significantly exceed the average busy hour traffic. Combined with the fact that the network will have busy sectors (often called data hotspots), it will be very difficult to accommodate data traffic spikes for every sector in a network all of the time.

1.3.1.2 Key access design considerations and trends for capacity and coverage

It must be noted that there is an increasing trend towards multi-cell design and seamless mobility across smart user spaces. Furthermore, networks become virtual in how they employ the physical network as suited to serve the intention/application and enhance the user's experience. To this end, network management, which is discussed at length in Chapter 3, becomes more dynamic, adaptive, and re-configurable (self-organizing). This is linked to access network design, deployment, and operation (such as performance optimization), and therefore is intimately linked to such wireless access considerations as coverage, capacity, and configuration plug and play (such as cell insertion and deletion), among others.

User access is becoming increasingly seamless across user spaces as shown in Figure 1-15.

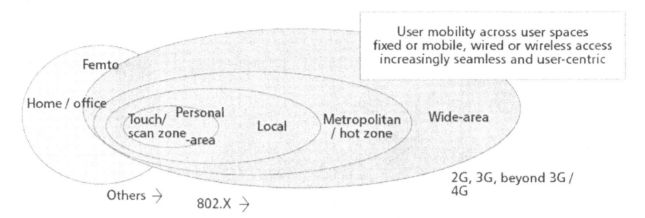

Figure 1-15: User's Seamless Mobility across Multiple Cells and Technologies

1.3.2 Mobility Management

1.3.2.1 Motivation

Mobility Management is responsible for supervising and controlling the mobile user terminal (or mobile station, MS) in a wireless network. The fact that a MS is not tethered and can move around freely presents an interesting set of challenges. Furthermore, the radio spectrum is a scarce resource that must be managed efficiently. Mobility management can be divided into registration and paging, admission control, power control, and handoff (also referred to as handover). Registration is to inform the network about the presence and location of the MS. Paging is the process by which the network alerts the MS to an incoming call or message. Admission control determines when the MS gets access to the network based on the priority of the request compared to network resource availability. Power control is necessary to keep interference levels at a minimum in the air interface and to provide the required quality of service. Handoff is needed in cellular systems to handle the mobility when the user is moving from the coverage area of one cell site to another, or to the service of another wireless technology. Figure 1-16 locates some of these features for a simplified radio access architecture.

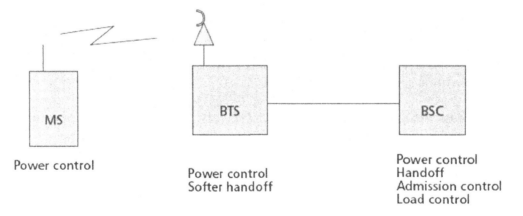

Figure 1-16: High Level Radio Access Architecture

Figure 1-17 shows the main states of the MS. Upon MS power up, the handset goes through an initialization state and acquires the preferred wireless network. Idle is the state the MS is in when not on an active call or connected to the network. Typically this is the state the MS is in the most. During this state the MS monitors overhead messages from the network or listens for incoming calls. System access refers to when the mobile attempts to access the network with the intent of setting up a traffic channel for a voice or data call. The connected or traffic state is when the MS has a dedicated connection to the network for the transfer of voice or data packets.

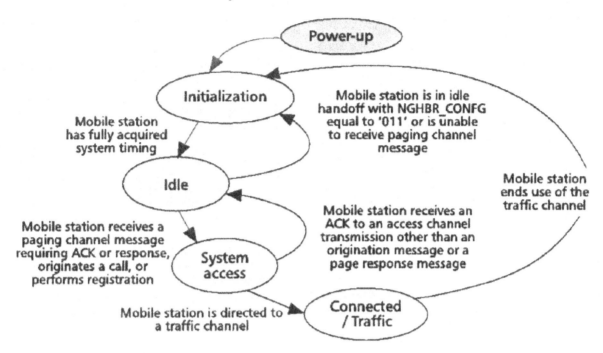

Figure 1-17: Main States of a Mobile Station

The concepts in this section are meant to provide perspective (and potentially examples) on mobility management considerations. It must be noted that differences do exist among various systems and advancements bring about new ways of doing things. Ultimately however, one can appreciate that a user device deals with powering up and down and moving between different modes and that it gets paged or registered and initialized, among other management functions.

1.3.2.2 Registration

Registration is the event by which a MS notifies the cellular system of its location, status, identification, and capabilities. The purpose of registration is to allow the network to efficiently page the MS when establishing a mobile-terminated call. Two types of registrations may exist:

i) Autonomous: triggered by some event or condition

    a. Power up registration – the MS powers on or switches serving system.

    b. Power down registration – the MS powers down (preventing unnecessary attempts to reach a user).

    c. Timer based registration – the MS registers when a timer expires. The timer is set by the network operator. This allows the system to de-register a MS that fails to register on power-down (i.e., moves out of coverage range).

    d. Distance based registration – the MS registers when the distance between the current base station and the base station where it last registered exceeds a specified threshold. This is particularly useful if the MS is not highly mobile.

    e. Zone based registration – the MS registers when it enters a new zone. A zone is defined by the network operator. Some technologies allow the MS to maintain a list of zones in which it is registered.

  ii) Non-autonomous: explicitly requested by the base station or implied, based on other messages sent to the MS

    a. Parameter change registration – the MS registers when a specific parameter (e.g., frequency band) has changed.

    b. Implicit registration – this occurs when the MS and base station exchange messages which convey sufficient information to identify the MS and its location.

    c. Ordered registration – the base station orders a MS to register (e.g., while on a traffic channel).

Not all registration methods are necessarily supported for a given network. It will depend on vendor implementation and operator configuration to optimize overhead signaling due to registrations, as the frequency of registrations can place a high load on reverse access channels.

1.3.2.3 Paging

When a MS is powered on, after going through initialization and network selection process, it goes into an idle state. In this state, it listens to the network for overhead messages containing network information or pages indicating that a call is being made to the MS. In this idle state, the MS is monitoring what is commonly referred to as a paging channel.

There is a link between the amount of pages sent and the quantity of registrations. A balance is needed and proper system design is required to efficiently page the MS for mobile-terminated calls. The more often the MS registers, the more precisely the network knows its location so that paging can be targeted to a smaller group of cells, reducing messaging that consumes precious wireless channel capacity.

1.3.2.4 Slotted mode

Typically a MS spends much of its time in the idle mode. In order to conserve battery life and maximize standby time for a mobile terminal, wireless technologies implement a sleep or slotted mode of operation. In slotted mode, a MS is able to power down some of its electronics and periodically wake up to check for new overhead messages or page messages indicating there is an incoming call. The network and MS must be synchronized so that pages for a specific MS are broadcast when it is awake. The longer the sleep period, the better the battery life, but this comes at the expense of a longer duration to page the MS to complete an incoming call. A sleep period of between 2–5 seconds is typically found in mobile networks.

1.3.2.5 Admission control

The air link in a wireless system is a shared resource and is intended to support a finite amount of capacity or users. If new users were indiscriminately allowed to join the network, at some point they would start to have a negative impact on the existing wireless connections that have been established with other users. For example, in CDMA based systems, as more users are allowed to establish connections, loading increases on the forward and reverse links. In turn the power levels for the existing connections must increase to overcome this loading increase. For those existing connections at the cell edge, the MS and/or the base station may already be at maximum power, so with the increased loading, the call quality cannot be guaranteed.

Admission control adds network intelligence to the call establishment process so that before adding a new connection or user, the system first ensures that it has sufficient resources so as to not affect existing customers. As these resources can be on the forward or reverse links, both are looked at separately and a call is only admitted if it passes both forward and reverse link admission control. Resources that can be part of admission control include:

- noise rise (reverse link systems);

- base transceiver station (BTS) power (forward link CDMA systems);

- codes (for CDMA systems);

- available frequencies or timeslots (FDMA or TDMA systems); and

- call processing resources.

For example, for the reverse link noise rise criteria, new calls from a mobile would not be admitted by the admission control algorithm if the resulting interference is predicted to be higher than a pre-defined threshold value.

For voice calls, if admission control rejects a new user, it is treated as a blocked call and the user must try again later. However, some technologies allow calls to be re-directed to a neighboring cell if it has capacity and is deemed to be able to handle the connection due to overlapping coverage. For data connections, it is possible for the network to downgrade the throughput of existing data calls to allow more users access to the network.

1.3.2.6 Power control

To achieve high capacity and quality in wireless systems, they must employ power control. The goal of power control is to minimize the transmission power on both the forward and reverse links to conserve system resources and also minimize interference to other users.

In CDMA based systems, with all mobiles using the same frequency assignment, reverse link power control is fundamentally needed to ensure that each mobile signal will be received at the cell site at the same level to deal with the well known near-far problem. Because mobiles in a given cell are always on the move, some will be close to the base station and some will be much further away. The closer mobiles would have stronger signals back to the base station and cause unnecessary interference on the reverse link unless their power is controlled. In addition, system capacity is maximized if the transmit power of each mobile is controlled such that it is received at the base station with the minimum required signal level, to keep the system noise floor as low as possible. On the forward link, there is a different type of problem. Mobiles that are near the cell edge need more power from the base station than those that are close to the base station, for similar performance.

Reverse link power control is made up of an open loop, a fast closed loop, and an outer loop. Reverse link loop power control is made up of both an open loop and a fast closed loop. Reverse link open loop power control is determined by the mobile. It measures the received power level from the base station and adjusts its transmit power accordingly. If it receives a strong signal, it can determine that the path loss going back to the base station is low, and it therefore lowers its transmit power. The required mobile transmit power can be determined by a calibration constant which factors in cell loading, cell noise figure, antenna gain, and power amplifier output.

Reverse link closed or inner loop power control is a function of the base station. The goal of the closed loop component is for the cell to provide rapid corrections to the mobile's open loop estimate to maintain the optimum transmit power level. The cell measures the relative received power level of each of its associated mobiles and rapidly compares it to an adjustable threshold. Each mobile can then be instructed to increase or decrease its power level. This closed loop corrects for any variation required in the open loop estimate to accommodate gain tolerances and unequal propagation losses between the forward and reverse links. In the case of cdma2000-1x, closed loop power control operates at 800 Hz on both the forward and reverse links. In Wideband CDMA (W-CDMA), this mechanism operates at 1500 Hz.

The reverse link outer loop power control serves to maintain communications quality by setting the target for the closed loop by periodically adjusting a signal-to-interference (SIR) target or setpoint based on the frame error rate (FER). If the FER is low, this wastes capacity and thus the SIR target is decreased. If the frame error rate is poor, this affects quality and the SIR target is increased.

Similar fast closed and outer loop algorithms are also implemented on the forward link for cdma2000-1x and W-CDMA. An example is shown in Figure 1-18.

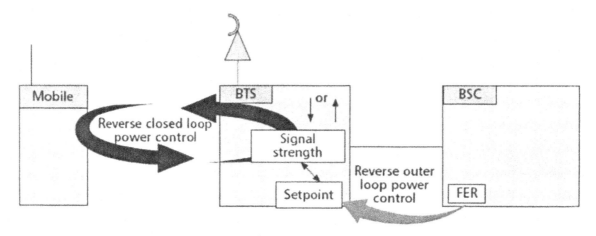

Figure 1-18: cdma2000-1x Reverse Link Power Control

1.3.2.7 Hand-off (inter-technology & intra-technology)

Hand-off (or handover) is one of the key features that enables a wireless network to support mobility. Hand-off is the ability to maintain communication between the MS and the network as the MS travels from the coverage area of one cell tower to another. As mentioned earlier, the two primary states of a MS are idle and connected mode and hand-offs are necessary for both.

In idle mode, the MS is monitoring the network for changes in network information or page messages for incoming calls. In order to receive these reliably, the MS should be communicating with the tower that can provide the best radio signal. This requires it to hand-off as the MS moves and the strongest signal comes from a new tower. Sometimes this process is also referred to as cell re-selection. When the MS is in idle state and communicating with a given tower, the network broadcasts information in the overhead messages about the surrounding or neighbor cells. The MS uses this information to periodically scan the strength and quality of the signals from neighboring towers, and if they are deemed to be better, the MS switches to the new tower to continue to listen to the network. This hand-off may trigger a registration criterion to be met so that the MS will update the network with its new location via a registration message. However, in many cases the network will not know or be concerned that the MS is monitoring a new tower.

In the connected mode, the MS has a voice or data connection with the network and reliable hand-off is even more important to maintain quality of service as the MS moves about. There are commonly two types of hand-off; soft hand-off and hard hand-off. Soft hand-off is also referred to as a "make before break" hand-off and hard hand-off is known as a "break before make" hand-off.

CDMA based technologies use a soft hand-off, which indicates that the MS is in communication with multiple base stations simultaneously, all with identical frequency assignments. This type of hand-off provides diversity on both the forward and reverse links at the boundaries between base stations. While connected, the MS is continually searching for the presence of neighboring sectors or cell sites by monitoring the quality of their pilot channels. If a pilot of sufficient strength is detected, the MS will send a message to the base station(s) with which it is communicating, containing information about the new pilot. If the base station decides that a soft hand-off should take place, it will check for and allocate resources at the new cell site and will send a hand-off message back to the MS directing it to perform a soft hand-off. Similarly, if a pilot with which the MS is in soft hand-off falls below a specific quality criterion, the MS will report this to the base station, and the network will remove this connection to the MS. With soft hand-off, as the MS moves from (say) cell A to B, cell B would be added to the MS as a hand-off leg before cell A is removed and hence this is described as a make before break hand-off mechanism. Softer hand-off is a special form of soft hand-off and applies to the situation where the MS is in hand-off with multiple sectors from the same cell site.

On the other hand, hard hand-off occurs when the MS transitions between base stations that operate on different frequencies or with different technologies. FDMA- and TDMA-based systems utilize hard hand-offs. In this situation the neighboring cell sites use different frequency assignments and the MS must retune itself to a new frequency assignment before it can continue the connection. Hence, this is known as a break before make hand-off. Hard hand-off can also apply to CDMA-based technologies when multiple frequency assignments are present (e.g., when a second carrier is laid on top of the first for capacity needs).

There is a fundamental difference in hand-off mechanisms between CDMA-based systems (cdma2000-1x or W-CDMA) and TDMA systems (GSM). TDMA uses discontinuous transmission, so there are gaps in time when the MS is not communicating with a base station. This provides an opportunity to make intersystem measurements with a single receiver on alternate frequency assignments or even technologies in the case of a GSM to UMTS hand-off.

On the other hand, CDMA-based technologies use continuous transmission and reception. Therefore, W-CDMA, as an example, introduces a compressed mode to create short gaps, on the order of a few milliseconds, in both the transmission and reception functions and provides the MS an opportunity to make GSM measurements.

1.4   Wireless Access Technology Standardization

1.4.1 Motivation

Wireless access standards are designed to specify users' connectivity to networks and access to services through a user terminal. In doing this, they define the essential functions and protocols, relying on and interacting with the user terminal capabilities at one end, and working with the core network at the other. An access technology is a transmission system to deal with communication through the wireless channels to achieve its connectivity and performance goals; it is designed as a multi-user sub-network with a multiple-access scheme, sharing of resources, and fast scheduling of simultaneous users.

As expected, standardized technologies aim to define interoperability and inter-working, a rich set of attributes, a graceful evolution path, and potentially strong product ecosystem. It must be noted that proprietary systems or system elements may also be introduced at different layers to provide differentiation, ease of implementation, or time-to-market advantage. The generations of wireless access technology, however, have increasingly been about standardized systems, building regional and global ecosystems, and enabling technology availability and user roaming. Typically, it has taken several years from the start of an access standard definition to its commercial launch, and naturally it typically stays in operation much longer. Recent years have seen explosive growth in wireless traffic, particularly given the increasingly mobile world, and have witnessed a proliferation of global deployments to provide high-speed wireless connectivity, communication, and content.

1.4.2 Design Goals and Technology Elements

It is intuitively appealing to think of a wireless access technology standard as one that has certain design attributes, in terms of:

- Connectivity, access;
- Mobility, coverage, roaming;
- Throughput, latency, performance, data symmetry;
- Efficiency; and
- Universality (technology availability, global ecosystem, user roaming).

Although these attributes are fundamental, there may be differences in how they are defined, depending on the system's role and its requirements:

- Low or high mobility;
- Personal-area, local-area, hot-zone/metro, wide-area, near-field (scan-zone);
- Satellite or terrestrial; line-of-sight (LOS) or non-LOS; and
- Indoor coverage extension or home network (e.g., Femtocell)

So far in this section, the goals and requirements, or attributes from a user or an application viewpoint, have been identified (i.e., what and why). An insight into technology elements and functional capabilities should then follow to address these goals and requirements in their phased evolution (i.e., how).

To meet particular system requirements, a number of technology elements may be expected to be outlined in a standard, and during its evolution through generations of access technology:

- Coding, modulation, spectral efficiency, round-trip transmission, receiver structure;

- Antenna technology, diversity, channel bandwidth and structure, FDD vs. TDD;

- Multiple-access mechanism, user-access scheduling; and

- Resource sharing and allocation, power management, topology and distribution architecture.

Although an access standard is generally not tied to any particular frequency band, the latter has significant impact on access technology planning and development. The implications of choosing one band vs. another include coverage, interference considerations, (indoor) signal penetration, line-of-sight constraints, user terminal ecosystem and availability, and potentially capacity.

We must bear in mind that there is no fixed definition for "access" technology and its boundaries. Different wireless networks are defined, some with different levels of architectural and functional details; moreover, the functions themselves evolve, while the entire network may be becoming more distributed and flatter, heteregenous, and with intelligence moved to the edges.

1.4.3 Technology Framework Definition and Standardization

The International Telecommunication Union (ITU) is the leading United Nations agency for information and communication technologies. It has three core sectors, radiocommunication, telecommunication standardization, and telecommunication development, in addition to organizing global telecom events. The radiocommunication sector (ITU-R) in particular has several Study Groups, on such topics as spectrum management, radiowave propagation, satellite services, terrestrial services, broadcasting services, and science services (http://www.itu.int/ITU-R).

A framework for the third generation (3G) network has been defined by ITU's International Mobile Telecommunications – 2000 (IMT-2000) family. The concept was born as early as the mid-1980s and the framework matured by 1999. This motivated global collaboration to define 3G standards with such design goals as flexibility, interoperability, affordability, compatibility, and modularity in mind. ITU-R Recommendation M.1457 [ITU11] identified five radio interfaces, while a sixth (based on WiMax, Worldwide Interoperability for Microwave Access) air interface was added to this family in 2007:

- IMT-DS (Direct Sequence) – W-CDMA/UTRA FDD

- IMT-MC (Multi-Carrier) – cdma2000

- IMT-TC (Time-Code) – UTRA TDD (TD-CDMA, TD-SCDMA)

- IMT-SC (Single Carrier) – EDGE

- IMT-FT (Frequency-Time) – DECT

- IMT-OFDMA TDD WMAN (Wireless MAN) – WiMAX

IMT-Advanced was subsequently defined through a set of requirements issued by ITU-R starting in 2008 for 4G mobile phone and Internet access services.

The industry players in telecommunications, computing, broadcasting, and applications, whether manufacturers, developers, service providers, or operators, have been involved for years in the development of wireless technology standards. With the growth of mobile communication, advances in technology, availability of new spectrum, and the need for universal cost-effective solutions to enable rich systems and ecosystems, broad terminal availability, and users' ability to roam, third-generation (3G) standards have been defined through global partnership projects (GPPs) involving standards bodies from different regions. Specifically, 3GPP and 3GPP2 have specified wireless access technologies (among others) for 3G and beyond.

The 3rd Generation Partnership Project, 3GPP, was established in 1998. The collaborating regional standards bodies, known as Organizational Partners, include the Association of Radio Industries and Businesses (ARIB), the China Communications Standards Association (CCSA), the European Telecommunications Standards Institute (ETSI), the Alliance for Telecommunications Industry Solutions (ATIS), the Telecommunications Technology Association of Korea (TTA), and the Telecommunication Technology Committee (TTC, Japan). The project is run by a Project Coordination Group, under which there are four Technical Specification Groups, each with a number of working groups (see Figure 1-19; more details can be found at http://www.3gpp.org).

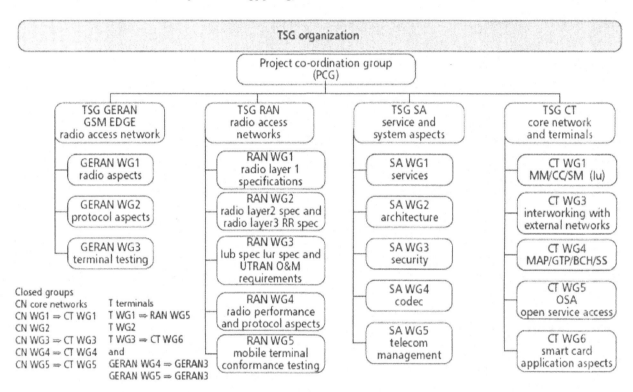

Figure 1-19: 3GPP Project Coordination and Technical Specification Structure

The scope of 3GPP has covered 3G systems based on evolved GSM core networks and the radio access technologies they support, Universal Terrestrial Radio Access (UTRA – both FDD and TDD), in addition to the maintenance and evolution of GSM technical specifications and reports, including evolved radio access technologies such as GPRS and EDGE. It has further defined systems beyond 3G and the phased evolution of IP networks, specifically Long-Term Evolution (LTE) and Evolved Packet Core (EPC). The work has been published in a number of Releases (e.g., Release 10 published in mid-2011). The technologies are described further in subsequent sections.

In parallel, 3GPP2 was born, inspired by the 3GPP partnership project concept, and out of the IMT-2000 initiative, with a focus on global specifications for systems (supported by ANSI/TIA/EIA-41 [ANS01]) towards 3G (cdma2000) and beyond (Ultra Mobile Broadband, UMB). 3GPP2 organizational partners include the regional standards bodies ARIB, CCSA, TIA (the Telecommunications Industry Association), TTA and TTC. Figure 1-20 shows the structure of the 3GPP2 Technical Specifcation Groups.

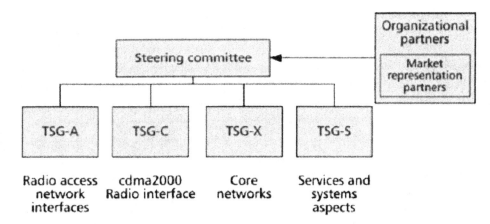

Figure 1-20: 3GPP2 Technical Specification Structure (Working Groups not Shown)

The IEEE has been involved in the creation of a broad range of standards in a wide range of areas within its scope (see http://standards.ieee.org). In particular, IEEE Project 802 (or 802 LAN/MAN) has developed standards with focus on the data link and physical layers, the bottom two layers in the Open Systems Interconnection (OSI) layered architecture and the corresponding sub-layer structure. The many working groups within the Project 802 family include those that focus on wireless technology standards, extending to personal-area networks (PANs), including:

- IEEE 802.11 – Wireless LAN;

- IEEE 802.15 – Wireless PAN;

- IEEE 802.16 – Broadband Wireless Access; and

- IEEE 802.20 – Mobile Broadband Wireless Access.

These, along with the working group for wireless sensor standardization, form the so-called IEEE Wireless Standards Zone. Each of these work streams covers a range of specifications that typically evolve in versions, revisions, or support of user requirements. For example, the 802.16 working group has specified broadband wireless access across all scenarios of fixed, nomadic, or high user mobility within a wide-area network. Furthermore, industry forums work extensively to further define, certify, and promote related technologies, variety of profiles, and products. Again as an example, the WiMAX Forum, with a number of working groups, certifies and promotes the compatibility and interoperability of broadband wireless products based on the harmonized IEEE 802.16/ETSI HiperMAN (High Performance Radio Metropolitan Area Network) standard (see http://www.wimaxforum.org).

1.4.4 Mobile Technology Generations and Nomadic Implementations

Figure 1-21 shows a simplified view of mobile cellular access technology evolution, in addition to highlighting nomadic broadband wireless access. Not all technologies or evolutionary steps are shown. Moreover, the context (e.g, Generation label) may not necessarily represent all scenarios accurately.

To provide an understanding of these technologies, the GSM evolution path (3GPP), Interim Standard IS-95 (cdmaOne) evolution path (3GPP2), IEEE 802.11 (WLAN), IEEE 802.16 (WiMAX), and OFDMA-based technologies beyond 3G are outlined in some detail in subsequent sections, in addition to an introduction to personal, home, and near-field communications.

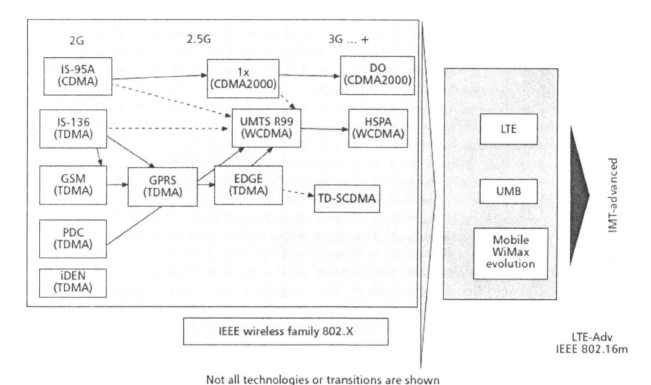

Not all technologies or transitions are shown

Figure 1-21: Evolution of Mobile Access Technologies

1.5   Digital Mobile Cellular Technologies – GSM to LTE

The preceding sections outlined wireless access technology design and spectrum considerations, multiple access schemes, mobility management, and standards definitions. In this section, a number of global wide-area mobile access technologies are discussed in some detail, including those of the second and third generations, based on TDMA and CDMA. OFDMA-based wide-area technologies are discussed later in the section on beyond 3G.

Again, it must be noted that the goal in this chapter has been to provide an end-to-end appreciation of wireless access technologies, with some details, particularly on global mobile and nomadic technologies. As such, it is neither inclusive of all technologies and scenarios, nor does it provide comparable details about competing or complementing technologies. Furthermore, every implementation has its own attributes and rationale, based on its own market, context/history, goals, and roadmap, with parameters which may vary from those represented in standards, or presented here. Similar concepts and mechanisms, however, apply to a variety of wireless access technologies, depending on applications and features, enabling technologies, architectural layers and details, and also ecosystem maturity.

1.5.1   3GPP Wireless Access Technologies

The GSM evolution path and wireless access standards specified by 3GPP are described in further detail in this section. The section starts with the widely implemented TDMA-based technologies of GSM, GPRS, (and briefly) EDGE, followed by UMTS Phase 1, and then High Speed Packet Access (HSPA) evolution. Long Term Evolution (LTE) is later discussed in some detail.

As also suggested in Figure 1-21 on the evolution of access technologies, the movement between different paths varies, based on needs and roadmaps. For example, operators with PDC or 3GPP2 technologies may have chosen to implement UMTS technologies, or a 3G operator may be in a position to choose any of the OFDMA-based technologies in its roadmap beyond 3G. As a significant example, LTE has gained global interest by virtually all mobile communication players.

1.5.2 GSM, GPRS, EDGE

In this section, we describe the evolution and enhancements of the GSM/UMTS standards and networks from GSM phase 1 to HSPA. GSM Phase 1 is a circuit-switched mobile network technology using TDMA, providing voice services and short-message service (SMS). The subsequent phases of GSM introduced packet services (GPRS) while keeping some fundamental principles such as TDMA radio transmission, the Mobile Application Part (MAP) signaling protocol for roaming, and the security features. UMTS phase 1 (often referred to as Release 99) has kept the network principles of GSM and GPRS but has a completely new radio access interface based on CDMA.

1.5.2.1 Main principles of GSM

GSM is a mobile digital technology developed in several phases. Though a second generation (digital) system (2G), the main principles of GSM phase 1 were set as early as 1987. It was primarily optimized to provide circuit switched voice services, although basic data services, notably SMS, were soon introduced. This section describes the radio interface of GSM and the main architectural principles of GSM phase 1.

GSM radio interface

The GSM radio interface is based on the FDD mode and TDMA with eight time slots per radio carrier. In other words, each uplink time slot is paired with a downlink time slot. Each radio carrier requires a 200-kHz uplink and a 200-kHz downlink. A time slot may be used for one of the following sets of logical GSM channels, although there are other possibilities: one TCH (Traffic Channel) full rate, two TCHs half rate, eight SDCCHs (Stand Alone Dedicated Control Channels), or one CCCH/BCCH (Common Control Channel, Broadcast Control Channel). The TCHs are used to transmit voice or data whereas the SDCCHs can only be used for signaling purposes or SMS transmission.

In Europe, the Middle East, Africa, and Asia, the GSM system generally operates in the following bands:

900 MHz (174 radio carriers):	Uplink: 880-915 MHz	Downlink: 925-960 MHz
1800 MHz (374 radio carriers):	Uplink: 1710-1785 MHz	Downlink: 1805-1880 MHz

In the Americas, on the other hand, the operating band is as follows:

1900 MHz (298 radio carriers):	Uplink: 1850-1910 MHz	Downlink: 1805-1880 MHz

It should be noted, however, that GSM is defined independently of the frequency resources meaning that it can operate in other bands as well.

A generic example will help explain what this means for a GSM operator. A typical spectrum allocation for a GSM operator may be 20 MHz which is sufficient for 100 carriers. A frequency reuse scheme will generally be used. As a result, around eight carriers may be available in each cell, which means 64 TDMA time slots per cell. Among these, four time slots may be used to accommodate the signaling traffic (for instance one timeslot for the CCCH/BCCH and three other timeslots to provide 24 SDCCHs). Therefore, 60 timeslots are available in each cell to accommodate voice traffic. If only Traffic Channel

Full Rate Speech (TCH/FS) is used, this allows 60 simultaneous voice calls. If Traffic Channel Half Rate Speech (TCH/HS) is used exclusively, 120 simultaneous voice calls are possible.

Protocol aspects

The signaling protocols of the radio interface are divided into a structure of three layers, which is similar to that of the DSS1 (Digital Subscriber Signaling System 1) protocols on the D channel in ISDN (Integrated Services Digital Network) and which is based on the OSI reference model.

According to the configuration requested by the mobile station, Layer 2 offers either a connectionless information transfer in unacknowledged mode (on point-to-multipoint or multipoint-to-point channels) or connection-oriented information transfer in acknowledged mode on a dedicated control channel. Layer 3 of the radio interface is subdivided into three sub-layers: Radio Resource (RR), Mobility Management (MM), and Connection Management (CM). The CM sub-layer comprises parallel entities: Supplementary Services handling, Short Message Service, and Call Control (CC).

GSM Phase 1 architecture

The radio coverage of a cell is provided by a Base Transceiver Station (BTS). Each BTS is linked to a BSC (Base Station Controller); a BSC and the BTSs which are linked to it constitute a BSS (Base Station Sub system). Each BSC is linked to a MSC (Mobile Switching Center), as shown in Figure 1-22. The interface between the BSS and the NSS (Network Sub-System) is called the A interface. Viewed physically, information flows between mobiles and BTSs, but viewed logically, the mobile communicates with entities in the BSS and in the MSC. Layer 1 and Layer 2 are handled by the BTS, the RR sub-layer is handled by the BSC and the MM and CC-sub-layers are handled by the MSC.

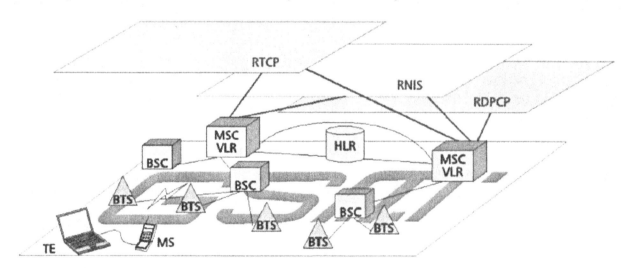

Figure 1-22: Basic Cellular (GSM) Network Architecture

There exist in fact slight exceptions to these general rules but this mapping essentially means:

- The BTS handles the radio transmission.

- The BSC organizes the allocation, release, and supervision of the radio channels, these actions being performed according to commands received from the MSC.

- The MSC handles the call establishment and call release, the mobility functions, and every aspect related to the subscriber's identity.

Schematically, the behavior is as follows: the MS makes a first access on the RACH (Random Access Channel, a multipoint to point channel) of the selected cell. In response to this, the BSC allocates the MS a first dedicated channel. After the MS has seized this radio channel, a dedicated link will exist between the MS and the BSC; the MS uses this link to send an initial message which includes its identity and the reason for the access (e.g., a requested service). Upon receipt of this message, the BSC establishes a signaling connection with MSC dedicated to this MS.

The core elements of a GSM network include the Mobile Switching Center (MSC), the Home Location Register (HLR), and the Visitor Location Register (VLR), as key functional elements. The MSC has an interface with the BSC, on the access side, and to the back-end and fixed networks, as indicated earlier. The MSC is a digital exchange entity, able to perform all necessary functions to handle the calls to and from mobile subscribers located in its area, and furthermore, it is able to cope with the mobility of the subscribers, using the HLR and VLR. The signaling exchanges between these entities are specified in the Mobile Application Part (MAP), which is the protocol used to provide roaming.

Core network technologies and architectures are covered in detail in Chapter 2.

SIM features in GSM

The GSM mobile station (user terminal) has two distinct elements: the mobile equipment which is able to connect with the network, and a chip card (namely the Subscriber Identity Module, SIM) which contains all subscriber-related data. There is a standardized interface between these two elements. Although, much of this is a back-end and networking concern, a brief overview is offered in here to complete the notion of access using the SIM. The data stored in the SIM include:

- The International Mobile Subscriber Identity (IMSI);

- The Ki key; which is linked to the IMSI and is allocated at subscription and stored unchanged; and

- The authentication security algorithm A3 and the key generation security algorithm A8.

These data are used for two security features, namely the authentication procedure and ciphering. The authentication procedure enables the network to validate the mobile subscriber identity, and thus protects the network against unauthorized use. When the MSC receives a mobile identity (IMSI) transmitted on the radio path, it triggers an authentication procedure. The network sends a random number RAND to the mobile station in order to check that it contains the Ki linked to the claimed IMSI. The Mobile Station applies algorithm A3 to RAND and Ki in order to compute the answer to be sent to the network.

The purpose of the ciphering procedure is to prevent an intruder from listening to what is transmitted over the radio interface. This protection covers both the signaling and the user data, for both voice and non-voice services. The layer 1 data flow transmitted on dedicated channels (SDCCH or TCH) is the result of a bit-per-bit addition of the user data flow and of a ciphering stream generated by the ciphering/deciphering algorithm. This algorithm uses both a ciphering key and the TDMA frame number. The ciphering key (Kc) is computed independently on both the MS side and the network side by the authentication procedure; algorithm A8 is used to derive Kc.

The SIM card is provided by the network operator; thus the network operator can control the security features even in a roaming situation. In addition, the SIM serves as a tool to support other features and services. See section 3.5.4 for more on the SIM and GSM security.

1.5.2.2 GPRS and EDGE – GSM evolution

After the success of GSM phase 1, it became necessary to define new features in GSM networks. These included service consistency even when roaming outside the home network, enhanced throughput to support data services, and the ability of the SIM to become an active device able to control the MS (SIM toolkit). These new features were developed to be compatible with legacy user devices, and their introduction was optional.

Introduction of packet mode in GSM

Data services were already defined in the first phase of GSM. These were circuit-switched data services which yielded the allocation of a TCH (one TDMA timeslot). The throughput was very low, typically 9.6 kb/s. Therefore it was highly desirable to increase the throughput of the radio interface. One method to achieve this for a given call is to utilize more than one TDMA time slot on the radio interface. Such a feature has been standardized as HSCSD (High Speed Circuit Switched Data). The main advantage of HSCSD is its simplicity. The GSM MSC and A interface are unchanged as long as the number of utilized time slots is equal or less than four, because in these cases a 64 kbits/s circuit is sufficient on the A interface. The drawback of HSCSD is its inefficient use of radio resources.

To solve this problem another service was designed, GPRS (General Packet Radio Service). GPRS is a set of GSM bearer services that provide packet-mode transmission and interworking with external packet data networks. GPRS allows the service subscriber to send and receive data in an end-to-end packet transfer mode without utilizing network resources in circuit-switched mode. The service aspects of GPRS are specified in GSM 02.60 [3GP99] and the technical realization is specified in GSM 03.60 [3GP98].

To accommodate sporadic transfers of large volumes of data, GPRS encompasses allocation and release mechanisms that optimize the use of the radio resources. During a data call, these resources (i.e., one or more time slots) are allocated only when data is being transmitted; they are then released, although the data-transfer session can be kept between the MS and the network. In addition, GPRS is well suited to asymmetric data transfer, typically with larger downlink speed. To take that into account, GPRS can allocate more time slots in the downlink than in the uplink.

GPRS was designed to allow a smooth sharing of radio resources between speech and data. In order to accommodate the data traffic, the network operator can decide for each radio carrier how many timeslots are allocated to GPRS traffic and how many timeslots are allocated to voice traffic. This sharing can be performed dynamically.

With a time slot, the following throughputs can be obtained over the radio interface according to the different coding schemes:

CS-1: up to 9.05 kb/s   CS-2: up to 13.4 kb/s   CS-3: up to 15.6 kb/s   CS-4: up to 21.4 kb/s

More than one time slot can be allocated for a data transfer. Typically, a GPRS MS may allocate up to four time slots in the downlink and two time slots in the uplink, thus allowing up to 85.6 kb/s in the downlink.

GPRS architecture

GPRS was designed to allow network operators to reuse the radio coverage deployed for voice. In addition, it was necessary to define two new functional entities on the core network side:

- The Serving GPRS Support Node (SGSN) is the node that controls the BSC serving the MS. It handles mobility, paging, security features, and interface with the BSC.

- The Gateway GPRS Support Node (GGSN) works with the packet data networks (fixed and mobile). It is connected to the SGSN through an IP network; it contains routing information for GPRS users.

These network entities have been defined to work with IP networks.

After the rollout of GPRS, a GSM network includes circuit-switched and packet-switched domains. The basic architecture is shown in Figure 1-23. For a detailed discussion of GPRS architecture, including the SGSN and GGSN, see section 2.3.3 in the next chapter.

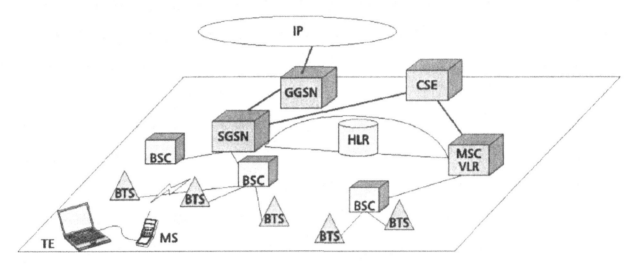

Figure 1-23: Evolved GSM Architecture (GPRS)

EDGE

EDGE (Enhanced Data Rates for GSM Evolution) uses more advanced modulation to increase the throughput at the radio interface. In a time slot, the following throughputs can be obtained over the radio interface, depending on the coding scheme: 28.8 kb/s, 32.0 kb/s and 43.2 kb/s. As in GPRS, a MS can be allocated more than one timeslot.

EDGE time slots are used with the GPRS architecture to provide packet services with increased data throughput; this is known as EGPRS (Enhanced GPRS). Further advancements have been proposed to enhance EDGE capabilities (EDGE$^+$).

1.5.3 UMTS Phase 1

UMTS (Universal Mobile Telecommunications System) was designed by a joint effort of the GSM community and other players as a Third Generation (3G) system, compatible with GSM to the extent possible and meeting the goals of IMT-2000 set by the ITU. To organize this common work, a new joint standardization forum was developed as a Partnership Project (3GPP).

The UMTS phase 1 specifications are based on the following key principles:

- A completely new radio interface using Wideband CDMA (WCDMA), and

- Reuse of the network, services and security principles of GSM and GPRS.

1.5.3.1 Radio interface

Whereas GSM radio technology is based on TDMA, the design of the UMTS radio interface is based on CDMA technology. A completely new radio interface was designed by ETSI and developed further by 3GPP. This radio interface is formed by two modes of operation in two different parts of the spectrum: FDD in paired-band configurations (for uplink and downlink) and TDD in an unpaired one.

The FDD component is built on a WCDMA concept which is based on direct-sequence CDMA with a 3.84 Mc/s chip rate and is designed both for flexibility for third generation services and for optimized GSM compatibility. The physical layer offers flexible multi-rate transmission capabilities and a service multiplexing scheme. An efficient support for packet access has been defined with a dual-mode packet transmission scheme supporting various multimedia services.

The TDD component is a TD-CDMA scheme, with both time and code multiplexing. It uses joint detection in the receiver on the uplink as well as on the downlink and requires neither high power control accuracy nor a soft handoff. This mode offers flexibility in downlink and uplink time slot allocation to meet asymmetric traffic requirements and it allows for the easy implementation of adaptive antennas. Basic system parameters such as carrier spacing, chip rate, and frame length are harmonized. FDD/TDD dual mode operation is thereby facilitated, providing a basis for the development of low-cost terminals.

Each UMTS WCDMA FDD radio carrier requires a 5 MHz downlink and a 5 MHz uplink. A UMTS WCDMA FDD radio carrier may transmit around 50 simultaneous voice calls in circuit mode. However, as mentioned earlier, the CDMA scheme provides trunking efficiency, with no fixed time or frequency slots per user, and therefore allows trade-offs between traffic load and performance attributes, such as throughput and the level of tolerable degradation. UMTS specifications provide the details. In Europe, the Middle East, Africa, and Asia, UMTS generally operates in the FDD 1920-1980 MHz band for uplinks and the 2110-2170 MHz band for downlinks, though other bands are possible (e.g., re-farming of 900 MHz). In the Americas, different frequency bands have been used notably the frequency bands around 850 MHz and 1900 MHz.

The split of the radio interface protocol stack into three layers is similar to GSM. Layer 1 is the physical layer with the new radio technology (WCDMA and TD-CDMA). Layer 2 is split into two sub-layers: the Media Access Control (MAC) sub-layer and the Radio Link Control (RLC) sub-layer. Layer 3 encompasses three sub-layers: RRC (radio resource control), MM (mobility management), and CC (call control). The FDD component of the WCDMA radio interface provides radio bearers for both circuit-switched (e.g., voice) and packet-switched communication, up to 384 kb/s (downlink) and 12 kb/s (uplink) in UMTS Phase 1. The radio interface protocol architecture is specified in UMTS 25.301 [3GP11a].

1.5.3.2 UMTS Phase 1 architecture

The architecture of the UMTS radio access network (UTRAN) is fundamentally similar to the GSM architecture, in terms of the Node B (base station) and Radio Network Controller (RNC) components. The functionality split, however, is somewhat different from that of GSM.

The core network of UMTS Phase 1 reuses the principles of GSM and GPRS. It comprises the MSC, VLR, HLR, SGSN, and GGSN (all introduced earlier) with a functional splitting quite similar to that of GSM. As in GSM, messages are exchanged directly between the mobile terminal and the core network, without any translation by the RNC or Node B. All UMTS radio resource control functionalities are located in the RNC, including handoff functions. This is different from GSM, where control of the handoff process is shared between the visited MSC and the BSC. The circuit domain is quite similar to that of GSM, but there has been more evolution in the packet domain, although it is based on GPRS. First of all, the GPRS interface between the SGSN and the BSC (the Gb interface) is completely changed. Indeed, a tunnel is established between the SGSN and the RNC. Unlike GSM, the transcoders are controlled by the MSC and not by the access sub-network. Lastly, the SIM is kept in UMTS and is known as the USIM (UMTS SIM).

UMTS specifications 23.101 [3GP11b] and 23.121 [3GP02] give the general UMTS architecture principles and requirements for UMTS Phase 1 (Release 99).

1.5.3.3 GSM/UMTS interworking

From the start, GSM/UMTS handoff and roaming were defined. The protocol for roaming (MAP) is common to GSM and UMTS. In addition, the UMTS Mobile Stations typically support GSM at a minimum, and generally support multiple modes and multiple bands. It was also decided to develop a core network common to GSM and UMTS, thus allowing GSM/UMTS operators to deploy a single core network enabling the different access technologies, as shown in Figure 1-24.

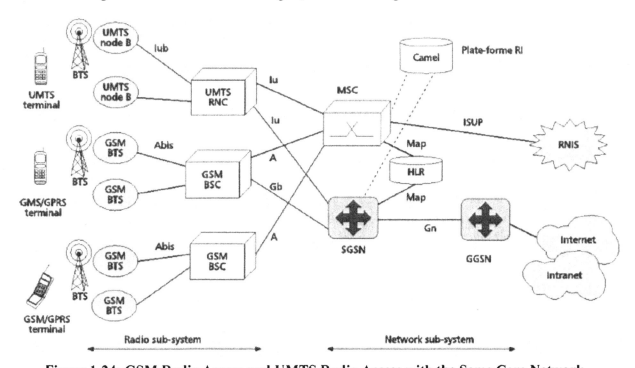

Figure 1-24: GSM Radio Access and UMTS Radio Access with the Same Core Network

1.5.4 HSPA (UMTS Evolution)

3GPP Release 5 defined the evolution of the UMTS access network, High Speed Downlink Packet Access, (HSDPA [3GP04]), with its significant enhancements to downlink throughput capabilities. In order to improve the data throughput in the uplink, High Speed Uplink Packet Access (HSUPA) was defined in the framework of 3GPP Release 6 [3GP06]. The different phases are generally defined with much consideration to the smooth and graceful evolution of phased implementations.

1.5.4.1 HSDPA (High-Speed Downlink Packet Access)

The growing importance of IP-based applications for mobile access made it necessary to evolve the UMTS radio interface to meet the requirements of new and expected applications:

- Packet data transmission,

- High-throughput with asymmetry between uplink and downlink, and

- Low time constraint and non-uniform quality of service requirements (traffic in bursts).

This led to the standardization of HSDPA by 3GPP in its Release 5, with two major improvements:

- The optimization of the spectral efficiency by a better allocation of radio resources, enabling a variety of speed profiles depending on the number of codes, modulation, and channel coding, reaching a defined peak data rate of 14.4 Mb/s; and

- The reduction of transmission delays.

HSDPA is an optimization of UMTS. The same frequency bands used for WCDMA Phase 1 may be used.

A fundamental goal of HSDPA, as is evident from its name, is to provide high throughput in the downlink. To achieve this, it uses methods known from GSM/EDGE standards, including link adaptation and the HARQ (Hybrid Automatic Repeat reQuest) retransmission algorithm. In addition, HSDPA exploits a new dimension of diversity that has not been exploited by other 3GPP systems: the multi-user diversity that takes advantage of the multiplicity of users to optimize the use of radio resources. This multi-user diversity is extracted using a fast scheduling algorithm, built in Node B. This scheduler selects (every 2 ms) the most appropriate MS (or user equipment, UE) to which data should be sent.

Radio interface

A simplified and basic functionality of HSDPA is shown in Figure 1-25. Node B estimates the channel quality of each HSDPA user on the basis of, for instance, power control, ACK/NACK ratio (i.e., ratio of packet transmission acknowledgements and non-acknowledgements), and HSDPA-specific user feedback (CQI, Channel Quality Indicator). Scheduling and link adaptation are then conducted at a fast pace that depends on the scheduling algorithm and the user prioritization scheme.

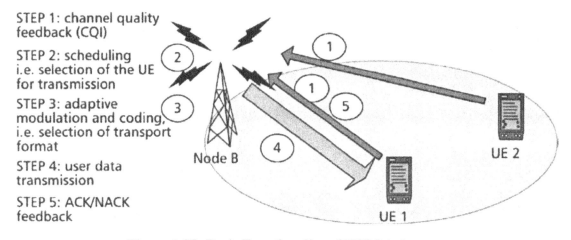

Figure 1-25: Basic Functionality of HSDPA Access

HSDPA relies on a new downlink transport channel, the High Speed Downlink Shared Channel (HS-DSCH). This channel is supported by new physical channels with distinct functions, on the uplink and downlink, for data traffic or signaling.

In HSDPA, two fundamental features of UMTS/WCDMA, variable Spreading Factor (SF) and fast power control, are disabled and replaced by adaptive modulation and coding, extensive multi-code operation, and a fast and spectrally efficient retransmission strategy. This allows selecting, for users in good radio conditions, a coding and modulation combination that provides better throughput, with the same transmitted power as for UMTS channels. To enable a large dynamic range for HSDPA link adaptation and maintain good spectral efficiency, a user may simultaneously use up to 15 codes of SF 16. Also, a new modulation is introduced, 16-QAM (Quadrature Amplitude Modulation), in addition to Quadrature Phase Shift Keying (QPSK). The use of more robust coding, fast HARQ, and multi-code operation removes the need for variable SF. Table 1-2 shows some possible modulation and coding schemes.

CQI Value	Transport Block Size	Number of Codes	Modulation
1	137	1	QPSK
2	173	1	QPSK
7	650	2	QPSK
10	1262	3	QPSK
15	3319	5	QPSK
16	3565	5	16-QAM
30	25558	15	16-QAM

Table 1-2: Applicable Modulation and Coding Schemes (MCS) for a Given Reported CQI

To allow the system to benefit from short-term radio channel variations, packet scheduling decisions are done in Node B. If desired, most of the cell capacity may be allocated to one user for a very short time, when its radio conditions are the most favorable. Thus, an MCS providing high user payload (i.e., a larger transport block size) can be used. The classical 10-ms frame of UMTS has been divided into five sub-frames, or TTIs (Transmission Time Intervals), of 2 ms. In the optimum scenario, scheduling tracks fast-fading of user equipment (UE), as shown in Figure 1-26.

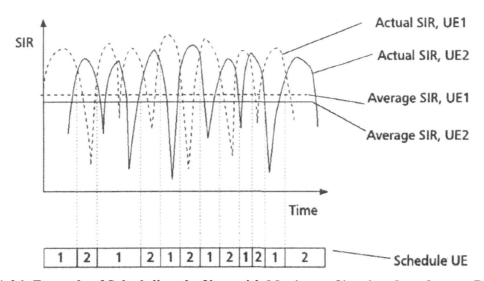

Figure 1-26: Example of Scheduling the User with Maximum Signal-to-Interference Ratio (SIR)

The choice of an appropriate scheduling algorithm is a crucial point for deploying HSDPA efficiently, since this algorithm has a direct effect on the throughput achievable by end users. Several categories of algorithms may be implemented, from fair resource sharing to efficient algorithms that aim at maximizing the cell throughput. A compromise is to use algorithms that aim at finding the middle ground between fairness and efficiency, at the expense of increased complexity.

The combining of physical layer packets (HARQ) basically means that the terminal stores the received data packets in soft memory and if decoding has failed, the new transmission is combined with the old one before channel decoding. The retransmission can be either identical to the first transmission (Chase Combining) or contain different bits compared with the channel encoder output that was received during the last transmission (Incremental Redundancy). With this incremental redundancy strategy, one can achieve both a diversity gain and an improved decoding efficiency.

Radio access network architecture

In 3GPP Release 5, the architecture of the radio access network with HSDPA remains the same as in Release 99, with Node B and the RNC. However, some upgrades were needed in nodes and interfaces to deploy HSDPA compared to UMTS Release 99. The major impact on access network nodes is that the HS-DSCH is terminated in Node B instead of the RNC. This means that a small part of the 3GPP Release 99 functionalities, initially located in the RNC, have been moved down to Node B: e.g., the HARQ functionality (as part of the MAC layer) and scheduling functionality, which are implemented in a new MAC entity. Therefore, software upgrades are necessary in Node B and the RNC. Concerning interfaces, the major issue is a dimensioning one: since HSDPA provides higher bit rates, interfaces (labeled Iu-PS and Iub) should be re-dimensioned to support a higher throughput than for UMTS Release 99.

Concerning the MS (or UE), enhancements were necessary. For example, the new HARQ functionality requires more buffering and computational power; moreover, advanced receivers are needed, as the traditional Rake receiver is not sufficient to decode more than five simultaneous codes efficiently.

A significant trend has been to move toward a flatter architecture to enable seamless mobility across different access networks. In 3GPP Release 7 (2007) and Release 8 (2008), enhancements of the architecture are introduced that enable deployment of a so-called flat architecture, by collapsing Node B and the RNC into a single node.

Physical layer performance

Cell throughput and the user throughput depend on several factors:

- Radio link conditions: A user with good radio link quality will be allocated a modulation and coding scheme (MCS) that will provide a higher throughput than if the link quality is poor.

- Scheduling algorithm: This has a major impact as discussed earlier in this section.

- Load of the cell: The user is scheduled more frequently and throughput is higher with less cell load.

- UE category: 12 categories of HSDPA-capable UEs have been defined by 3GPP, according to their capacity to simultaneously decode several codes, the modulations they can support, and the periodicity at which data can be scheduled. The maximum achievable bit rate for a UE varies depending on these parameters, from 1.8 Mb/s to 14.4 Mb/s (the theoretical peak rate).

- Power allocated to the HS-DSCH: The power allocated to HSDPA channels may be set to a fixed value or may be the remaining available power after power is allocated for Release 99 dedicated and control channels.

- Number of codes allocated to HS-DSCH: This depends on the UE category but also on the operator's configuration. For example, fewer codes are available if a carrier is shared with a Release 99 Dedicated Channel (DCH).

1.5.4.2 HSUPA (Enhanced DCH)

HSUPA is actually identified in the 3GPP specifications as Enhanced DCH (E-DCH) or enhanced Uplink. It is a feature added to UMTS in 3GPP Release 6 [3GP06] to handle high data rate packet services in the uplink, with a maximum (target) data rate of 5.8 Mb/s.

The motivation for development of HSUPA was fundamentally the same as HSDPA, but in this case addressing the need for enhanced uplink speeds. HSUPA introduces better handling of radio and network resources and also transmission time in the uplink communication.

Contrary to HSDPA, HSUPA is not based on a shared channel, but rather on an optimization of the Release 99 uplink DCH, by the use of traditional DCH features:

- A new enhanced dedicated transport channel E-DCH;

- Power control to adapt E-DCH to a changing environment;

- Transport format selection according to the current buffer status and available power;

- A soft handoff that benefits from inherent uplink macro-diversity; and

- A predictable bit rate.

Moreover, some HSDPA-like enhancements are introduced:

- MAC functions are moved from the RNC to Node B to reduce the Round Trip Time (RTT).

- Short uplink TTI is introduced (2ms as an option; 10ms as default), which also reduces RTT.

- Improved re-transmission (HARQ) is used to benefit from combining received packet versions.

- Multi-code operation and low spreading factor are supported to increase the peak data rate, with the definition of several UE categories corresponding to various multi-code and SF configurations.

- Node B scheduling is applied to control the uplink interference level, the cell capacity, and the UE QoS:
  - o Node B dynamically allocates scheduling grants to *all* users in the cell, according to their rate requests, QoS requirements, current uplink cell load, and the maximum uplink target cell load.

The process of scheduling is illustrated in Figure 1-27. It is, by nature, less efficient than in HSDPA, as the entity that performs the scheduling (Node B) is not the one that allocates the resources (the UE).

One UE has at most one E-DCH transport channel that can be mapped over a variable number of physical channels. There are other physical channels in the uplink and downlink for traffic and signaling (control).

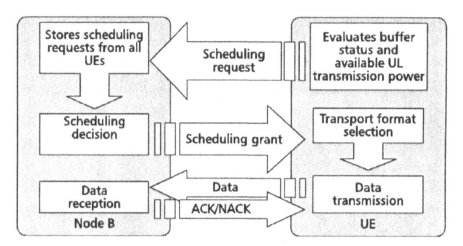

Figure 1-27: Simplified View of HSUPA Node B Uplink Scheduling

Macro-diversity

As in Release 99 and contrary to HSDPA, soft and softer handoffs are supported for the E-DCH channel. This exploits the available uplink macro-diversity to improve the radio link transmission for users at the cell edge and is particularly necessary to provide seamless mobility to delay-sensitive services. However, this feature also adds some complexity to the Node B scheduler and the HARQ algorithm. The number of HSUPA physical channels to monitor for UE is considerably higher with soft handoff on E-DCH.

In summary, HSUPA aims to increase uplink transmission throughput, decrease delays, and improve coverage, compared to Release 99. HSUPA introduces a new uplink dedicated transport channel called Enhanced Dedicated Channel (E-DCH). E-DCH is very similar to DCH, with some improvements to support a shorter sub-frame, and techniques such as HARQ and Node B scheduling. Because of the uplink configuration, HSUPA is more complex than HSDPA, and more signaling consuming. The HSUPA scheduler controls the radio resource less tightly than does the HSDPA scheduler.

USIM and UICC

It is possible to access a GSM network using a SIM card, as indicated earlier, and a UMTS network using a USIM card. A UICC (Universal Integrated Circuit Card) can provide both capabilities, in addition to an increasing number of other applications, within embedded or removable memory. The ETSI Smart Card Platform (SCP) project and the ISO develop and maintian standards and uniform platforms.

1.5.4.3 HSPA+

There is a broad deployment of HSPA technology around the world today, mostly based on Release 6 or beyond, allowing high speed access in both the downlink and the uplink.

3GPP has continued to define further enhancements to the HSPA standards, specifically through the recommendations in Release 7 to define HSPA+ versions. The principal additional advancements include:

- Use of higher-level modulation (64 QAM) in addition to 16 QAM and QPSK;
- Use of Multiple Input Multiple Output (MIMO) antenna technology; and
- Concatenation of two or more transmission carriers (particularly DC or Dual-Carrier).

These advanced technologies have allowed HSPA+ deployments around the world, enabling high speed wireless access at defined peak rates of 21, 28, 42 (or more) Mb/s.

1.5.5 Long Term Evolution (LTE)

3GPP started work on the evolution of 3G in 2004. A set of high level requirements was identified, including reduced cost per bit, increased service provisioning, flexibility in the use of existing and new frequency bands, simplified architecture, open interfaces, and reasonable terminal power consumption.

A feasibility study of UMTS Long Term Evolution (LTE) was subsequently started. The objective was to develop a framework for the evolution of 3GPP radio-access technology toward a high-data-rate, low-latency, and packet-optimized radio access. The study focused on the following aspects:

- The radio-interface physical layer (downlink and uplink), to support flexible transmission bandwidth up to 20 MHz, by introducing new transmission schemes and advanced multi-antenna technologies;

- Radio interface layer 2 and 3 and signaling optimization;

- UTRAN architecture, identifying the optimum network architecture and the functional split between Radio Access Network (RAN) network nodes; and

- RF-related concerns.

A set of basic requirements resulted from the first part of the study:

- Peak data rate – an instantaneous downlink peak data rate of 100 Mb/s within a 20 MHz downlink spectrum allocation (5 bps/Hz) and an instantaneous uplink peak data rate of 50 Mb/s (2.5 bps/Hz) within a 20 MHz uplink spectrum allocation;

- Control-plane latency – a transition time of less than 100 ms from a camped state to an active state and a transition time of less than 50 ms from a dormant state to an active state;

- Control-plane capacity – at least 200 users per cell supported in the active state for spectrum allocations up to 5 MHz;

- User-plane latency – as low as several milliseconds;

- User throughput – an average downlink user throughput per MHz that is three to four times that of HSDPA and average uplink user throughput per MHz that is two to three times that of Enhanced Uplink (Release 6);

- Spectral efficiency – the same improvement factors as for throughput;

- Mobility – performance optimized for low mobility but maintaining high mobility across the cellular network at speeds greater than 120 km/h (even up to 500 km/h depending on the frequency band);

- Coverage – the above throughput, spectrum efficiency, and mobility targets met for 5 km cells, and with a slight degradation for 30 km cells;

- Enhanced Multimedia Broadcast Multicast Service (eMBMS);

- Paired and unpaired spectrum arrangements;

- Spectrum flexibility and scalability;

- Co-existence and inter-working with existing 3GPP radio access technologies;

- Architecture and migration – a single E-UTRAN architecture and optimized backhaul communication protocols;

- Radio Resource Management requirements for enhanced end-to-end QoS support, efficient support for transmission of higher layers, and load sharing and policy management across different radio access technologies; and

- Complexity optimization (in terms of a limited number of options and no redundant mandatory features).

A number of multiple access candidate schemes have been considered, taking into account the objective of supporting both FDD and TDD operation. OFDMA was selected for the downlink, while single carrier frequency division multiple access (SC-FDMA) was chosen for the uplink. OFDMA is an attractive solution for the downlink, not only for its increased spectral efficiency but also for its low implementation complexity attributes. This becomes even more crucial with the use of MIMO and the inherent orthogonality in transmission that eliminates intracell interference. While OFDMA could have been an attractive multiple access scheme for the uplink, OFDM signals have a high peak-to-average power ratio (PAPR). A large PAPR requires a large back-off in the power amplifier and decreases the coverage area, resulting in poor cell edge performance. To address the PAPR issue and at the same time preserve the advantage of the "orthogonality" of OFDM, and therefore combine the benefits of OFDMA and single carrier transmission, an OFDMA scheme was chosen, called single carrier FDMA (SC-FDMA).

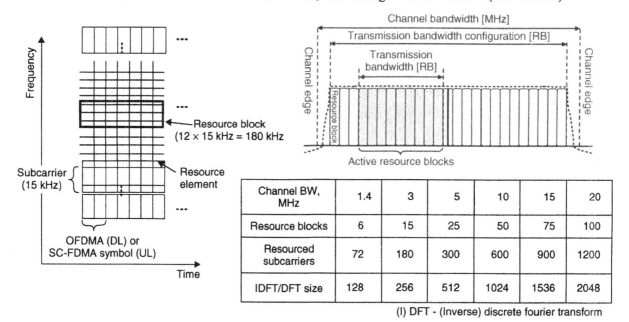

Channel BW, MHz	1.4	3	5	10	15	20
Resource blocks	6	15	25	50	75	100
Resourced subcarriers	72	180	300	600	900	1200
IDFT/DFT size	128	256	512	1024	1536	2048

(l) DFT - (Inverse) discrete fourier transform

Figure 1-28: LTE Transmission Channel Structure Formed of Resource Blocks

The resource block structure includes 12 subcarriers each, as shown in Figure 1-28, thus providing a great deal of flexibility and efficiency, and therefore both high throughput and support of variety of channel bandwidth (i.e., a large link budget for a variety of data streams, agnostic to channel size, etc.). Note that the discrete Fourier transform (DFT) size is chosen to be greater than the total resourced subcarriers with the difference set to zero. See the 3GPP standards (Release 8 and beyond) for details.

Adaptive modulation and coding are used in the downlink to adapt user data rates and QoS requirements to the channel quality. Various coding schemes have been considered, such as duo-binary turbo, rate compatible/quasi cyclic low density parity check (RC/QCLDPC), concatenated zigzag LDPC, turbo single parity check (SPC) LDPC, and shortened turbo codes through insertion of temporary bits. These all vary in performance merits for different block sizes and complexity assumptions.

To address the interference-limited scenarios resulting from universal frequency reuse, interference mitigation can be deployed in several forms:

- Inter-cell interference randomization, by deploying cell-specific scrambling, cell-specific interleaving, or random frequency hopping;

- Interference avoidance, by means of fractional frequency reuse – users are allocated among multiple classes according to their distance from the cell center, and different bandwidth allocation patterns are assigned to different user classes; and/or

- Interference cancellation by means of multiple antenna processing.

Multiple antenna techniques are considered in 3GPP LTE for increased data rates and improved link quality through the use of spatial multiplexing and diversity gains. As these schemes are deployed in combination with the packet scheduler, performance optimization needs to be studied, taking into account both spatial and multi-user diversity gains at a system level (multi-cell) sense, in order to include realistic interference modeling. In the single user MIMO case, for example in the per antenna rate control (PARC) scheme, spatial multiplexing is implemented by transmitting parallel data streams to a single user in order to improve the link rate, while in a multi-user MIMO case, multiple spatial streams are transmitted to different users to increase the system throughput. Space division multiple access (SDMA) can be realized through sectorization, multi-user beam forming, or multi-user MIMO precoding. Decisive factors in the realistic assessment of candidate MIMO approaches are the system level performance under realistic channel and interference modeling assumptions and the required channel state information, overhead signaling, and pilot design constraints.

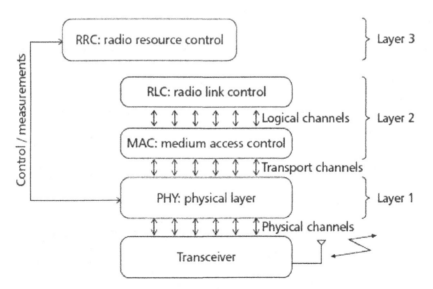

Figure 1-29: A Model of the Layers in LTE and a Schematic View of Channel Mappings

In summary, LTE is a broadband wireless access technology (with a flexible set of capabilities) defined by 3GPP in Release 8 (2008), and enhanced in subsequent releases, to enable high data rates, high capacity, and reduced latency, among other features, with significant enhancements in spectral and resource efficiencies. It supports both TDD and FDD schemes, and enhanced broadcast/multicast, with transmission bandwidths ranging from 1.25 MHz to 20 MHz. The layer mapping is shown in Figure 1-29. OFDMA is used in the downlink and single-carrier FDMA in the uplink. In addition, such elements as sophisticated antenna systems (e.g., 2×2, 2×4, etc. MIMO) and coding/modulation schemes (e.g., QPSK, 16QAM, 64QAM) will allow an evolution and/or variety of LTE (product) capabilities. LTE is expected to support service continuity and hand-off across a variety of 3GPP and non-3GPP access technologies.

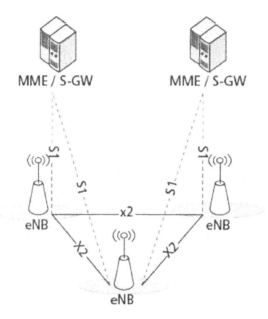

Figure 1-30: LTE/EPC End-to-End Simplified View

LTE is defined as part of an end-to-end all-IP network which can be simplified into two nodes, the LTE access functions within an enhanced Node B (eNB) and the Evolved Packet Core (EPC), as shown in Figure 1-30. The EPC includes a bearer gateway (with a Serving GW facing the eNBs) and a Mobility Management Entity (MME) for control and management functions. Key interfaces defined in the standards are those between eNBs (labeled X2) and between an eNB and a MME (labeled S1).

1.6   3GPP2 Radio Access Standards Evolution

Figure 1-31 shows a high-level view of 3GPP2 radio access technology evolution. Some details on cdmaOne to cdma2000 evolution (and beyond) are provided in this section.

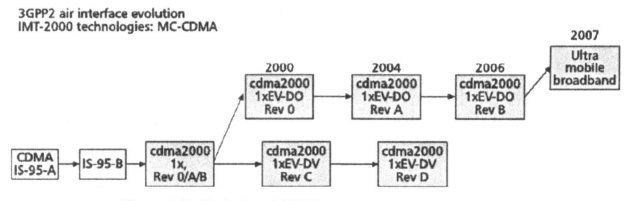

Figure 1-31: Evolution of 3GPP2 Access Technology Standards

1.6.1 cdmaOne and cdma2000-1X

cdmaOne is the first commercial CDMA network based on ANSI-95 standards. cdma2000-1X doubled the spectral efficiency as shown in Figure 1-32. The spectral efficiency here is defined by the sector capacity (in Mb/s) for both the forward link (base-to-mobile station) and the reverse link (mobile-to-base station) divided by the total required bandwidth (MHz). The spectral efficiency was significantly enhanced by definition of the evolved version (data-optimized) cdma2000-1xEV-DO Rev 0, and subsequently Rev A.

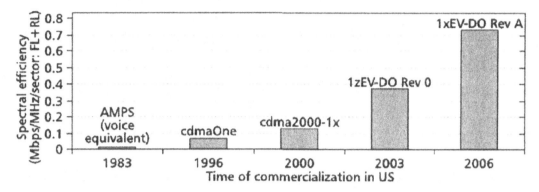

Figure 1-32: Evolution of Spectral Efficiency in 3GPP2

A cdmaOne network has pilot, sync, and paging control channels on the forward link and an access control channel on the reverse link. These control channels support forward- and reverse-link traffic channels. The forward-link channels are orthogonally covered by a Walsh function and then spread by a quadrature pair of pseudo-random noise (PN) sequences at a fixed chip rate of 1.2288 Mchips per second. Note that in the spread spectrum concept, the high rate spreading signal is defined by "chips", distinct from the information symbols. After the spreading operation, the in-phase (I) and quadrature (Q) signals are applied to the inputs of the I and Q baseband filter. The (QPSK) modulated signals are then sent over the air as shown in Figure 1-33.

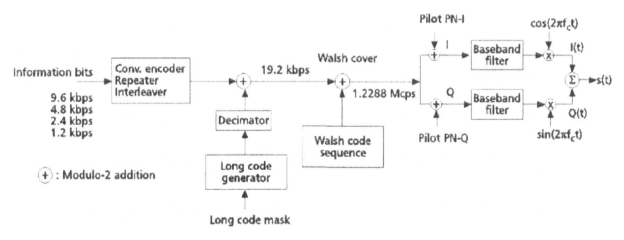

Figure 1-33: Block Diagram of Basic cdmaOne Forward-Link Transmission

The pilot channel is an unmodulated all-zero signal transmitted continuously by each CDMA base station to provide mobile stations with timing for initial system acquisition, phase reference for coherent demodulation, and signal strength for handoff decisions. The sync channel is a modulated spreading signal to transmit the sync channel message, which conveys key information to mobile stations, such as protocol revisions, system identification number (SID), system time, pilot PN sequence offset index, and others. The paging channel sends control information to CDMA mobile stations that have not been assigned to a traffic channel. Forward traffic channels are used by the base station to pass user data and signaling messages to mobile stations while on a call. Forward traffic channels can operate at different rates depending on the service options being supported.

When the mobile station is turned on, it will first complete pilot acquisition, and change its de-spreading to the sync channel to receive sync channel messages. Mobile stations use sync information to synchronize their system timing to the CDMA system time. Mobile stations then perform system registration initialization and begin to monitor the paging channel. Once a mobile station is paged or wants to access the network, the mobile station will switch to the traffic channel to start communication.

The reverse-link transmission is achieved by using 42-bit PN codes, referred to as long codes. Each subscriber uses a unique long code. The reverse-link transmission is illustrated in Figure 1-34.

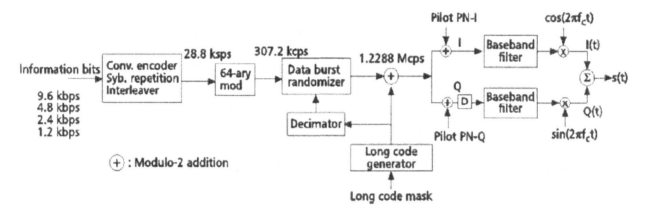

Figure 1-34: Block Diagram of Basic cdmaOne Reverse-Link Transmission

Access channels are used by mobile stations that are not engaged in a call to send messages to the base station. Typical messages sent include registration, call origination messages, and responses to requests or orders received from the base station. Reverse traffic channels are used by mobile stations to transmit user data and signaling information to the base station while on a call. The reverse traffic channels can operate at different rates depending on the service options being supported.

cdma2000-1x maintains the basic cdmaOne traffic channel structure and expands it to forward and reverse fundamental channels. It introduces forward and reverse supplemental channels, as well as associated forward and reverse dedicated control channels. Variable data rates in cdma2000 are achieved through variable spreading rates (factors) from 4 to 256 times instead of the fixed 128 times in cdmaOne. Both the forward and reverse link support up to a 153.6 kb/s physical link data rate. A reverse link pilot channel is added to support a coherent receiver on the reverse link. Powerful convolution and turbo codes are specified that double the voice capacity compared to cdmaOne, as well as supporting 144 kb/s packet data applications. The control channels are also upgraded to Enhanced Access Control Channels and Forward and Reverse Common Control Channels with higher rates and shorter frames. A quick paging channel and a broadcast channel are introduced to improve call setup time and battery life.

Power control is essential for a CDMA system to control interference and ensure reliable system operation. Link level radio power management consists of forward link power control, reverse link power control, and soft/softer handoff. The link level power management determines the sector capacity. The two-fold increase in capacity from cdmaOne to cdma2000-1x was largely due to the introduction of fast forward link power control. The forward link power control update rate increased from around 50 Hz in cdmaOne based on control messages, to a dedicated control channel at 800 Hz in cdma2000. Reverse link power management has both open-loop and closed-loop power control. The open loop allows a mobile station to set its transmit power based on the strength of the received power on the forward link, while closed loop supports power control to a desired reverse link traffic channel E_b/N_o, directed by the base station. These mechanisms ensure that both the mobile station and the base station transmit at the minimal required power level.

Soft and softer handoffs are also defined in cdma2000 to ensure make-before-break communication (as discussed earlier). Soft handoff refers to a handoff between two adjacent sectors belonging to different cell sites. Softer handoff is a handoff between two adjacent sectors of the same cell site. Although soft handoff improves the voice quality and handoff performance, it ties up multiple channel receivers, and decreases the capacity.

cdma2000-1x has also introduced various bearer service profiles to support voice-only, data-only, and mixed voice and data services over traffic channels. Medium access layer, link access layer, and upper layer protocols provide an efficient way to transmit bursty packet data traffic with differentiated QoS capability. New network elements were also introduced to address data mobility, IP address, packet data routing, authentication, authorization, and accounting as illustrated in Figure 1-35.

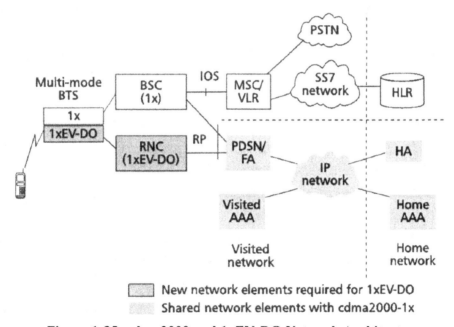

Figure 1-35: cdma2000 and 1xEV-DO Network Architecture

1.6.2 cdma2000-1xEV-DO (Rev 0 and Rev A)

cdma2000-1xEV-DO is designed for a high data rate, flexible latency, and high quality applications. The 1xEV-DO high-speed packet data network is an integral part of the cdma2000 family of standards. It has the same spectrum bandwidth as that of cdma2000-1x and is spread by the same PN sequences as cdma2000-1x. The hybrid mode supports a smooth and seamless handoff between 1xEV-DO and cdma2000-1x. An integrated 1xEV-DO and cdma2000-1x network architecture is shown in Figure 1-35.

This architecture provides a spectrally efficient means for both delay-sensitive (e.g., conversational) and delay-tolerant (e.g., data) services. The integrated network can share packet data network elements, such as the Packet Data Serving Node (PDSN), Foreign Agent (FA), Authentication, Authorization and Accounting (AAA), and Home Agent (HA). Only a Radio Network Controller (RNC) and dedicated 1xEV-DO base station modem cards are needed to upgrade a deployed cdma2000-1x network.

A cdma2000-1xEV-DO Rev 0 network is dedicated to packet data traffic, and is optimized for maximum data throughput. The system takes advantage of the burst- and delay-tolerant nature of data to increase the throughput efficiency. High spectral efficiency in 1xEV-DO is achieved on a shared full power and full code forward link fat pipe through Adaptive Modulation and Coding (AMC), fast scheduler, and HARQ.

The forward traffic channel is a common shared channel for all access terminals and their control signals. The modulation and coding are adaptive to the channel carrier-to-interference ratio (CIR). Accordingly, QPSK, 8-PSK, and 16 QAM modulations with various turbo coding are used to achieve data rate ranges from 38.4 kb/s to 2.457 Mb/s in a 1.25 MHz channel as shown in Table 1-3. The transmission data rate to each terminal is adapted to the RF channel condition and decided by the terminal. Therefore, users under good RF conditions will transmit at higher data rates than users in poor RF conditions. The Data Rate Control (DRC) channel on the reverse link provides fast feedback of the data rate and the best serving sector at a 1.667 ms slot interval. Fast cell selection supports virtual soft hand-off. Only the best serving sector is transmitting to the terminal at a time, reducing interference and hardware resource consumption.

Data Rate (kb/s)	Number of Slots	Code Rate	Modulation	CIR (dB)
38.4	16	1/5	QPSK	-11.5
76.8	8	1/5	QPSK	-9.7
153.6	4	1/5	QPSK	-6.8
307.2	2	1/5	QPSK	-3.9
614.4	1	1/3	QPSK	-3.8
307.2	4	1/3	QPSK	-0.6
614.4	2	1/3	QPSK	-0.8
1228.8	1	1/3	QPSK	1.8
921.6	2	1/3	8-PSK	3.7
1843.2	1	1/3	8-PSK	3.8
1228.8	2	1/3	16-QAM	7.5
2457.6	1	1/3	16-QAM	9.7

Table 1-3: Modulation and Coding Schemes - 1xEV-DO Rev 0

The fast scheduler is enabled by channel state information feedback from the DRC channel as well. The scheduler takes advantage of rapidly changing fading channel characteristics, and serves users when they have better than average channel CIR. This achieves multiple user diversity gain. The transmission time and duration are decided by the base station based on both the history of the transmission and the usage pattern. This adaptive data scheduler can support fairness and differentiation among users.

cdma2000 1xEV-DO Rev A is designed for versatile multimedia services, both delay-sensitive and delay-tolerant applications. Major improvements over EV-DO Rev 0 are increased data rates, improved reverse link, and Multiple Flow Packet Application to support delay-sensitive applications with more symmetric data rate requirement, such as voice and video telephony.

EV-DO Rev A increases the forward link data rate to 3.072 Mb/s by increasing the physical layer packet size from 4096 bits at 2.457 Mb/s in Table 1-3 to 5120 bits per slot. Reverse link data rates are expanded from 4.8 kb/s to 1.843 Mb/s. Higher data rates and better packing efficiency improve the forward and reverse link spectral efficiency, and enable VoIP capability, among other enhancements.

EV-DO Rev A reverse link performance is improved through HARQ and two transmission modes, as compared to EV-DO Rev 0, which has a similar reverse link structure as cdma2000-1x. HARQ achieves higher effective data throughput by earlier termination of multiple sub-packet transmissions. HARQ exploits power control imperfections due to radio channel and interference variations by using a-posteriori channel feedback. Two transmission modes are High Capacity (HiCap) and Low Latency (LoLat) modes. HiCap has a termination target of four sub-packets, while LoLat has a termination target of two to three sub-packets. HiCap achieves higher sector capacity gain with lower required bit-energy-to-noise-density ratio. LoLat minimizes transmission delay and achieves a lower packet error rate.

A multiple flow packet application supports spectral efficiency and latency trade-offs for each application flow. A service provider can deliver various multimedia applications with distinctive flow characteristics for each Radio Link Protocol (RLP) flow over EV-DO Rev A. EV-DO Rev A base stations manage radio resources to meet user throughput, latency, and packet loss requirements and ensure end-to-end QoS. Forward-link scheduler, reverse-link traffic power and medium-access control, call-admission control, backhaul QoS support and overload control, together form an integrated and effective resource management system.

Forward link QoS is managed by a RF scheduler that has a queue for each RLP flow. It assigns the packets in accordance with the needs of each flow QoS profile to perform Expedited Forwarding (EF), Assured Forwarding (AF), and Best Effort (BE) resource allocation. Voice flow receives the highest delay- and jitter-sensitive EF priority to ensure voice quality. Background file download is allocated BE.

Reverse link QoS is controlled by a token bucket scheme to ensure that average and peak resources utilized by each reverse traffic channel are less than or equal to the limitations imposed by the RAN. The limitations are specified by Traffic-to-Pilot ratios (T2P). LoLat has higher T2P ratios for the first two to three sub-packets. HiCap has the same T2P ratios for all 4 sub-packets as illustrated in Figure 1-36.

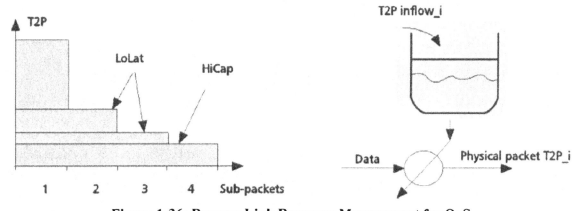

Figure 1-36: Reverse Link Resource Management for QoS

EV-DO Rev A has improved physical, MAC, and higher layer network efficiency and latency performance for multimedia services. The spectral efficiency for packet data applications has achieved 0.72 Mb/s/MHz/sector. EV-DO can also support voice user capacity of 35 Erlangs using a 1.25 MHz bandwidth, and it can also support a wide range of multimedia services.

1.6.3   Ultra Mobile Broadband

The 3rd Generation Partnership Project 2 (3GPP2) standards body engaged in a similar activity to the 3GPP LTE effort, aiming at evolving the cdma2000 high-rate packet data (HRPD) 3G system into the next generation. EV-DO Rev C standard, now known as Ultra Mobile Broadband (UMB), was specified and released in 2007. It is an OFDMA solution, using sophisticated control and signaling mechanisms, radio resource management (RRM), adaptive interference management, and advanced antenna techniques, such as MIMO, SDMA (Space Division Multiple Access), and beam-forming.

UMB aims to address a large cross-section of advanced mobile broadband services by economically delivering voice and multimedia. UMB supports inter-technology hand-offs and seamless operation with existing cdma2000 systems.

The basic target requirements for UMB are:

- High-Speed Data – peak rates of 288 Mb/s (with 2x20 MHz FDD, 4x4 MIMO) download and 75 Mb/s (with 2x20 MHz FDD, 1x2 MIMO) upload;

- Increased Data Capacity – ability to deliver both high-capacity voice and broadband data in all environments - fixed, pedestrian, and fully-mobile in excess of 300 km/h;

- Low Latency – an average latency of 14.3 ms over-the-air to support Voice over IP (VoIP), push-to-talk, and other delay-sensitive applications with minimal jitter;

- Increased VoIP Capacity – up to 1000 simultaneous VoIP users within a single sector, 20 MHz of bandwidth in a mobile environment without degrading concurrent data throughput capacity;

- Coverage – large wide-area network coverage areas equivalent to existing cellular networks, with either ubiquitous coverage for seamless roaming or non-contiguous coverage for hot zone applications;

- Mobility – robust mobility support with seamless handoffs;

- Converged Access Network – an advanced IP-based RAN architecture being developed by 3GPP2 to support multiple access technologies and advanced network capabilities, such as enhanced QoS, with fewer network nodes and lower latencies;

- Multicasting – support for high-speed multicast of rich multimedia content; and

- Transmission Bandwidth – deployable in flexible bandwidth allocations between 1.25 MHz and 20 MHz.

The UMB air interface supports both TDD and FDD modes and several coding schemes, including turbo and LDPC (optional). OFDMA is used for uplink and downlink traffic, while for uplink control channel transmission both OFDMA and CDMA are used. Orthogonal access schemes (such as OFDMA) eliminate intracell interference but require centralized assignment of resources to users, thus requiring additional signaling and possibly adding to the overall latency. On the other hand, non-orthogonal access schemes can take advantage of "soft capacity" and do not require explicit signaling for resource allocation. However, they involve more complex receivers and control methods to achieve a similar link performance as orthogonal schemes. For this reason, for the UMB uplink control channel (which is by nature bursty, delay-sensitive, and with small payloads), a hybrid scheme is used, composed of both OFDMA and a non-orthogonal multi-carrier CDMA (MC-CDMA) component. Both uplink and downlink support a variety of modulation mechanisms.

The UMB downlink supports two MIMO schemes.

- Single codeword (SCW) MIMO with closed loop rate and rank adaptation: A single packet is encoded and sent on a number of spatial beams for rate adaptation. The initial UMB design will be based on SCW MIMO.

- Multi-codeword (MCW) MIMO with per-layer rate adaptation: Multiple packets are encoded separately before spatial multiplexing.

UMB also supports adaptive interference management for increased cell-edge data rates. Furthermore, flexible resource allocation is optimized for low overhead signaling through the classification of assignments as "persistent" or "non-persistent". Persistent assignments eliminate bandwidth request latency for delay-sensitive applications, whereas non-persistent assignments are best suited for best-effort traffic. Seamless handoffs and advanced QoS mechanisms enable enhanced performance and support for real time services.

1.7 IEEE and Other Wireless Access Technologies

Although some access technologies compete with each other as optimal choices to serve an application, given the other considerations, their complementary nature must be appreciated. This complementary nature across user domains and sensitivity to the application and user requirements, within a cooperative and dynamic communication environment, is gaining increasing momentum and significance.

1.7.1 Wireless Local Area Network (WLAN)

The WLAN protocol is defined by the IEEE 802.11 standard which was originally published in 1997 and later updated with a major revision in 2007. There are various versions of WLAN as denoted by the letter next to 802.11, all of which are amendments to the original IEEE 802.11 standard. Within the IEEE 802.11 family, the following standards and amendments exist.

- IEEE 802.11 – The WLAN standard, originally 1 Mb/s and 2 Mb/s, 2.4 GHz RF and infrared (IR) (1997). All others below are Amendments to this standard. Recommended Practices 802.11F and 802.11T were withdrawn/cancelled.

- IEEE 802.11a – 54 Mb/s, 5 GHz standard (1999)

- IEEE 802.11b – Enhancements to 802.11 to support 5.5 and 11 Mb/s (1999)

- IEEE 802.11c – Bridge operation procedures; included in IEEE 802.1D (2001)

- IEEE 802.11d – International (country-to-country) roaming extensions (2001)

- IEEE 802.11e – Enhancements: QoS, including packet bursting (2005)

- IEEE 802.11g – 54 Mb/s, 2.4 GHz standard (backwards compatible with b) (2003)
- IEEE 802.11h – Spectrum Managed 802.11a (5 GHz) for European compatibility (2004)
- IEEE 802.11i – Enhanced security (2004)
- IEEE 802.11j – Extensions for Japan (2004)
- IEEE 802.11-2007 – New release includes amendments a, b, d, e, g, h, i, and j. (2007)
- IEEE 802.11k – Radio resource measurement enhancements (2008)
- IEEE 802.11n – Higher throughput improvements using MIMO antennas (2009)
- IEEE 802.11p – WAVE—Wireless Access for the Vehicular Environment (2010)
- IEEE 802.11r – Fast BSS transition (FT) (2008)
- IEEE 802.11s – Mesh Networking, Extended Service Set (ESS) (2011)
- IEEE 802.11u – Interworking with non-802 networks (2011)
- IEEE 802.11v – Wireless network management (2011)
- IEEE 802.11w – Protected Management Frames (2009)
- IEEE 802.11y – 3650–3700 MHz Operation in the U.S. (2008)
- IEEE 802.11z – Extensions to Direct Link Setup (DLS) (2010)
- IEEE 802.11mb – Maintenance of the standard; will become 802.11-2011 (2011)
- IEEE 802.11aa – Robust streaming of Audio Video Transport Streams (anticipated 2012)
- IEEE 802.11ac – Very High Throughput <6 GHz; potential improvements over 802.11n: better modulation scheme (expected ~10% throughput increase); wider channels (80 or even 160 MHz), multi user MIMO (anticipated 2012)
- IEEE 802.11ad – Very High Throughput 60 GHz (anticipated 2012)
- IEEE 802.11ae – QoS Management (anticipated 2011)
- IEEE 802.11af – TV Whitespace (anticipated 2012)
- IEEE 802.11ah – Sub 1GHz (anticipated 2013)
- IEEE 802.11ai – Fast Initial Link Setup

Today's commercial WLAN products are mostly based on the 802.11a, b, g, and n protocols. While 802.11a was the first wireless networking standard, the first standard broadly implemented was 802.11b, followed by 802.11g. IEEE 802.11n (next-generation) has the objective of speeds above 100 Mb/s using such advanced technologies as MIMO, and with flexibility in the choice of modulation scheme and operational frequency (2.4 or 5 GHz). Standards c–f, h, and j are service amendments and extensions or corrections to previous specifications. 802.11i was introduced to fix the security weaknesses of the WEP (Wireless Encryption Protocol) that came with the original 802.11 standard.

1.7.1.1 Protocol description

As with other 802.x protocols, 802.11 deals with the data link and physical layers of the OSI (Open System Interconnection) protocol stack. The data link layer functions are covered by the MAC (Media Access Control) protocol of the 802.11 standard. Figure 1-37 shows the relevance of 802.11 in the OSI protocol stack.

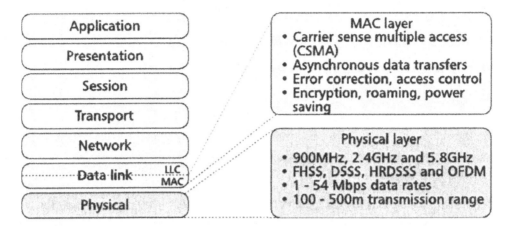

Figure 1-37: The IEEE 802.11 Protocol Stack

1.7.1.2 MAC layer

Following are some of the major functions of the MAC layer.

- It utilizes CSMA/CA (Carrier Sense Multiple Access with Collision Avoidance) as its basic access mechanism. The main aspect of this method is that it does not transmit if the medium is thought to be occupied by other users, but only transmits when the medium is believed to be available for communication. Upon sensing that the medium is free for a specified time called DIFS (Distributed Inter Frame Space), the transmitting station is allowed to transmit. The receiving station will then check the CRC (Cyclic Redundancy Check) of the received packet and send an ACK (Acknowledgement). The receipt of an ACK by the sender station will ensure that no collision has taken place. If the sender does not receive any ACK within a given time frame, a retransmission will occur. The packet will be discarded by the sender if the number of retransmissions reaches a set maximum limit.

- The 802.11 standard provides a mechanism for "virtual carrier sense" with an acknowledgement-based signaling protocol. A station willing to transmit a packet first sends a short packet called a Request To Send (RTS) which will contain the source, destination, and duration of the following transaction. If the receiving station is free, it will send a response with another short packet called a Clear To Send (CTS) that denotes the sender can now transmit its message. All stations receiving either the RTS or the CTS will set their Network Allocation Vector (NAV) for the given duration and use this information (along with the physical carrier sense) to sense the medium.

- A simple send-and-wait algorithm is used to address the fragmentation and reassembly functions. The transmitting station is not allowed to send any new fragment until it receives an ACK for the sent fragment or it decides that the fragment was retransmitted too many times.

- 802.11 uses exponential back-off algorithm to resolve contention between multiple stations attempting to access the medium at the same time. This method requires each station to wait for a random number of slots before reattempting to access the medium.

1.7.1.3 Physical layer

Following are some of the highlights of the 802.11 Physical Layer.

- An 802.11 protocol is allowed to operate in the unlicensed 915 MHz, 2.4 GHz, and 5 GHz Industry, Science and Medical (ISM) bands.

- The 802.11a protocol operates on 5 GHz Unlicensed National Information Infrastructure (U-NII) band with 12 non-overlapping channels. It uses a 52 subcarrier OFDM-based modulation.

- Both the 802.11b and 802.11g protocols use the 2.4 GHz ISM spectrum with three non-overlapping channels. Since ISM spectrum is free, 802.11b and 11g equipment may occasionally suffer interference from other household devices such as microwave ovens and cordless telephones.

- The 802.11b protocol uses Direct Sequence Spread Spectrum (DSSS) modulation while 802.11g uses an OFDM-based modulation.

- The 802.11n protocol builds on previous 802.11 standards by adding MIMO to the physical layer. MIMO uses multiple transmitter and receiver antennas to improve the system performance. 802.11n operates in both the 2.4 GHz and 5 GHz spectrum.

Table 1-4, with data taken from Wikipedia [Wik11], briefly compares physical layer characteristics.

Protocol	Release Date	Operating Frequency (GHz)	Realistic Throughput (Mb/s)	Maximum Data Rate (Mb/s)	Modulation Technique	Range Radius (Meters)	
						Indoor	Outdoor
Legacy	1997	2.4	0.9	2		~20	~100
802.11a	1999	5	23	54	OFDM	~35	~120
802.11b	1999	2.4	4.3	11	DSSS	~38	~140
802.11g	2003	2.4	19	54	OFDM	~38	~140
802.11n	2009	2.4 and 5	74	248	DSSS or OFDM	~70	~250

Table 1-4: Physical Layer Characteristics of Various 802.11 Protocols

1.7.1.4 Network architecture

Figure 1-38 shows a typical reference architecture of an 802.11-based WLAN. The network is subdivided into regions called Basic Service Sets (BSS) which are controlled by a radio base station called the Access Point (AP). The boundary of each BSS is defined by the range of the AP for providing wireless connectivity to the MSs within that BSS. The interconnection between multiple BSSs is typically handled by an Ethernet-based backbone network. However, the BSSs can be connected via a wireless backbone network as well. The interconnected BSS area is called an Extended Service Set (ESS) in the standard.

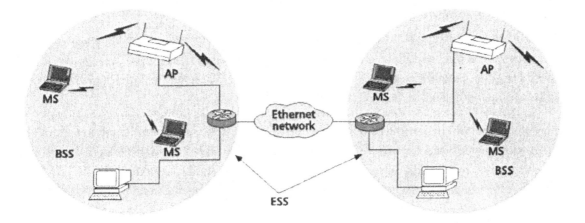

Figure 1-38: IEEE 802.11 WLAN Network Reference Model

A network architecture where the MSs directly talk to each other and are not connected to a wired network (e.g., the Internet and/or a private IP-based network) is called an Independent BSS (IBSS) as shown in Figure 1-39. These MSs can only talk to each other and thereby build their own LAN (Local Area Network).

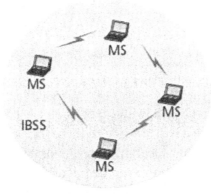

Figure 1-39: An IEEE 802.11 Ad-Hoc Network

In an IBSS, a MS can only communicate with another MS if they are in direct radio contact with each other. There is no central node (e.g., an AP) to which all the MSs can connect, and the connections between the MSs are established on an as-needed basis. In such a case, the IBSS is also called an Ad-hoc Network, which is usually small and short-lived (relative to a BSS and/or an ESS). For example, in a conference setting, the presenter and/or the participants could exchange their files without having to log onto the conference center's private network.

IEEE 802.11 has had a great and growing success. The emerging and future wireless access paradigms include a heteregenous environment of multiple technologies and cells, with WiFi playing a significant role.

1.7.2 Wireless Personal Area Networks (WPAN)

Wireless Personal-Area Networks (WPANs) enable wireless connectivity and communications between devices or sensors within a few meters to tens of meters of each other. They typically operate at low power and have a range of capabilities from low to high data rates. Examples include the Infrared Data Association (IrDA), Bluetooth, Zigbee, and Ultra-wideband (UWB).

The sub-working groups within IEEE 802.15 cover a range of capabilities, from simple and low-rate (e.g., Zigbee) interaction of devices and sensors (802.15.4), to high data rate definition (802.15.3) for such applications as short range multimedia imaging or vehicular communication systems. The 802.15.1 has been a cooperative effort with the Bluetooth Special Interest Group (Bluetooth SIG, formed in 1998, www.bluetooth.org). Many devices, particularly mobile user terminals, are equipped with Bluetooth.

The IEEE 802.15 WPAN task groups are listed in Table 1-5 [IEE11].

IEEE 802.15 WPAN Task Group	Scope
802.15.1	WPAN / Bluetooth
802.15.2	Co-existence
802.15.3	High-Rate WPAN
802.15.4	Low-Rate WPAN
802.15.5	Mesh Networking
802.15.6	Body Area Networks (BAN)

Table 1-5: IEEE 802.15 WPAN Task Groups

UWB is a technology that uses a large bandwidth, potentially shared with other users, to transmit information at low power. It is intended to provide a range of applications, notably high-data-rate personal-area networking (PAN). While technical, spectrum, and regulatory concerns have been debated, and the question of co-existence is expected to be further addressed or coordinated, UWB has received a flurry of attention, research, and innovation, as a technology with a great deal of promise.

The parameters of importance in wireless PAN naturally include such attributes as transmit power, range, data rate, spectrum, co-existence/interference, configuration/set-up, pairing, and mesh networking.

1.7.3 WiMAX

WiMAX is a broadband wireless technology that is designed to support both delay-sensitive (conversational or interactive) and delay-tolerant (non-real-time) applications. It has made significant advancements as a candidate for broadband wireless applications.

The Mobile WiMAX air interface is based on the OFDMA protocol as specified in IEEE 802.16-2009 [IEE09]. The mobility enhancements of mobile amendment IEEE 802.16e provide improved multi-path performance in non-line-of-sight environments and support for scalable channel bandwidth operation from 1.25 to 20 MHz (as outlined by the WiMAX Forum in February 2006 [WiM06a, WiM06b]).

The Mobile Technical Group (MTG) in the WiMAX Forum has defined several mobile WiMAX profiles that describe various capacity-optimized and/or coverage-optimized configurations. Initial mobile WiMAX profiles, however, will cover the 5, 7, 8.75, and 10 MHz channel bandwidths for licensed worldwide spectrum allocations in the 2.3 GHz, 2.5 GHz, and 3.5 GHz frequency bands.

The Network Working Group (NWG) of the WiMAX forum defined the higher-level networking specifications for mobile WiMAX systems beyond the IEEE 802.16 standard. The combined effort of IEEE 802.16 (http://www.ieee802.org/16) and the WiMAX Forum (http://www.wimaxforum.org) has helped define the end-to-end system solution.

- Mobile WiMAX supports target peak data rates of up to 63 Mb/s on the downlink and up to 28 Mb/s on the uplink in a 10 MHz channel. This requires the use of MIMO antennas, flexible sub-channelization schemes, efficient MAC frames, and enhanced Modulation and Coding Schemes (MCS).

- The dynamic allocation of RAN resources based on service flows is another advanced feature of mobile WiMAX. Service flows enable the RAN to utilize its resources efficiently based on such QoS requirements as low latency.

- Mobile WiMAX supports disparate allocations of spectrum based on particular requirements and is designed to scale within a variety of channelization schemes between 1.25 MHz and 20 MHz.

- Mobile WiMAX provides a security solution by implementing Extensible Authentication Protocol (EAP)-based authentication, Advanced Encryption Standard – CTR mode with CBC-MAC (AES-CCM)-based authenticated encryption, and other message authentication codes. (See Chapter 3 for detailed information on security codes and protocols.)

- Mobile WiMAX supports Subscriber Identity Module/Universal Subscriber Identity Module (SIM/USIM) cards, smart cards, digital certificates, and username/password associated with relevant EAP methods for credential management.

- Since fast handoff is a critical requirement for supporting delay-sensitive applications, mobile WiMAX has an optimized handoff scheme with a mute time of 50 milliseconds.

1.7.3.1 Network architecture

Figure 1-40 shows the network reference architecture for mobile WiMAX. It depicts the key normative reference points R1-R5. Each of the entities, MS, ASN (Access Service Network), and CSN (Connectivity Service Network) represent a grouping of functional entities; each of these functions may be realized in a single physical device or may be distributed over multiple devices. The ASN consists of all the radio components within the network access provider domain while the CSN consists of core network components (e.g., AAA server, user database, inter-working gateway, etc.). As shown, both the visited and home NSPs (Network Service Providers) have been accommodated in the network architecture via the R5 interface, while R2 defines a specified interface between a NSP and user terminal (subscriber or mobile station). While the ASP (Application Service Provider) or the Internet is shown connected to the CSN, the interface between them is left open at this time and is assumed to be implementation dependent.

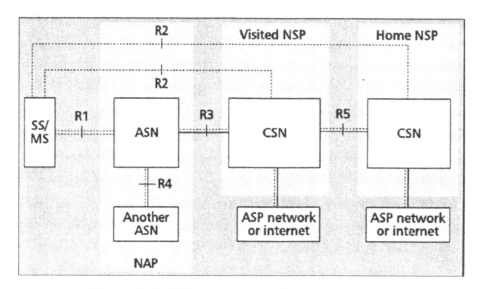

Figure 1-40: WiMAX Network Reference Model

1.7.3.2 Protocol description

The WiMAX physical layer (OFDMA PHY) supports sub-channelization in both the downlink and the uplink. The minimum frequency-time resource unit of sub-channelization is one slot, which is equal to 48 data tones (sub-carriers).

There are two types of sub-carrier permutations for sub-channelization; diversity and contiguous. The diversity permutation draws sub-carriers pseudo-randomly to form a sub-channel. It provides frequency diversity and inter-cell interference averaging. The contiguous permutation groups a block of contiguous sub-carriers to form a sub-channel. The permutation scheme enables multi-user diversity by choosing the sub-channel with the best frequency response.

In general, diversity sub-carrier permutations perform well in mobile applications while contiguous sub-carrier permutations are well suited for fixed, portable, or low mobility environments. These options enable the system designer to trade off mobility for throughput.

The MAC layer of IEEE 802.16 supports both busrty data traffic for bandwidth intensive applications and lower-bandwidth continuous traffic for delay-sensitive (conversational) applications. The resources allocated to the user device by the MAC scheduler can vary from a single time slot to the entire frame allowing a large range of throughput variation. Additionally, the MAC scheduler has the ability to allocate the resources ahead of sending each frame by including that information in the message before each frame.

QoS in the MAC layer of mobile WiMAX is implemented via service flow as shown in Figure 1-41. Service flow is a unidirectional flow of packets (identified by the term SF ID) that contains specific QoS parameters associated with that WiMAX "connection" (a unidirectional logical link between the peer MACs at the mobile and base station, identified by the term Connection ID or CID). The QoS parameters associated with the service flow include a traffic flow "classifier" and the "QoS category."

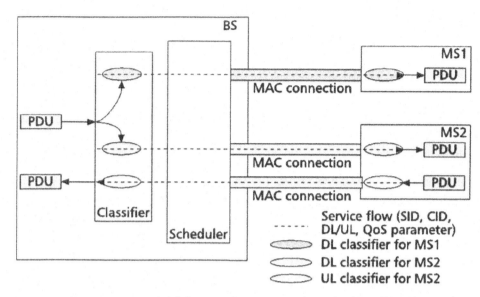

Figure 1-41: MAC Layer QoS Model for Mobile WiMAX

As shown, a protocol data unit (PDU, or basic message), whether in uplink (MS to BS) or downlink transmission, is sent to the classifier associated with the service flow, where it is translated into a specific scheduler priority according to its QoS category. The scheduler then transmits that PDU based on its priority in relation to the available resources at that specific instant.

Mobile WiMAX supports a wide range of data services and applications with varied QoS requirements. These are summarized in Table 1-6.

QoS Category	Proposed Applications	QoS Specifications
UGS Unsolicited Grant Service	VoIP and PTT	Maximum sustained rate Maximum latency tolerance Jitter tolerance
rtPS Real-Time Packet Service	Streaming audio or video	Minimum reserved rate Maximum sustained rate Maximum latency tolerance Traffic priority
ertPS Extended Real-Time Packet Service	Voice with activity detection (VoIP)	Minimum reserved rate Maximum sustained rate Maximum latency tolerance Jitter tolerance Traffic priority
nrtPS Non-Real-Time Packet Service	File Transfer Protocol (FTP)	Minimum reserved rate Maximum sustained rate Traffic priority
BE Best-Effort Service	Data transfer, web browsing, etc.	Maximum sustained rate Traffic priority

Table 1-6: Mobile WiMAX Quality of Service Categories

1.7.3.3 Scheduling

The mobile WiMAX MAC scheduling service provides optimized delivery of both delay-sensitive and delay-tolerant applications over broadband wireless access. The main aspects of this service are:

- Fast Data Scheduler – The MAC scheduler, located at each BS, allocates the available RF resources for bursty data traffic under time-sensitive channel conditions. Each PDU has well-defined QoS characteristics (associated with its SF) that allows the scheduler to provide over the air (OTA) transmission priority.

- Scheduling for the downlink and uplink – The scheduling service is provided in both directions. Bandwidth requests through a ranging channel, piggyback requests, and polling methods are used to support uplink bandwidth demands. Downlink resource allocation is performed by the downlink scheduler based on the service flow priority against the available resources.

- Dynamic Resource Allocation – The MAC scheduler allows frequency- and time-based resource allocation in both directions on a-per frame basis. A MAP message delivers the resource allocation information at the beginning of each frame, allowing for finer resource management per frame in response to the variations in traffic and/or channel conditions. Furthermore, each resource allocation can consist of one slot or the entire frame.

- QoS-Oriented Scheduling – The scheduler provides QoS-based data transport per MAC connection as indicated earlier. The scheduler can dynamically allocate/de-allocate RF resources in both the downlink and the uplink based on the QoS parameters associated with each connection.

- Frequency-Selective Scheduling – The MAC scheduler is capable of operating on various types of sub-channels. While using an Adaptive Modulation and Coding (AMC)-based continuous permutation method, the sub-channels may vary in their attenuations. The frequency-selective scheduling can allocate the strongest sub-channel available to a mobile user. Additionally, the frequency-selective method can increase system capacity with a slight increase in Channel Quality Indicator (CQI) overhead over the uplink.

1.7.3.4 Mobility management

Two important aspects of mobility management addressed by mobile WiMAX are power consumption and seamless handoff.

- Power Management – Mobile WiMAX supports two modes, sleep and idle, within a connection state, for efficient power consumption. Sleep mode is a connection state in which the MS turns off its communication with the BS for pre-negotiated periods of time. Idle mode, on the other hand, is a mechanism for the MS to be periodically available for downlink transmission without the re-registering with the BS. Idle mode allows the MS to be reachable in a highly mobile environment.

- Handoff – Three handoff methods are supported within the 802.16e standard – Hard Handoff (HHO), Fast Base Station Switching (FBSS), and Macro Diversity Handover (MDHO). Of these, HHO is mandatory while FBSS and MDHO are optional. The WiMAX Forum has developed several techniques for optimizing HHO within the 802.16e framework. These improvements have been developed with the goal of keeping Layer 2 handoff delays to less than 50 ms.

    o In HHO, the MS must go through the entire registration process when it changes the BS.

    o In FBSS, the MS maintains continuous RF communication with a group of BSs that are called its "active set". The MS is also assigned an "anchor" base station within its active set. The anchor base station manages the uplink/downlink signaling and traffic bearer. Transition from one anchor base station to another is performed based on the signal strength measured by the MS on its CQI channel without exchanging any explicit HO (Hand Off) signaling messages. An important requirement of FBSS is that the data destined for the MS are received by all base stations that are in its active set.

    o In MDHO, the signaling and bearer communications over the RF resources are performed between the MS and all of the BSs that are in its active set. This is done for both uplink and downlink transmission.

1.7.3.5 Security

Mobile WiMAX exploits advanced technologies to provide security. These include mutual device/user authentication, flexible key management protocol, strong traffic encryption, control and management plane message protection, and security protocol optimizations for fast handoffs. The reader is again referred to Chapter 3 for a detailed discussion of wireless security.

1.7.4 Broadcast/Multicast Distribution over Cellular

Broadcast/Multicast services such as mobile TV and video applications are increasingly common multimedia content applications enabled by evolving wireless systems. The distribution systems may be:

- Satellite ("S") and/or terrestrial ("T"), or
- Cellular network and/or overlay access.

These systems leverage sophisticated technologies for such functions as coding, modulation, error correction, media formatting and presentation, content protection and management, time-slicing, radio-networking, and data-casting, among others. Typically frequencies below 800 MHz have high capabilities for non-line-of-sight reach and indoor penetration for an application such as over-the-air TV, although higher frequencies are also used, e.g., for high-capacity direct access. An end-to-end block diagram indicating the wireless-access component is shown in Figure 1-42.

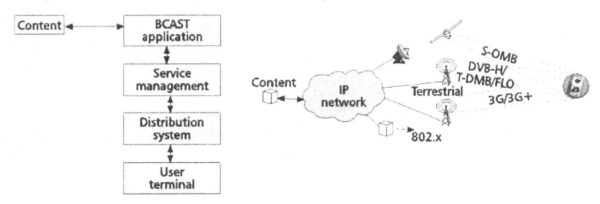

Figure 1-42: Simplified End-to-End Broadcast/Multicast System

The Open Mobile Alliance has worked to develop standards for service enablers, including unified broadcast/multicast service enabler standards, with adaptation to a number of distribution systems.

Examples of mobile broadcast/multicast distribution systems, integrated within or overlaid on top of the mobile cellular systems, include but are not limited to:

- MBMS, 3GPP Multimedia Broadcast/Multicast Service and its enhanced version eMBMS;
- BCMCS, 3GPP2 Broadcast and Multicast Services;DVB-H, Digital Video Broadcast – Handheld;
- DMB (Digital Multimedia Broadcasting;
- MediaFLO, Forward-Link Only Technology – Qualcomm;
- ATSC M/H; and others.

1.7.5 Femtocells

A femtocell may be viewed as a local (e.g., home) coverage area, serving a limited number of users, but as a somewhat independent extension of a wide-area cellular network. The fundamental design goal is for the cell to have as many features of the wide network as possible, with minimal impact on it (e.g., in terms of performance, capacity, efficiency, etc.).

The macro-cellular network features and attributes which may be supported (within assigned policy) include:

- Services at comparable quality;
- Radio interface(s), signaling;
- Mobile terminals (governed by authorization policy and admission control); and
- Mobility management, handoff, operational management, security, and others.

The overall architecture is intuitively simple. Each femtocell has a base station at low power and local (home) coverage serving authorized mobile devices, which may move in and out of the cell. The femtocell home network is connected through existing (IP) broadband connection to the Internet and a cellular network as shown in Figure 1-43.

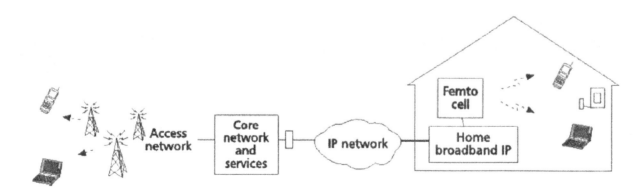

Figure 1-43: A Femtocell as an (Independent) Extension of a Wide-Area Network

The macro-cellular core network must be capable of supporting a large number of femto cells, again without noticeable compromise in efficiency, performance or security.

1.7.5.1 Femtocell specification in 3GPP

In 3GPP, femtocells are defined as customer-premises equipment that connects a 3GPP mobile device over a 3GPP air interface to a mobile operator's network using a broadband IP backhaul [3GP11c]. In addition, femtocells can be deployed by the end user using a consumer-grade backhaul connection. This is facilitated by a standardized management interface that allows the mobile operator to remotely configure all the necessary operating parameters.

3GPP started work on defining the specifications for femtocells starting with Release 8 for both UMTS and LTE. The work covers the network architecture, radio, security, testing, and network management aspects. 3GPP specifications for Release 8, Release 9, and Release 10 include femtocell functionality.

1.7.5.2 Femtocell architecture

The UMTS femtocell architecture (see Figure 1-44) includes the femtocell access point and femto gateway, respectively termed the Home Node B and Home Node B Gateway in 3GPP specifications. The interface between these two network elements is called the IuH.

The Home Node B contains the Node B and most of the RNC functions – the radio interface, radio resource management, medium access control (MAC), etc. The Home Node B Gateway aggregates the connections from multiple Home Node Bs to the core network and also presents the core network with an Iu-PS and/or Iu-CS interface to Packet Switched or Circuit Switched network, respectively.

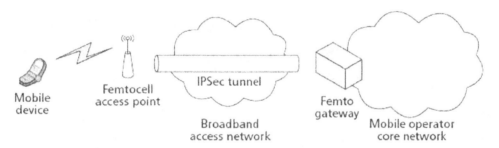

Figure 1-44: Generic View of Femtocell Architecture

An LTE femtocell has a similar architecture, the only differences being that the femto gateway is an optional element and the interface between the femtocell access point and the mobile operator core network is called S1.

1.7.5.3 Security aspects

Security is one of the most important aspects of femtocell architecture. As the femtocell access point itself is not within a secured mobile network operator's premises, it is more susceptible to additional security threats. In addition, the mobile network operator needs to ensure that any communication via the femtocell access point can be done in a secure manner. There is also the need to ensure that the mobile operator's network is protected from threats originating from a rogue femtocell access point.

To that end, 3GPP designed the interface between the femtocell access point and the mobile operator core network using an Internet Protocol Security (IPSec) tunnel that terminates at the Security Gateway in the mobile operator's core network (see Figure 1-45). All communication between the femtocell access point and the mobile operator core network is secured by this tunnel.

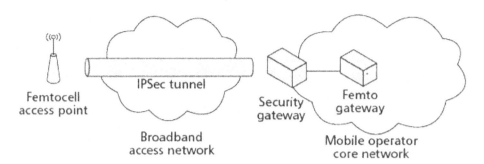

Figure 1-45: Interface between the Femtocell Access Point and the Mobile Network

The Security Gateway also performs mutual authentication of the femtocell access point, for example, when the femtocell access point initially connects with a mobile operator's core network.

1.7.5.4 Femtocell connectivity with the mobile operator core network

For a femtocell access point to provide services, it must first be connected to a mobile operator's core network. The first point of contact for the femtocell access point is the Security Gateway, where mutual authentication is performed. Once this process is complete, an IPSec tunnel is established between the femtocell access point and the Security Gateway. Then the femtocell access point is registered on the mobile operator's network. At this point, the femtocell access point will be configured with the appropriate radio parameters by the operator's management system.

The femtocell access point can also detect the configuration and parameter settings of the macro network directly by "listening" to the information broadcast by the macro network on the 3GPP air interface in order to set its own suitable radio parameters.

1.7.5.5 Access control

There are three modes of operation for a 3GPP femtocell:

- Open – access given to any subscriber allowed access to the mobile operator's network;

- Closed – access given to a defined set of subscribers, typically subscribers of the same mobile operator's network; and

- Hybrid – access given to any subscriber allowed access to the mobile operator's network; however, preferential treatment is given to a defined set of subscribers.

1.7.5.6 Mobility

For mobile devices in the idle state, the mobile device will scan for any available cells. When a femtocell is detected as the best cell, the mobile device will check to see if it is allowed access to that femtocell before selecting the femtocell to camp on. If the mobile device is not allowed access, then the mobile device will camp on the next best cell, if it is available, and so on.

For mobile devices in the connected state, when the device moves between macro network coverage and femtocell coverage, handover will be performed. In the case where the target cell is a femtocell, checks will be made prior to handover to ensure the mobile device is allowed access to the femtocell. For the case where the macro network and femtocell are both UMTS, this is a hard handover. Handover is also supported when the mobile device moves between different femtocells.

1.7.5.7 Network management

For femtocells, a standardized management interface is extremely important, as it allows for the remote configuration of radio parameters and the collection of performance counters and fault management information. The Broadband Forum's procedures have been used as the basis for 3GPP femtocell network management.

1.7.5.8 End user perspective

The main requirement for the end user is that the femtocell access point must allow zero-touch installation that needs no intervention by the end user, apart from physically plugging in the power cord and cable for IP connectivity.

1.7.6 Near-Field Communication and RFID

Radio-Frequency Identification (RFID) is a mechanism for reading data stored on tags over distances typically within centimeters to potentially tens of meters. There has been an increasing interest in RFID, particularly for enterprise applications such as inventory tracking. RFID tags may be passive, semi-passive, or active. Passive ones do not have a power source and are activated by the tag reader.

Near-Field Communication (NFC) allows wireless connectivity over a short range, typically within centimeters, enabling such applications as secure and reliable mobile payment and ticketing (in a crowd of users and in the presence of other applications). NFC extends the capabilities of RFID and is compatible with it. It can co-exist with WPAN, e.g., to configure or pair with it. Despite some overlap between the two, the distinctions between NFC and WPAN are self-evident in terms of attributes such as range, power, data rate, set-up time, topology, and applications. As with other access mechanisms, however, their complementary nature must be appreciated to be effectively leveraged as applicable.

It is worth noting that mobile (2D) barcodes are also finding a significant market in mobile commerce, among other applications, typically using optical (camera) scanning of information (e.g., a URL).

1.7.7 Satellite Systems

Satellite communication has advanced in the course of the last several decades. There is obviously a great deal of science, technology, engineering, and operations involved in the design and structure, launch and propulsion, orbit selection, power, weight, tracking, and control, among other aspects. From a communication perspective, similar concepts apply as have been discussed so far, and also as are covered in Chapter 4. These include transmission and reception concepts such as modulation, coding and detection, antenna technologies, operational frequencies, and receiver structures in terms of sensitivity, noise, filtering, amplification, equalization, figure of merit, etc.

There have of course been different systems intended for different satellite services, with Mobile Satellite Service (MSS) being a notable example. Wireless access technologies, as we have seen, can be any of a broad range of enablers including satellite communication or a hybrid satellite/terrestrial system. Not only are services such as connectivity, communication, and content increasingly and seamlessly made possible by a variety of mechanisms and potentially in a multi-technology environment, but the mobile standards we have discussed (such as LTE) may be deployed in multiple ways.

1.8   Wireless Access Technology Roadmap: Vision of the Future

1.8.1 *Motivation and Design Objectives*

Wireless technology evolution has attracted a great deal of interest in research, standardization, industry forum initiatives, innovations, and regulatory fora, in an effort to shape design goals and specify technical requirements to realize the vision and potential requirements of ubiquitous connectivity, communication, and content, to serve an increasingly mobile world society, as represented by Figure 1-46. In the same framework, a number of global and regional initiatives continue to define the roadmap and enable the future communication space.

Design goals for the systems beyond 3G/4G include:

- High data rate (user experience, speed) and capacity (traffic volume);

- Low latency (real-time and near-real-time), high performance, constant connectivity;

- Spectral efficiency;

- Cost/resource efficiency;

- Flexibility (e.g., resource allocation), adaptability (e.g., interference management), seamless mobility, scalability (e.g., spectrum);

- Enhanced broadcast/multicast and end-to-end QoS; and

- Co-existence and inter-working with other access technologies.

Figure 1-46: Enabling Connectivity, Communication, and Content on the Go

Some of these goals (such as seamless mobility and inter-working, end-to-end QoS, etc.) are also tied to the evolution of the core packet network, which in parallel is maturing towards a heterogeneous all-IP flat (or at least flatter) architecture. Core network evolution is the subject of Chapter 2.

Wireless access technologies beyond 3G leverage such sophisticated, efficient, and high-capacity technologies or mechanisms as OFDMA, MIMO, and advanced coding/high-level modulation (e.g., 16 or

64 QAM), among others, to meet their goals. There is a flurry of activities to define the requirements and technologies for the next 5-10 years and beyond.

1.8.2 Technology Roadmap Long-Term Initiatives
1.8.2.1 ITU IMT-Advanced

"IMT-Advanced" has been developed by the ITU in consideration of evolving marketplace needs for future wireless communications, along with the required technologies, spectrum, and other issues associated with the worldwide advancement of IMT-2000 systems. This evolving definition for next generation (fourth generation) systems has a baseline defined by ITU-R Recommendation M.1645 [ITU03]. This baseline was also taken into account at the World Radiocommunication Conference (WRC) in 2007 for the identification of additional frequency spectrum for mobile and wireless communications.

IMT-Advanced goes beyond IMT-2000 to provide advanced mobile multimedia services, with significant improvement in performance and user experience across both fixed and mobile networks. The key features of IMT-Advanced include:

- A high degree of commonality, flexibility, and efficiency;

- Superior user experience and user-friendly applications and devices;

- Inter-working of access technologies and global roaming; and

- Enhanced peak data rates to support advanced services and applications – envisioned peak aggregate user data rates of ~100 Mb/s for high user mobility and ~1 Gb/s for low user mobility (new nomadic/local-area wireless access).

The objectives of IMT-Advanced were addressed in 3GPP Release 10 and continue to be addressed in subsequent releases. Figure 1-47 shows the evolving capabilities of IMT-2000 and IMT-Advanced.

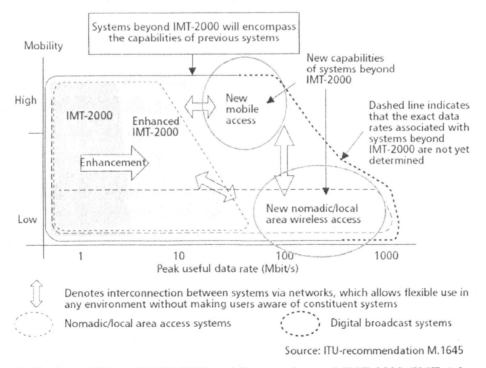

Figure 1-47: Capabilities of IMT-2000 and Systems beyond IMT-2000 (IMT-Advanced)

1.8.2.2 Next Generation Mobile Networks

The Next Generation Mobile Networks (NGMN) initiative consists of a group of mobile operators (Members), who along with other stakeholders in the mobile ecosystem collectively comprise the group of Partners. NGMN intends to provide a coherent vision for technology evolution beyond 3G (specifically an LTE/EPC roadmap) for the delivery of broadband wireless services in the decade beyond 2010.

The vision of NGMN is to "provide a platform for innovation by moving towards one integrated network for the seamless introduction of mobile broadband services. The initial objective of the alliance was the commercial launch of a new experience in mobile broadband communications and to ensure a long and successful cycle of investment, innovation and adoption of new and familiar services that would benefit all members of the mobile ecosystem" [NGM11]. In the phase beyond the introduction of LTE, NGMN continues to guide standardization and the overall ecosystem with respect to the requirements of future networks such as interoperability, converged operational excellence and network management, multi-mode multi-band (user device) technologies, spectrum and band requirements, TDD/FDD dual and seamless support, etc. A number of publications, including a white paper on technical achievements, provide more details.

1.8.2.3 Wireless World Research Forum

The Wireless World Research Forum (WWRF) is a global organization which was founded in August 2001. It has a large number of members from across the globe, representing all sectors of the mobile communications industry and the research community. The objective of the Forum is to formulate visions on strategic future research directions in the wireless field, among industry and academia, and to generate, identify, and promote research areas and technical trends for mobile and wireless system technologies.

Forum activities aim towards the creation of a shared global vision for the future of wireless, intended to drive research and standardization. In addition to influencing regional and national research programs, WWRF members contribute to the work done within the ITU, UMTS Forum, ETSI, 3GPP, 3GPP2, IETF, and other relevant bodies regarding commercial and standardization issues derived from the research work. As part of the effort to shape a common vision for the future of wireless, the WWRF has published a number of white papers, addressing the technical areas of the Working Groups and Special Interest Groups, as well as several generations of the WWRF "Book of Visions" or System Concept document [WWR11].

1.8.2.4 Other initiatives

A large number of research and innovation initiatives exist, particularly in the areas of future radio, future networks, future Internet, smart networks, machine-to-machine, etc., all with involvement from academia, industry, governments, regional commissions and bodies, and global forums.

1.8.3 Technology Trends and Final Words

Communication systems have fundamental attributes in terms of transmission, reception, channel performance, networking, session management, etc. Wireless systems in turn have similar considerations with respect to radio engineering, antennas, operational frequencies, and receiver sensitivity and performance, among others. There is variety, however, in terms of design goals (and constraints), enabling technologies, performance objectives, and applications. In mobile cellular communications, there is a great deal of consideration in terms of radio engineering, capacity and coverage, spectrum and band aspects, mobility and user management, networking, and other aspects. The enabling technologies are becoming more user-centric, more adaptive, reconfigurable, and dynamic, creating a certain level of abstraction such that the intention is served in a smart user space with context and proximity awareness, but somewhat transparent to the technology used. In this context, there is also an increasing trend towards multi-cell and heterogeneous networks.

There is a broad range of wireless access technologies: terrestrial, satellite-based or hybrid; fixed, nomadic, or mobile; point-to-point and line-of-sight or non-line-of-sight; mesh, relay, or ad hoc; and a possible combination of femto, personal, local, or wide-area networks. New mobile networks are already IP-centric and content-centric. This is a new era. In addition, the functions of different elements have evolved as shown in this chapter, including more intelligence moved to the edges. Scalability and flexibility in the design and deployment will be essential characteristics in order to achieve adaptability to varying conditions, related to the radio environment, the usage scenario, and the network conditions.

Personalization, adaptive and reconfigurable devices, ambient awareness, and multi-modal "natural" user interaction in smart user spaces with "enhanced" reality, among others, are trends for vision, research, and innovation. Such technologies as cognitive radio, self-organizing networks, and smart beamforming, coupled with advanced technologies in other areas, such as sensing, robotics, nano-engineering, and others, will enable this vision of future user-centric, context-aware, ubiquitous, and intention-based communication across all user spaces.

Finally, it must be noted that identification of design objectives and requirements, and definition of technology innovations and solutions, will not sufficiently succeed without adequate modeling and performance evaluation methodologies and systems. These concepts are not new but the environment in which wireless access technologies operate has evolved remarkably and thus demands new modeling and evaluation criteria, in the presence of new user interaction paradigms, dynamic user interfaces and access mechanisms, traffic growth and patterns, mixed services, interference sources, and time-varying communication parameters.

Wireless technologies co-exist with non-wireless ones (particularly optical fiber). Furthermore, a variety of wireless mechanisms and technologies potentially cooperate. Similar fundamental concepts apply, although differences exist, as discussed in this chapter, not only in different environments but in technology evolution and with new paradigms, towards an increasingly seamless and dynamic communication environment to serve and empower the mobile world.

1.9 References

[3GP98] 3rd Generation Partnership Project, Technical Specification Group Service and Systems Aspects, *Digital cellular telecommunication system (Phase 2+); General Packet Radio Service (GPRS); Service description; Stage 2*, Specification TS 03.60, V7.9.0, 1998.

[3GP99] 3rd Generation Partnership Project, Technical Specification Group Services, *Digital cellular telecommunications system (Phase 2+); General Packet Radio Service (GPRS); Service description, Stage 1*, Specification GSM 02.60, V8.1.0, 1999.

[3GP02] 3rd Generation Partnership Project, Technical Specification Group Service and Systems Aspects, *Architectural Requirements for Release 1999 (Release 1999)*, Specification TS 23.121, V3.6.0, 2002.

[3GP04] 3rd Generation Partnership Project, Technical Specification Group Radio Access Network, *High Speed Downlink Packet Access (HSDPA); Overall description; Stage 2 (Release 5)*, Specification TS25.308, V5.7.0, Release 5, 2004.

[3GP06] 3rd Generation Partnership Project, Technical Specification Group Radio Access Network, *FDD Enhanced Uplink; Overall description; Stage 2 (Release 6)*, Specification TS25.309, V6.6.0, 2006.

[3GP11a] 3rd Generation Partnership Project, Technical Specification Group Radio Access Network, *Radio Interface Protocol Architecture (Release 10)*, Specification TS 25.301, V10.0.0, 2011.

[3GP11b] 3rd Generation Partnership Project, Technical Specification Group Service and Systems Aspects, *General Universal Mobile Telecommunications System (UMTS) architecture (Release 10)*, Specification 23.101, V10.0.0, 2011.

[3GP11c] 3rd Generation Partnership Project, Technical Specification Group Service and Systems Aspects, *Service requirements for Home Node B (HNB) and Home eNode B (HeNB)*, Specification TS 22.220, V11.3.0, 2011.

[ANS01] American National Standards Institute, Telecommunications Industry Association, and Electronic Industries Alliance, *Cellular Radio-Telecommunications Intersystem Operations*, Standard ANSI/TIA/EIA-41-E, 2001.

[IEE09] Institute of Electrical and Electronics Engineers, Working Group on Broadband Wiureless Access Standards. *Air Interface for Fixed and Mobile Broadband Wireless Access System*, Standard 802.16-2009.

[IEE11] Institute of Electrical and Electronics Engineers, www.ieee802.org/15, IEEE 802.15 Working Group for WPAN.

[ITU03] International Telecommunication Union, Radiocommunication Sector, *Framework and overall objectives of the future development of IMT-2000 and systems beyond IMT-2000*, Recommendation ITU-R M.1645, 2003.

[ITU11] International Telecommunication Union, Radiocommunication Sector, *Detailed specifications of the terrestrial radio interfaces of International Mobile Telecommunications-2000 (IMT-2000)*, Recommendation ITU-R M.1457-10, 2011.

[NGM11] Next Generation Mobile Networks, http://www.ngmn.org, 2011.

[Wik11] Wikipedia, *IEEE 802.11*, http://en.wikipedia.org/wiki/IEEE_802.11, 2011.

[WiM06a] WiMAX Forum, *Mobile WiMAX – Part I: A Technical Overview and Performance Evaluation*, 2006

[WiM06b] WiMAX Forum, *Mobile WiMAX – Part II: Competitive Analysis*, 2006.

[WWR11] Wireless World Research Forum, http://www.wireless-world-research.org, 2011.

1.10 Suggested Further Reading

Although not specifically cited in this chapter, the following references are suggested by the authors as sources of useful background and/or detailed information.

3rd Generation Partnership Project, Technical Specification Group Radio Access Network, *Feasibility Study for Enhanced Uplink for UTRA FDD*, Technical Report TR25.896 V6.0.0, 2004.

3rd Generation Partnership Project, Technical Specification Group Radio Access Network, *Physical channels and mapping of transport channels onto physical channels (FDD) (Release 5)*, Specification TS25.211 V5.8.0, 2005.

3rd Generation Partnership Project, Technical Specification Group Radio Access Network, *Physical layer procedures (FDD) (Release 5)*, Specification TS25.214, V5.11.0, 2005.

3rd Generation Partnership Project, Technical Specification Group Radio Access Network, *Requirements for Evolved UTRA (E-UTRA) and Evolved UTRAN (E-UTRAN)*, Technical Report TR25.913 V7.3.0, 2006.

3rd Generation Partnership Project 2, Technical Specification Group C: Radio Access, *cdma2000 High Rate Packet Data Air Interface Specification*, Technical Report 3GPP2 C.S0024-A, Version 2.0, 2005.

A. Alexiou, et al., *Duplexing, Resource Allocation and Inter-Cell Coordination-Design Recommendations for Next Generation Systems*, Wireless Communications and Mobile Computing, Wiley Online Library, vol. 5, no. 1, 2005.

A. Alexiou and D. Falconer, *Challenges and Trends in the Design of a New Air Interface*, IEEE Vehicular Technology Magazine, vol. 1, no. 2, 2006.

C. Antón, P. Svedman, M. Bengtsson, A. Alexiou, and A. Gameiro, *Cross-Layer Scheduling for Multi-User MIMO Systems*, IEEE Communications Magazine, vol. 44, no. 9, 2006.

R. Bachl, P. Gunreben, S. Das, and S. Tatesh, *The Long Term Evolution Towards a New 3GPP Air Interface Standard*, Bell Labs Technical Journal, vol. 11, no. 4, 2007.

S. Das, et al., *EV-DO Revision C: Evolution of the cdma2000 Data Optimized System to Higher Spectral Efficiencies and Enhanced Services*, Bell Labs Technical Journal, vol. 11, no. 4, 2007.

G. Nair, J. Chou, T. Madejski, K. Perycz, P. Putzolu, and J. Sydir, *IEEE 802.16 Medium Access Control and Service Provisioning*, Intel Technology Journal, vol 08, no. 3, 2004.

R. Tafazolli (editor), *Technologies for the Wireless Future, Wireless World Research Forum (WWRF)*, Volume 1, John Wiley and Sons, 2005.

R. Tafazolli (editor), *Technologies for the Wireless Future, Wireless World Research Forum (WWRF)*, Volume 2, John Wiley and Sons, 2006.

Telecommunications Industry Association, *Mobile Station – Base Station Compatibility Standard for Wideband Spread Spectrum Cellular Systems*, Standard ANSI/TIA/EIA-95-B, 1999.

H. Yagoobi, *Scalable OFDMA Physical Layer in IEEE 802.16 Wireless MAN*, Intel Technology Journal, vol. 08, no. 3, 2004.

J. Yang, S-H. Shin, and W. C.Y. Lee, *Design Aspects and System Evaluations of IS-95 based CDMA System*, Proceedings of ICUPC, 1997.

J. Yang and J. Lin, *Optimization of Power Management in a CDMA Radio Network*, Proceedings of IEEE Vehicular Technology Conference, 2000.

J. Yang, *Performance and Deployment of a Mobile Broadband Wireless Network based on IS-856 (EV-DO)*, Proceedings of the 2004 World Wireless Congress, 2004.

J. Yang, *Multimedia Services over 1xEV-DO Mobile Broadband Network*, Proceedings. of SATEC, October 2006.

Chapter 2

Network and Service Architecture

2.1   Introduction

In a typical wireless network, a core network sits behind the radio access network, which is responsible largely for functions related to radio resource management. The core network, on the other hand, deals with switching/routing, mobility management, network quality of service (QoS), service provisioning and service management. This chapter focuses on the core network; the radio access network has been covered in Chapter 1.

There have been two major threads in the development of wireless networks. First came voice-centric cellular networks from the telephony world. The second thread, which rose to prominence quickly after the introduction of IEEE 802.11-based wireless local area networks (WLANs), encompasses wireless data networks from the data communications world. The success of WLANs in the late 1990s has led to various concepts of integrating cellular and WLAN technology [Var03], and it has been argued that rather than competing, WLAN and cellular technologies can complement each other. Also fostering integration is the convergence of cellular and WLAN technology in 3G wireless standards. As the specifications of 3GPP (3rd Generation Partnership Project) and 3GPP2 evolve, they have both been moving towards more convergence, e.g., towards an all-IP network with the creation and development of the IP Multimedia Subsystem (IMS) [Won03].

Exploring these developments will take up a significant portion of this chapter. We will also, however, discuss alternative wireless network architectures, service architectures and teletraffic analysis. By service, we mean a particular user application, which may be enabled by a particular technology platform or viewed as a customer solution. Examples include short message service (SMS), push-to-talk, and location based services. By service architecture, we mean how different technology components, platforms, and service enablers are put together to compose and deliver services.

Last but not least, teletraffic analysis and engineering, which are useful for network planning and capacity planning, are discussed briefly.

This chapter covers:

- Traditional voice-centric cellular networks;

- More data-centric, IP-centric network architectures, e.g., those used in WLANs;

- The latest converged network architectures that include voice, video, and data services; IMS is an important example of such an architecture. IMS also provides a platform for service and application development;

- Protocols and technologies for carrying voice and video over IP;

- (Higher layer) services and service architectures; and

- Alternative network architectures such as ad hoc networks and mesh networks.

2.1.1 Assumed Knowledge

It is assumed that the reader is familiar with the following general knowledge about networking (see Chapter 7):

- Circuit switching vs. networking switching;

- Basic differences between telephone networks and data communications networks;

- Network layering concepts, including protocol stacks;

- TCP/IP (Transmission Control Protocol/Internet Protocol) basics, including addressing, packet forwarding, hosts vs. routers, routing tables, etc.; and

- Queuing theory basics (for the topic of teletraffic analysis later).

2.2 Circuit-Switched Cellular Network Architecture

We focus here on the basic, essential network elements in a traditional voice-centric cellular network. Since the terminology varies slightly from system to system, we illustrate with the Global System for Mobile Communications (GSM) system, the most widely deployed 2nd-generation cellular system. While the 1st-generation cellular systems, first deployed in the 1970s, were analog, 2nd-generation cellular systems like GSM, IS-95 CDMA (Interim Standard 95 Code Division Multiple Access), and JDC (Japanese Digital Cellular) are digital. Third-generation (3G) systems like UMTS (Universal Mobile Telecommunications System) and cdma2000 evolved from 2nd-generation systems, but have gradually become more IP-based, converged networks (to be discussed in greater detail later in this chapter).

2.2.1 Basic Network Elements

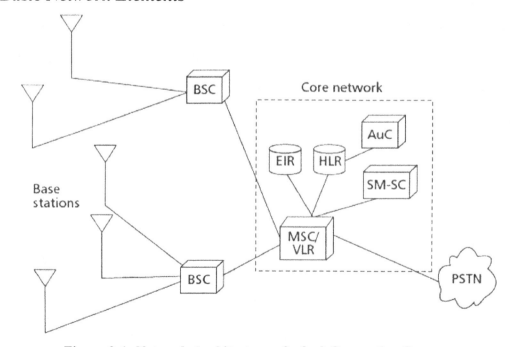

Figure 2-1: Network Architecture of a 2nd-Generation System

The essential elements of the core network of a 2nd-generation cellular system are shown in Figure 2-1. We focus in this chapter on the elements in the core network, whereas, Base station controllers (BSC) and base stations are involved in radio resource management, as discussed in Chapter 1. The core network contains a number of elements that together support telephony services, mobility management (roaming, etc.), security, short message service, and so on. Although we have chosen to illustrate GSM and its nomenclature, other 2nd-generation cellular networks are similar.

The *Mobile Switching Center* (MSC) is like a normal switch in a telephone network but with additional functions to support mobility management. Thus, it understands standard Signaling System 7 (SS7) signaling between telephony switches and supports the SS7 mobile application part (MAP). SS7 MAP uses the services of SS7 transaction capabilities application part (TCAP) to communicate. The use of SS7 MAP will be illustrated through a couple of examples shortly. An MSC typically needs to be connected to other networks, such as the public switched telephone network (PSTN), and suitable inter-working functions are needed for this. A service provider may have multiple MSCs in its network, especially if it is large. When an MSC plays a role as the first MSC to be reached by calls from other public land mobile networks (PLMNs), it takes on the gateway MSC role.

The *Home Location Register* (HLR) and *Visitor Location Register* (VLR) are databases that keep track of the location of mobile subscribers. However, they also contain other subscriber information, such as subscription information and security data. The subscriber information for each mobile subscriber is typically stored in an HLR in the subscriber's home network. This would be the network of the provider with which the mobile subscriber has a subscription for mobile phone services. When a subscriber is roaming (in the network of another service provider), a VLR in the new network will temporarily store some subscriber information, plus certain security data, to speed up authentication (how this works is explained in more detail in Chapter 3).

To support roaming, the HLR and VLR also together store enough information to allow roaming mobile stations to be located. When roaming, a mobile station registers its location in a process known as registration (introduced in Chapter 1). Even when idle, a mobile station will send location updates as it moves between location areas (the concept of location areas will be introduced shortly), allowing the network to keep track of its location. The location information is updated and tracked in the HLR and VLR as follows: the HLR points to the current network in which the mobile station is roaming, and the VLR contains more fine-grained information on its location. The VLR is often physically integrated with an MSC, and, hence, in Figure 2-1, the MSC and VLR are combined into one network element, the MSC/VLR. It is, however, also possible to have separate MSCs and VLRs, e.g., when a service provider has multiple MSCs in its network, and they all share the same VLR.

The *Authentication Center* (AuC) is central to the security functions in the network. Security is covered in more detail in Chapter 3.

The *Short Message Service Center* (SM-SC) handles short message services, which do not need an end-to-end connection to be established. Messages are first sent to an SM-SC, and then on to their destination. The SM-SC can store messages for awhile when the destination mobile station is unreachable.

The *Equipment Identity Register* (EIR) is a database containing information on the identity of stolen mobile equipment to prevent calls from stolen, unauthorized, or defective mobile stations.

2.2.2 Call Flow Examples

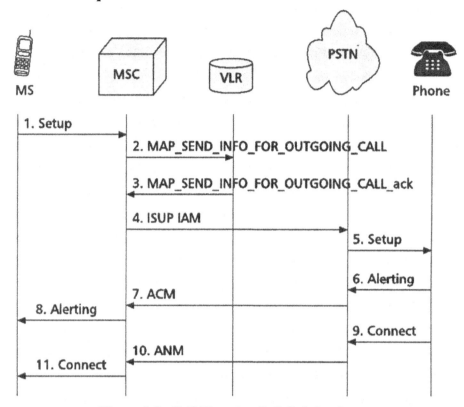

Figure 2-2: Call Flow for Call Origination

We will next illustrate how the elements of the core network work together. The first example is a call origination from a mobile station (MS). The call flow is shown in Figure 2-2, although with details omitted such as the radio network signaling and radio resource allocation. Instead, this is all lumped together under step 1, "Setup," which results in the MSC being requested to set up the call origination. SS7 signaling is then used within the mobile network, for example, to query the VLR. Then, if the MS is allowed to originate calls and has enough of a balance (if prepaid), etc., an ISUP IAM (ISDN User Part Initial Address Message) message is sent to the PSTN. This is a standard initial address message in SS7 to set up circuits. While the destination phone is being alerted (by ringing), an address complete message (ACM) is sent back to the MSC so the MS can be notified. When the phone is off-hook, an answer message (ANM) is sent.

In Figure 2-3, on the other hand, we see the more complicated case of call delivery to a roaming MS. Call delivery is more difficult than call origination because the MS needs to be located, whereas if the MS initiates call origination, the network receiving the signals from the MS knows where it is. For call delivery to an MS, the signaling first goes to the MS's home network because the rest of the network does not know where the MS is. As always when entering a mobile network, it enters through a gateway MSC (GMSC). This MSC can query the HLR of the MS using MAP_SEND_ROUTING_INFORMATION. As mentioned earlier, the HLR does not have complete location information for the MS so the GMSC cannot rely only on the HLR information to complete the call to the MS. Instead, the HLR knows which VLR to contact, and it does so to obtain a crucial number, the roaming number. This roaming number is a temporary number that the GMSC can use to set up the next leg of the voice circuit towards the MS. The GMSC then forwards the ISUP IAM to the MSC in the roaming network serving the MS.

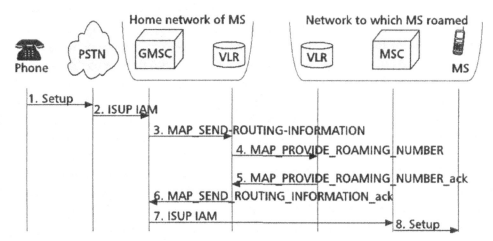

Figure 2-3: Call Flow for Call Delivery to a Roaming Mobile Station

Once the voice circuit connection set-up has reached all the way to the MSC serving the MS, a final challenge remains – to find the base station serving the MS. The MSC may not know precisely where the MS is if the MS is idle, in which case it only sends location updates to the network whenever it crosses location area boundaries. Typically, a location area would comprise multiple cells. The usual procedure for locating the MS more precisely within the location area is known as paging. Thus, the MSC would initiate paging for the MS, i.e., paging messages for the MS would go out on all the base stations within the location area. Finally, the MS responds to the paging message at its current serving base station, and the MSC sets up the last leg of the voice circuit, completing the call delivery.

The size of a cellular system's location areas must be set carefully. However, the smaller the location area, the more often an idle MS must perform location updates. In the extreme of a location area with only one cell, the whole purpose of the location area is defeated; it is, after all, meant to reduce the frequency that the MS has to update the network. On the other hand, the larger the location area the more cells would be involved in paging during call delivery.

2.3   TCP/IP in Packet Switched Networks

TCP/IP is the predominant set of data network protocols in the world today. In the 7-layer Open Systems Interconnection (OSI) model of protocol layering, Transmission Control Protocol (TCP) is a layer 4 (transport layer) protocol, whereas Internet Protocol (IP) is a layer 3 (network layer) protocol. The conjunction of TCP and IP in TCP/IP is due to the close historical relationship between TCP and IP, and even today TCP is often used as the transport protocol over IP. However, as we will see in section 2.4, TCP is not as suitable for traffic as VoIP (Voice over Internet Protocol), and alternatives like UDP (User Datagram Protocol) and RTP (Real-time Transport Protocol) are used.

TCP/IP was developed in the 1970s, based in part on 1960s research funded by the Advanced Research Projects Agency (ARPA). The name "internet" came about because one of the main goals of the network layer, and hence the Internet Protocol, was to inter-connect networks of different types, using different layer 2 technologies. This concept of inter-networking proved very successful, and many universities and research institutions joined the Internet (a network using TCP/IP technology) in the 1980s. With the arrival of the World Wide Web (WWW) and the hypertext transfer protocol (HTTP) in the 1990s, the Internet started being used not only by researchers and students, but also by the general public.

Two main reasons that TCP/IP is important for wireless networks are:

- By the late 1990s, the IEEE 802.11 wireless LAN (WLAN) standard was introduced, and the use of WLAN technology exploded, especially for people with laptop PCs who appreciated the convenience of being able to connect to the Internet while on the go.

- Cellular systems have been evolving from voice-centric to more IP-centric, as will be seen in section 2.5.

2.3.1 Basics of TCP/IP Protocols

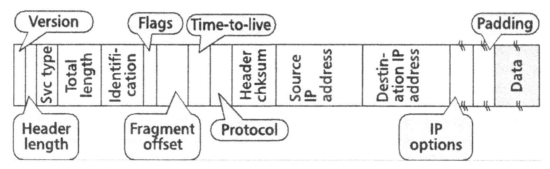

Figure 2-4: An IP Header

The IP header shown in Figure 2-4 contains all the information about a packet that a router needs to know in order to process it. More details on the IP header and the router processing of the IP header can be found in books on TCP/IP [Com06].

User requirements have changed dramatically since TCP/IP was introduced over 30 years ago, and Internet usage has expanded. The current version of IP is known as IPv4. A new version known as IPv6 [Dee98] has been created, and is gradually being deployed around the world. Meanwhile, features continue to be added to IPv4, including mobility support (see section 2.3.2).

IP hosts, the end points of an IP network, are often the source or destination of IP packets. IP routers, on the other hand, are intermediate nodes with multiple interfaces that forward incoming packets on outgoing interfaces based on the contents of the packet headers and routing tables. The contents and interpretation of routing tables can be found, for example, in [Com06]. Several protocols have been developed to exchange routing information between routers so routing tables can have some degree of automated updates. These protocols include router information protocol (RIP) [Mal98], open shortest path first (OSPF) [Moy98, Col99] and Intermediate System to Intermediate System (IS-IS) [ISO02].

2.3.2 IP and Mobility

TCP/IP was designed before all the contemporary uses for it came into existence. Thus, its use for multimedia (with its different quality of service requirements), in wireless mobile networks, and so on, was not even imagined. With the rise of wireless networks, and especially wireless data networking, mobility support has become increasingly important.

Mobility should be distinguished from portability. Portability is the concept that a mobile wireless terminal can move and reconnect at each new attachment point with a new IP address. However, if portability is supported, certain applications will not work whenever the terminal moves if the applications expect the IP address to be unchanged. For example, for server applications, clients would typically perform a Domain Name System (DNS) look-up to find the server's IP address. With portability and IP address changes (of the server), the clients would have trouble finding the server. Furthermore, TCP sessions expect the IP address to remain the same. Changes in IP address will change the TCP checksum, for example.

Mobility support requires, therefore, the mobile terminal to keep its "permanent" IP address while moving, and at the same time still be reachable by other IP hosts. Mobile IP [Per02] solves this problem by allowing the mobile terminal to keep a permanent IP address as it moves. However, it relies on the existing IP routing fabric, and so for routing purposes, another IP address is required, known as a care-of address. This is another IP address routable to the mobile terminal's current location.

To allow IP packets addressed to the permanent IP address to be routed to the care-of address, Mobile IP introduces an agent, the home agent, in the mobile terminal's home network that intercepts packets addressed to the mobile terminal's permanent address, and forwards them to the care-of address. The packet forwarding is through a packet encapsulation procedure where the original packet is encapsulated inside a new packet addressed to the care-of address, and tunneled to the care-of address. At the foreign network, another agent, the foreign agent, unencapsulates the packet and delivers it to the mobile terminal. The home agent knows the current care-of address of the mobile terminal because of the Mobile IP registration procedure. In this procedure, the mobile terminal registers its care-of address with its home agent, which can then associate the two addresses. The operation of Mobile IP is illustrated in Figure 2-5.

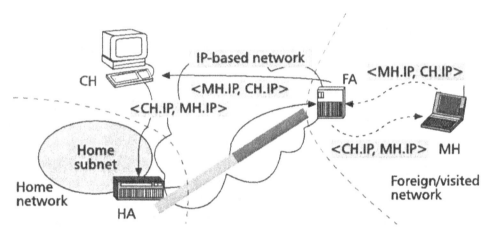

Figure 2-5: Basic Operations of Mobile IP

IPv6 also provides mobility support, which is sometimes called Mobile IPv6 [Joh04]. It is similar in many ways to Mobile IP, though Mobile IPv6 has some enhancements. These include:

- Support of route optimization;

- Support of fast handoffs;

- Reduced header overhead (using the IPv6 routing header); and

- No need for foreign agents.

Furthermore, building on top of Mobile IPv6, not just host mobility is supported, but network mobility as well. Network mobility (NEMO for short) is specified in Internet Engineering Task Force (IETF) Request for Comments RFC 3963 [Dev05].

2.3.3 IP and General Packet Radio Service (GPRS)

Mobile IP and Mobile IPv6 can provide mobility support in networks where the underlying link or bearer technology (that transports the IP packets) does not itself provide mobility support. A notable exception is General Packet Radio Service (GPRS). Two key networks elements in GPRS, serving GPRS support node (SGSN) and gateway GPRS support node (GGSN), act like routers but with the functionality to support the users' mobility. Hence, GPRS uses its own mobility protocols, not mobile IP or mobile IPv6. We will discuss GPRS mobility management shortly.

GPRS has already been introduced in Chapter 1. Recall that GPRS is the extension to GSM to provide better packet data services than what the original GSM system could support. The innovative use of GSM time slots for GPRS allows the support of variable data rates and higher data rates than a single GSM channel supports. There are also a variety of innovations on the network side, which is our focus in this chapter. The GPRS network elements are added onto the GSM network elements, as shown in Figure 2-6. Thus, we see the familiar GSM network elements as well as the SGSNs and GGSNs. The base stations (also known as Base Transceiver Systems, BTSs) and BSCs need to be upgraded to support GPRS functions such as dynamic resource allocation over the air.

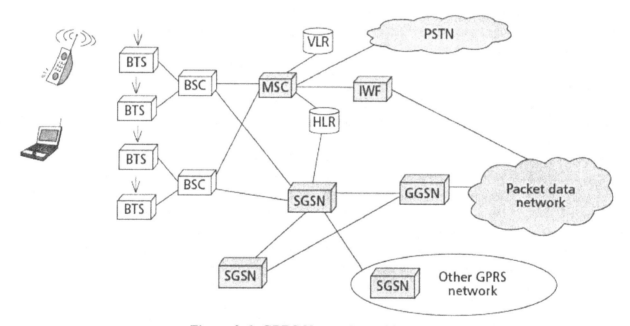

Figure 2-6: GPRS Network Architecture

One of the most interesting design choices in GPRS was to design it to be a "general" packet service, not specific to any one network protocol. The importance of this design choice can be seen in the name – General Packet Radio Service (GPRS); the first word of the name indicates this choice. This means that it is not just an IP packet service, but GPRS can support many packet services over the same network. The initial GPRS specifications mentioned IP and X.25 as two initial examples of packet data protocols that would be supported by GPRS. There are pros and cons to being a generalized packet radio service. One advantage is the obvious flexibility of being able to support multiple protocols. Disadvantages include:

- Less efficient packets – there is significant overhead in GPRS packets to allow it to support multiple protocols. For example, when IP packets are being transported over GPRS, IP is used both internally within GPRS and also at the user level.

- More signaling overhead – the concept of PDP (packet data protocol) is needed, to distinguish between multiple possible packet data protocols that can be transported over GPRS; this involves various additional signaling to set up, e.g., the PDP context activation.

To use GPRS, then, a mobile needs to perform not only the GPRS attachment procedure, but also one or more PDP context activations. Besides being needed to distinguish between multiple possible packet data protocols that can be used with GPRS, PDP contexts can also be used for multiple QoS profiles for one user, e.g., one for signaling, one for voice over IP, etc. The PDP context activation occurs between the mobile and the GGSN.

Internal to the GPRS network, i.e., between GGSN and SGSN, GPRS networks transport IP packets using the GPRS Tunneling Protocol (GTP) [3GP11a]. This rides on top of the internal IP network. Thus, if IP packets are going over GPRS, this results in IP over IP (plus GTP), and is thus less efficient than it might be if designed differently. At the user level, the GGSN acts as the first IP hop, i.e., the gateway router.

Mobility management in GPRS parallels the original GSM mobility management. Just as with the original mobility management, the mobile could be in idle mode or actively in a call; the GPRS terminal could also be in "standby" or "ready" state. The standby state corresponds to the idle mode of GSM, and during the standby state, the GPRS terminal is tracked to within a "Routing Area". A Routing Area is analogous to a Location Area in GSM and also consists of multiple base stations. When the GPRS terminal becomes active, in "ready" state, then it is tracked as it moves from base station to base station. GPRS adds one more state, besides ready and standby states. This is the idle state, and it corresponds to when the terminal is on but not GPRS attached.

When the mobile moves around and changes between SGSNs, the GRP tunnel will be changed from the GGSN to the latest SGSN. Thus, Mobile IP is not used for packet re-routing as the mobile terminal moves around.

The initial introduction of GPRS to GSM sets up a parallel network, a packet switched (PS) domain, that is parallel to the circuit switched (CS) domain of the original GSM. There has been a process of evolution since the introduction of GPRS, and it is only recently, with LTE (Long Term Evolution, see section 2.6) that the CS domain goes away. Meanwhile, support for VoIP has been added, in the form of the IMS (section 2.5).

2.3.4   3GPP2 and the Packet Data Serving Node (PDSN)

Interestingly, the evolution of CDMA/cdma2000 networks towards better support of packet data and IP in particular took a different path from that of GSM evolution. Unlike GSM, which created the "elaborate" GPRS system that could handle multiple packet data protocols, the CDMA evolution went for a simpler approach that reused standard IETF-defined IP-based protocols heavily.

The PDSN (packet data serving node) is analogous to the GGSN of GPRS, and it is the first IP hop in CDMA networks. Like the GGSN, it plays multiple roles.

- It is the termination point for Point-to-Point Protocol (PPP) sessions between the terminal and the PDSN (unlike GPRS, which has its own protocols for PDP context activation, and then GTP tunnels, 3GPP2 re-used the widely deployed PPP for its IP network).

- There are two mobility options, namely "simple IP" and "mobile IP". With simple IP, a new PPP session is established with a new PDSN when the mobile moves, whereas with mobile IP, the PDSN has Mobile IP Home Agent (HA) functionality, so Mobile IP is used.

2.4   VoIP/SIP for IP Multimedia

Voice over Internet Protocol (VoIP) is one of the emerging technologies developed recently in the telecommunication industry. Compared to traditional circuit-switched voice service in the public switched telephone network (PSTN), VoIP promises a major reduction in capital and operational costs of the high-capacity packet network infrastructure. With VoIP it is also possible to create new revenue-generating services in a converged high-speed packet network. In particular, VoIP has been successful in wireline networks, with affordable commercial VoIP services available in the market. A set of new signaling and media-transport mechanisms have been defined. For example, Session Initiation Protocol (SIP) is used for call signaling control. Real-time Transport Protocol (RTP) is used for media delivery. Quality of Service (QoS) strategies and mechanisms are considered in the design of the service platform and network infrastructure.

VoIP in wireless has drawn much interest recently because of the evolution to an all-IP architecture in converged wireless and wireline networks. Initial VoIP efforts were focused on wireless local area networks (WLANs) and fixed wireless since the achievable data rate is comparable to that of the wireline network. Recent advances in cellular high-speed packet networks have incorporated innovative techniques to improve spectrum efficiency and enable multimedia applications. Seamless integration and roaming of VoIP service across different wireless networks, i.e., WLAN, wireline VoIP, and cellular voice, will make VoIP not only a reality but a mainstream application in wireless networks.

2.4.1   Protocols for VoIP Application
2.4.1.1 Session Initiation Protocol – SIP

SIP [Sch08] is an application-layer control (signaling) protocol for creating, modifying, and terminating sessions with one or more participants. These sessions include Internet telephone calls, multimedia distribution, conferences, presence, events notification, and instant messaging. SIP is specified in IETF (Internet Engineering Task Force) RFC 3261 [Ros02] from the Multiparty Multimedia Session Control (MMUSIC) working group. SIP was adopted by both 3GPP and 3GPP2 standards as the main signaling protocol for multimedia over IP, and as a key element in the IMS architecture (see section 2.5). SIP has been widely used as a call control signaling protocol for VoIP applications in non-wireless networks.

SIP is a request-response and text-based protocol, very similar to HTTP (the protocol for WWW) and SMTP (the protocol for email). It is designed to be independent of the underlying transport protocol. SIP can run on top of TCP, UDP (User Datagram Protocol), or SCTP (Stream Control Transmission Protocol), which makes it very flexible and easy to implement. SIP consists of a set of messages that are used to set up and tear down a session. A simple example of a call setup and termination procedure using SIP (between user A and user B) is as follows (see Figure 2-7):

   1. The caller, user A, sends an INVITE message to initiate a call to the called SIP client, user B. The INVITE message consists of basic call setup information, such as the called and caller addresses, proxy information and vocoder (voice coder) options.

   2. The called party responds with a SIP TRYING message to acknowledge the call setup request.

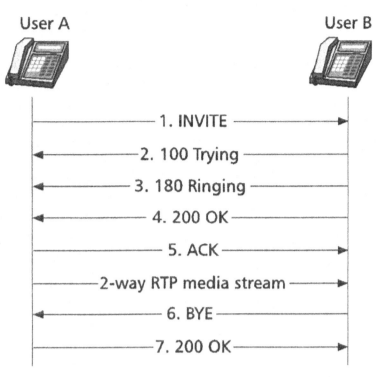

Figure 2-7: A Basic SIP IP Phone-to-IP Phone Call Flow

3. The called party sends a RINGING message to the caller, where the appropriate ring back to the called can be generated.

4. A connection is established when the called party goes off-hook, and a media stream enabled. The called party generates an OK message to indicate the state change.

5. The caller connects the call and acknowledges the connection with an ACK message. The media stream for a two-way voice call is now established.

6. When the called party hangs up and the call is over, the called party sends a BYE message. The caller disconnects the call and acknowledges the termination with an OK message, and the call is terminated.

Meant for IP-based networks, SIP design provides a set of call-processing functions and features similar to those in the current PSTN. SIP emphasizes on-call setup and signaling and does not define these PSTN-like features directly. It must work with other protocols and build these features in such network elements as proxy servers and gateways. An example of SIP network elements is the IMS architecture, described in section 2.5. An example of a generated SIP INVITE message is shown in Figure 2-8.

INVITE sip: UserB@ieee.org SIP/2.0
Via: SIP/2.0/UDP 200.101.200.005:5060
To: <sip:UserB@ieee.org>
From: User A <sip:UserA@1.1.1.1>
Call-ID: 1234567890
CSeq: 1 INVITE
Contact: <sip:UserA@1.1.1.1>

Figure 2-8: An Example of a SIP INVITE Message Format

2.4.1.2 Real-time Transport Protocol – RTP

The Real-time Transport Protocol (RTP) [Sch10] is a standard protocol for transporting multimedia applications over the Internet. It was developed by the Audio-Video Transport working group of the IETF and first published in RFC 1889 in 1996, and then in RFC 3550 in 2003 [Sch03]. It defines a packet format for delivering audio and video packets before passing them to the transport layer. RTP can be used for transporting common voice formats, such as PCM (Pulse Code Modulation), ITU standard voice codec G.711 and others, GSM codec, audio formats such as MPEG (Moving Picture Experts Group) audio (MP1, MP2, MP3), and video formats such as MPEG, H.263, H.264, etc. It can also be used for transporting proprietary audio and video formats. RTP has been widely implemented for streaming multimedia applications.

RTP usually runs on top of UDP because real-time applications have stringent delay requirements. The sender takes a media packet or data chunk and encapsulates it into an RTP packet. It then encapsulates the RTP packet into a UDP packet and passes it to the IP layer. The receiver extracts the RTP packet from the UDP packet, then extracts the media packet or data chunk from the RTP packet and passes it to the media layer for decoding and play-out.

As an example, consider the use of RTP to transport a G.711 voice stream. The G.711 encoder generates a compressed voice stream at 64 kb/s (an 8-kHz sample frequency × 8 bits per sample). Assume that the application layer collects the encoded data every 20 ms, or equivalently, 160 bytes of data, and passes it to the RTP layer. The sender encapsulates the 160 bytes into an RTP packet with a header. The RTP header is usually 12 bytes, including payload type (7 bits), sequence number (16 bits), time stamp (32 bits), synchronization source identifier (SSRC, 32 bits), and miscellaneous fields (9 bits), as shown in Figure 2-9. The 172-byte RTP packet (a 160-byte payload plus 12-byte header) is passed to the UDP socket interface and sent to the receiver. The receiver extracts the RTP packet from the UDP socket interface. The application layer extracts the voice payload from the RTP packet and uses the header information to decode and play out the voice stream.

Misc.	Payload type	Sequence number	Time stamp	SSRC
9 bits	7 bits	16 bits	32 bits	32 bits

Figure 2-9: RTP Header Format

RTP is used purely for media transport over the Internet. It can be viewed as a sub-layer of the transport layer. It provides a mechanism for easy interoperability between networked multimedia applications. Note, however, that RTP does not provide any QoS guarantees of timely delivery of the applications. The RTP encapsulation is visible only at the end points, and invisible to routers or switches along the path. RTP is often used in conjunction with other protocols and QoS mechanisms to provide VoIP service.

2.4.1.3 Robust Header Compression (RoHC) protocol

VoIP packets are often carried over RTP/UDP/IP, with an uncompressed header of about 40 bytes per voice frame. The voice codec used in cellular systems often has a very low bit rate, from 4 kb/s to 12.2 kb/s, to save bandwidth in the air interface. Consider a 22-byte voice frame (equivalent to an 8 kb/s full rate voice codec). The RTP/UDP/IP header overhead is more than 65% (~40/62) of the packet. The radio application protocol stack adds additional overhead. Without any header compression, the overhead becomes dominant and has a significant impact on the air interface capacity. Therefore, a header-compression scheme becomes an essential component in providing VoIP in wireless.

Robust header compression (ROHC) [Bor01] is a specification of a highly robust and efficient header-compression scheme. Developed by the ROHC working group of the IETF, it was published in RFC 3095 in 2001. It specifies a framework supporting four profiles, i.e., uncompressed, RTP/UDP/IP, UDP/IP, and ESP/IP (Encapsulating Security Payload/IP). ROHC offers the ability to run over different types of links by operating in different modes based on the availability of a feedback channel and how much feedback information can be provided. With ROHC compression of RTP/UDP/IP headers, the average header size is between 1 and 4 bytes. Within a voice stream, most of the packets can carry the header information in only a single byte. The protocol stack with header compression is shown in Figure 2-10.

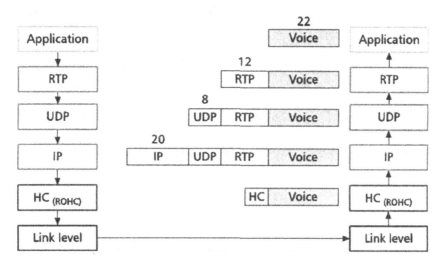

Figure 2-10: A Protocol Stack with Header Compression

Basically, header compression relies on exploiting the redundancy in the headers. RTP, UDP, and IP headers have significant redundancy in both intra- and inter-packets of a media stream. An intra-packet has identical or deducible information. For example, the source and destination address in the IP header and the source and destination port in the UDP header do not change for the same media stream. They only need to be transmitted once at the media stream's start. The payload type in the RTP header only needs to be sent once as well.

An inter-packet has incremental differences in the same stream. For example, the time stamp in the RTP header increases by a fixed amount with every increase in the sequence number. The time stamp need not be sent to the ROHC decoder at every talk spurt along with the sequence number. It must be sent only at the beginning of a new talk spurt after a silence compression.

Note that ROHC is not used as an end-to-end compression scheme, since routers need the full IP header information for routing purpose. It is often used in the radio-access network and improves the air interface capacity for VoIP applications.

The ROHC compression scheme is different from earlier header compression standards, such as IETF RFC 1144 (compressing TCP/IP headers) and RFC 2058 (compressing RTP/UDP/IP headers). The earlier header compression standards were developed to reduce transmission overhead over low-speed serial links. The new ROHC standard defines a compression scheme that performs effectively over links with high packet loss rates, such as wireless links. In addition to header compression, payload compression can reduce the amount of bandwidth required. Voice codec technology can compress the VoIP packet payload but at the cost of decreased overall voice quality. The tradeoff between voice quality and bandwidth requirements needs careful consideration.

Two new RFCs were published recently to clarify and simplify the ROHC specifications. RFC 5795 [San10] explicitly defines the ROHC framework and the uncompressed profile separately. The definition of the framework does not modify or update the definition of the framework specified by RFC 3095. RFC 5225 [Pel08] defines a second version (ROHCv2) of the ROHC profiles defined in RFC 3095, RFC 3843, and RFC 4019. The ROHCv2 profiles introduce a number of simplifications to the rules and algorithms that govern the behavior of the compressor and decompressor. For example, they specify a three-level decompressor state machine, but do not specify the compressor state machine in much detail. They also define robustness mechanisms to address the situations that packets can be lost and/or reordered on the ROHC channel. Finally, the ROHCv2 profiles define their own specific set of header formats, using the ROHC formal notation.

2.4.2 *Quality of Service (QoS) for VoIP in Wireless*

One of the most important requirements for Internet telephony is to provide Quality of Service guarantees, with service classes defined for different applications. For example, a common set of QoS definitions include conversational, streaming, interactive, and background classes. QoS metrics usually include delay and delay jitter, bandwidth, and reliability, which vary for different applications. Most stringent is the delay requirement for conversational voice applications, with the end-to-end packet delay less than 200 ms. Packet loss rate should be less than 1%. The mean opinion score (MOS) is often used to provide a numerical measure of human speech. The score is generated by using subjective tests to obtain a quantitative indicator of speech quality. There are five scores ranked as 1 (bad), 2 (poor), 3 (fair), 4 (good), and 5 (excellent). Acceptable voice quality requires a MOS score of at least 3.5. A specification of MOS is part of the Perceptual Evaluation of Speech Quality (PESQ) which is standardized by ITU. PESQ is better suited to measure the quality of VoIP than an earlier framework, Perceptual Speech Quality Measure (PSQM).

Providing end-to-end QoS guarantees for VoIP is extremely challenging because the wireless infrastructure is heterogeneous and operated by different service providers. The common practice is to provide QoS mechanisms at each network segment so that the overall end-to-end QoS can eventually be met. Those mechanisms include how to minimize the latency, assess the delay jitter, calculate the bandwidth requirements, and compensate for the packet loss. The methods differ for different types of wireless networks.

In cellular networks such as 3G/4G wireless communication systems and WiMAX, at the radio access network (RAN) side, the QoS design includes flow classification, a packet scheduler over the air interface, backhaul buffer management and control, call admission control and overload control, and header compression for improved efficiency. At the core network side, the QoS design includes support of DiffServ (Differentiated Services) for priority routing and handling, Multiprotocol Label Switching (MPLS) for ensured bandwidth and service guarantee and IMS session control for multimedia applications. The end-to-end QoS mechanism includes user authentication and negotiation, de-jitter play-out buffer design, and end-to-end delay improvement.

The IEEE 802.11 wireless LAN has been widely deployed as the wireless access technology in office buildings, in public hot-spots, and in homes. The early version of 802.11 standards does not define specific QoS mechanisms. VoIP and other streaming multimedia services can only be offered as best effort without any QoS guarantees. To remedy this problem, the 802.11e standard [IEE05], as an amendment to 802.11, defines a set of QoS enhancements for delay-sensitive applications such as VoIP. 802.11e improves the original 802.11 medium access control layer by defining a new coordinating function: the hybrid coordinating function (HCF). Within the HCF, two methods of channel access are defined: HCF Controlled Channel Access (HCCA) and Enhanced Distributed Channel Access (EDCA). Both EDCA and HCCA define Traffic Categories (TC) to differentiate QoS classes. In EDCA, high priority traffic has a higher access probability than low priority traffic. The levels of access priorities are called access categories (AC). In addition to access priority, other QoS features are specified as well, such as packet transmission scheduling and admission control. These QoS enhancements are beneficial to VoIP and other steaming multimedia applications in WLANs.

2.5   Packet-Switched Mobile Networks and IMS

With over four billion users in about 25 years, cellular telephony has been a phenomenal success. The goal of first-generation analog cellular systems was to provide voice telephony to mobile users (though voiceband data was feasible to a certain extent). Second-generation digital cellular systems introduced short message service (SMS) and circuit-switched data. SMS became very popular and enjoyed wide acceptance, especially among GSM users in Europe and the Far East. Even after data capability was introduced in second-generation systems, voice telephony continued to be the major application.

Circuit-switched connection is very inefficient for data services when it comes to scarce radio bandwidth resources. The bursty nature of data traffic means that the channel bandwidth is too narrow when the user has data to send and, when there is no data, the channel is idle most of the time. The advantage of packet switching is well known when a large number of users with bursty data traffic share a common medium. GSM introduced a packet overlay network in the core mobile network, as well as a packet protocol over the air interface to make better use of radio spectrum and network resources. Known as General Packet Radio Service (GPRS – see section 2.3.3), it was a key step in the evolution towards packet-switched core networks and all-IP mobile networks.

2.5.1   Evolution to Packet-Switched Core Networks and All-IP Mobile Networks

As third-generation cellular systems evolved, supporting higher speed data over the air interface became a key requirement. The ITU, which defined the high-level requirements for third-generation wireless, set forth the data rates to be supported in mobile, pedestrian, and fixed environments. In the meantime, the emerging Internet was changing the way people accessed information, and its widespread popularity made wireless data services even more appealing.

Telecommunications has seen two major trends lately. The growth in voice services has slowed, resulting in voice revenues that are flat or even decreasing. Data traffic is growing rapidly and with it a major shift towards packet-switched networks. In the early days of telephony, when voice dominated traffic, data services could be supported in the voice-centric network. With the growth in data traffic, network providers were left with the choice of either maintaining two separate networks (one for voice and one for data) or finding ways to support voice traffic in data networks. Since maintaining two different networks is not economically attractive, evolution towards a single packet-switched core network began in earnest.

The trend towards packet networks in cellular systems started with the introduction of the Internet protocol (IP) in the core network, as shown in Figure 2-11. Traditional mobile switching centers (MSCs) are decomposed into MSC servers, providing the call-control functions of the MSC, and into media gateways (MGWs) that provide conversion between circuit-switched voice and packet voice. Interworking with legacy circuit-switched systems is achieved with the help of additional signaling gateways that convert signaling over IP transport to signaling over SS7 networks. Even though the core network transports all the signaling and voice traffic over an IP network, call control for voice services continues to use existing protocols, and voice over-the-air interfaces use circuit-switched connections. The previous interface between the MSC and BSC, Iu-CS (Circuit Switched), was split into user and signaling parts, to support the decomposition of MSCs into MSC servers and MGWs. On the PSTN side, the SGW (Signaling Gateway) performs transport conversion between the IP network side and the PSTN transport protocols.

Note also that mobile stations are now called UE (User Equipment). Base stations are called Node B; and the BSCs have been replaced with RNCs (Radio Network Controllers).

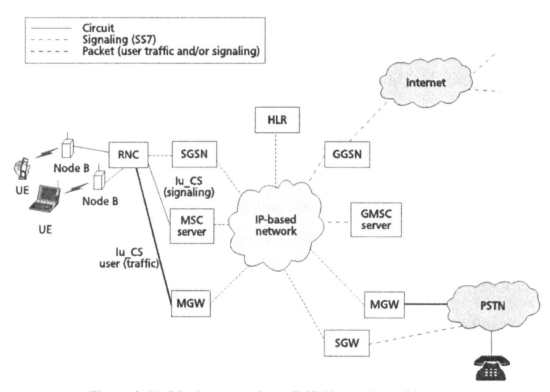

Figure 2-11: Moving toward an all-IP Network Architecture

2.5.2   The Need for IMS and Its Requirements

As cellular systems evolved to 3G, voice services remained as the major application. Even though 2.5G and 3G cellular technologies provided higher bandwidth access, packet-switched services did not take off as expected. The "killer application" for data continued to elude everyone, even though native connection to the Internet is available through wireless access, and the access bandwidth increased with new air-interface technologies. Service providers realized that providing higher bandwidth access alone was insufficient and rich data services were required. The voice-centric cellular infrastructure had limitations and could not take full advantage of IP-based services. It became apparent that a new approach was needed to stimulate IP-based services. IP multimedia subsystem (IMS) [Won05] emerged as the new approach that could eventually stimulate IP services in wireless networks. As cellular technology is evolving to 4G where end-to-end connections will be purely packet based (with no circuit-switched connections), the need for supporting voice over IP networks becomes even more important.

The vision of IMS is to offer an open architecture with a common control/session layer that promises the best of the fixed, cellular, and Internet worlds. IMS also attempts to replace the vertical stovepipe model, where each technology has its own horizontal layered architecture of core network, call control, and supplementary services and applications. In this architecture, different access technologies (DSL [Digital Subscriber Line], cable, cdma2000, UMTS, WiMAX, WLAN) can share a common IP transport layer, common control/session layer, and a common pool of applications. By merging wired and wireless networks, IMS [Won03] brings (a) the broadband capability of fixed networks, (b) the convenience of mobility from cellular networks and (c) a rich user experience of new and ever-expanding applications from the Internet. The objectives and requirements for IMS include [3GP11b]:

- Enabling operators to offer IP multimedia services;
- Enabling a mechanism for negotiating QoS;
- Supporting interworking with circuit-switched (CS) networks (PSTN, cellular) and the Internet;
- Allowing roaming;
- Providing home network control;
- Allowing rapid service creation;
- Supporting third-party development of IP-based services;
- Providing access independence; and
- Relying on IETF-approved standards.

With access bandwidth increasing as new air-interface technologies are introduced and native connection to the Internet is available through wireless access, the need for IMS and the role of the operators have been questioned in many quarters. The argument is that subscribers should be allowed to try third-party applications and use third-party service providers for IP multimedia services. Though there may be some benefit to this approach, it may not be in the best interests of both operators and subscribers.

Offering bandwidth as a commodity and letting users choose their own Internet services will shut network operators out from lucrative new service offerings. Indiscriminate use of Internet applications over cellular networks could also impact the integrity of the cellular network and quality of service. However, operators cannot anticipate all the applications users may want and develop them because this is not their area of expertise. In addition, such an approach could stifle the application development seen in the Internet domain. IMS offered a new approach where the operators could provide the session control while developing some applications themselves. At the same time, third parties would be allowed access to some network services and interfaces, and could develop the majority of applications.

2.5.3 IMS Architecture and Network Elements

The IMS architecture, shown in Figure 2-12, is a collection of functions linked by standard interfaces. The architecture includes the IMS terminal, commonly referred to as user equipment (UE), IP connectivity access networks (IP-CAN), one or more SIP servers, collectively known as call state control functions (CSCFs), one or more databases, application servers, nodes to interwork with legacy circuit-switched networks, and media resource functions. The elements in the cloud are the nodes of the IMS core, which interface with other IMS networks. The transport plane is a core IP network of routers. In IMS, signaling and media planes are completely separate. The only entity that handles both is the UE.

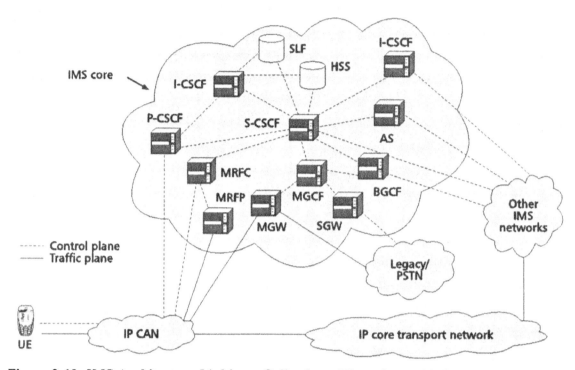

Figure 2-12: IMS Architecture Linking a Collection of Functions with Standard Interfaces

The UE can be a voice, data, or multimedia terminal. An IP Connectivity Access Network (IP-CAN) provides IP-based connectivity between a UE and the IMS core network for signaling and between a UE and the IP core transport network for user traffic. When the UE is turned on, it first attaches to the IP-CAN to obtain an IP address and to discover the first point of contact in the IMS core network. The IP-CAN assigns the UE an IP address as well as gives the contact address in the IMS core. The IP address may be static or dynamic depending on the capabilities of the network. Examples of IP-CAN are GPRS, W-LAN, cdma2000 packet network, Evolved Packet System (EPS), WiMAX, and cable or DSL access.

Home subscriber server (HSS) is a central database for user-related information. It is an evolution of the home location register (HLR) from the cellular world and contains user information such as subscription, location, security, and user profile to support multimedia sessions. IMS allows more than one HSS in a network. Subscriber location function (SLF) is a database that maps a user address to a given HSS. Networks with only a single HSS do not need an SLF. When the IP-CAN is either GPRS or EPS, the HSS is the central database that authenticates and authorizes the UE in the IP-CAN as well as in the IMS domains. When using other IP-CANs (e.g., WLAN, WiMAX), an Authentication, Authorization, and Accounting (AAA) server may act as an intermediary. In this case, the HSS still needs to include the profile for the user and the authentication key; however, the AAA server interacts with the HSS to obtain authentication credentials and to authenticate the UE in the IP-CAN. Additionally, this allows authorization and accounting to be properly handled. The communications with the AAA server could be done using an AAA protocol like RADIUS or Diameter, which are discussed in detail in Chapter 3.

Call state control function, or CSCF, is a key function in IMS. These are SIP servers and they process SIP signaling in IMS. There are three types of CSCFs:

- The proxy CSCF (P-CSCF), the first point of contact between the UE and IMS network, acts as an outbound/in-bound SIP proxy server. Each IMS network includes a number of P-CSCFs, each serving a number of IMS terminals. The P-CSCF is usually located in the visiting network when the mobile is roaming.

- The serving CSCF (S-CSCF) is the central node in the IMS. In addition to SIP server functions, the S-CSCF also performs session control and enforces the policy of the network operator. The S-CSCF also acts as a SIP registrar binding the IP address of the terminal to the public address. The S-CSCF is always located in the home network, which may include a number of S-CSCFs.

- The interrogating CSCG (I-CSCF) is a SIP proxy at the edge of an administrative domain. Its address is listed in the DNS and allows CSCFs to route the SIP messages from one IMS network to another via an I-CSCF.

The signaling gateway (SGW), the media gateway control function (MGCF), and the media gateway (MGW) comprise the PSTN gateway and allow IMS terminals to make and receive calls from other CS networks. The MGW converts media from packet to CS and vice versa. The MGCF converts between SIP messages and messages understood by the CS networks. In addition, the MGCF controls the MGW. The SGW performs the lower-layer protocol conversion between the IP and CS domains for the messages between the MGCF and a switch in the CS domain. The breakout gateway control function (BGCF) determines where in this or in some other IMS network the PSTN breakout should occur, and decides which MGCF to involve.

The media resource function (MRF) provides the ability to play announcements, to mix media streams (for conference services), and to transcode between different codecs. The MRF includes a signaling component (MRFC) and the processing component (MRFP).

An application server (AS) is a SIP node that hosts and executes IMS services. There are three types of application servers. A SIP application server hosts and executes IMS services on SIP. IMS-specific services are likely to be developed using SIP ASs. To use existing applications developed for Open Service Access (OSA) and for GSM, a service capability server (SCS) is included to convert the protocols so that these ASs look to the IMS networks like SIP ASs.

2.5.4 IMS Procedures

IMS is designed to use IETF protocols as much as possible. The most widely used protocol is the session initiation protocol (SIP). It is used between the UE and the P-CSCF, between CSCFs, between the S-CSCF and the AS, between the S-CSCF and the BGCF, between the BGCF and the MGCF and between the S-CSCF and the MRFC. Diameter (see section 3.5.1.2) is another protocol used in IMS. Diameter is used between the CSCFs and databases like the HSS and the SLF, as well as between the AS and the databases. Diameter is used for authentication, authorization and accounting (AAA) in IMS. Media Gateway Control Protocol H.248 is used between the MGCF and the media gateway, as well as between the MRFC and the MRFP. The MGCF uses stream control transmission protocol (SCTP) between the MGCF and the SGW. Real-time protocol (RTP) is used for end-to-end delivery of real-time data. RTP provides time stamps, among other things, and allows the receiver to play out the media at a proper pace.

Like mobile phones, IMS terminals also need to register before they receive services from the network. It is during the initial registration that the network assigns an S-CSCF for the subscriber that establishes a path between the UE and the S-CSCF for IMS call control. Before the terminal can register, it needs to gain access to the IP-CAN and acquire an IP address. Depending on the IP-CAN, the IP address may be obtained as part of getting IP connectivity (in GPRS) or separately (in Dynamic Host Configuration Protocol [DHCP]). The terminal needs to discover the P-CSCF and get the IP address of the P-CSCF. In GPRS, the PDP context-activation response provides the P-CSCF address, otherwise DHCP and DNS are used. Registration involves many subprocedures. It binds the public user identity to a contact address, home network authenticates the subscriber, the subscriber authenticates the network, and the UE and the P-CSCF establish a security association for further signaling. Once registration is completed, the terminal stays in an idle mode until an active session needs to be established.

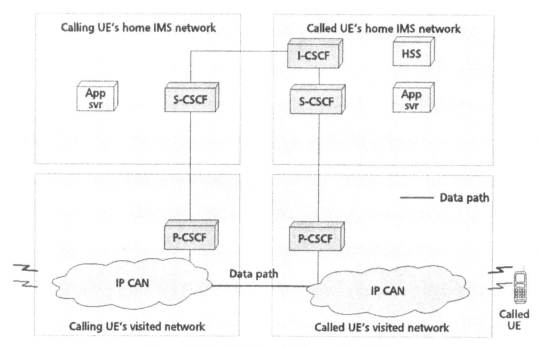

Figure 2-13: Call Establishment in IMS

Figure 2-13 shows the steps involved in the most general session establishment case where both calling and called party are roaming outside their home networks. The left side of the figure shows the visiting and home networks for the caller; the right side shows the visiting and home networks for the called user.

Session establishment starts when the calling UE sends a SIP INVITE message to the P-CSCF in the visiting network. The INVITE includes the SDP describing the parameters of the session and the public user identity of the called user. The P-CSCF processes the INVITE request and forwards the SIP INVITE to the S-CSCF in the user's home network. The S-CSCF performs session control and uses the filter criteria to invoke applications from application servers. An example of the use of AS could be to check the call-barring list for this user. There could potentially be a chaining of application invocations.

Using the public user identity of the called user, the S-CSCF resolves the SIP URI (Uniform Resource Identifier) and the IP address of the I-CSCF in the home network of the called user and forwards the INVITE. The I-CSCF queries the HSS and gets the address of the S-CSCF serving this user. The S-CSCF performs service control and invokes the AS service, if required. The S-CSCF then forwards the INVITE to the P-CSCF in the visited network. Session establishment continues with the negotiation of session parameters and preconditions and provisional acknowledgments.

When the called user answers the session, a 200 OK is returned to mark the successful completion of the session setup in response to the original INVITE message. The calling UE returns an ACK as the final response. When the session setup is complete, the data path is established between the calling user's IP-CAN and the called users IP-CAN. Both IP CANs in the illustration are in the visited networks and the data path may traverse intermediate IP networks. Signaling and media planes in IMS are completely separate. Also, signaling in IMS always traverses the home network. This is unlike a traditional cellular system where, for a call made from the visited network, the signaling need not traverse the home network.

To be viable, IMS needs to interwork with legacy circuit-switched (CS) networks like PSTN and existing cellular systems. When a call to a phone in the CS network is made by an IMS user, the SIP INVITE arrives at the S-CSCF of the calling user. The address analysis will indicate that the call is destined for a user on the CS network and the S-CSCF forwards the INVITE to the BGCF, which determines to which MGCF the INVITE should be forwarded. The MGCF converts SIP signaling to a signaling protocol that the CS network can understand and forwards the signaling to the switch on the CS network via the SGW signaling gateway. The MGCF also involves an MGW which performs media conversion between packet voice and circuit-switched voice. In the reverse direction, for a call originating from the CS network and destined for a user on the IMS, the CS network forwards the signaling via the SGW to the MGCF, which then converts the signaling to SIP INVITE and forwards the INVITE to the S-CSCF for further signaling.

Interworking between the IMS and CS network requires translation of SIP signaling to SS7 signaling and vice versa, and transmission of SS7 messages between the MGCF and the SGW. IMS resides in the IP domain, and the transport between MGCF and the SGW is over IP networks. Thus, SS7 messages must be transported over IP networks. To transmit these messages reliably requires acknowledged and error-free transmission. While UDP cannot ensure reliable acknowledged transmission, TCP is too restrictive, and that may affect the performance (e.g., delay requirements). Therefore, the IETF formed a working group known as SIGTRAN to define specifications for reliable transport of CS network protocols (e.g., SS7 and ISDN) over IP networks. It uses stream control transmission protocol (SCTP) to provide transparent transport of message-based signaling protocols. The scope of SCTP includes definition of encapsulation methods, end-to-end protocol mechanisms and IP capabilities that support the functional and performance requirements for signaling. SCTP provides the following functions [Ste07]:

- Acknowledged error-free non-duplicated transfer of user data;
- Data fragmentation to conform to discovered path Maximum Transmission Unit (MTU) size and sequenced delivery of user messages within multiple streams, with an option for order-of-arrival delivery of individual user messages;
- Optional bundling of multiple user messages into a single SCTP packet; and
- Network-level fault tolerance through support of multi-homing at either or both ends of an association.

As with CS interworking for voice calls, IMS needs to interwork with the Internet. This means that the S-CSCF may need to exchange SIP messages with a SIP server located outside IMS. These external clients may not support one or more of the SIP extensions required for IMS. In this case, the SIP user agents within IMS may fall back to basic SIP. Operators may also decide to restrict session initiation with external SIP clients not supporting extensions.

Another aspect is IP version interworking. When IMS was originally designed, IPv6 was expected to be widely available by the time IMS was deployed. SIP and associated protocols also had problems with the network address translation (NAT) traversals required in an IPv4 network. Hence, 3GPP (originally) decided to use only IPv6. However, though IMS is ready to be deployed, IPv6 has not taken off and IPv4 and NATs are becoming ubiquitous. Thus, 3GPP allowed IPv4 in early IMS, with dual-stack implementations (IPv4 and IPv6) allowed in IMS terminals and nodes. Two elements, IMS-application-level gateway (ALG) for SIP interworking and transition gateway (TrGW) for RTP interworking have been added to the IMS architecture.

2.5.5   IMS as a Platform for Convergence

Though IMS was conceived and is being standardized in 3GPP, it has become a platform for fixed mobile convergence (FMC) where mobile networks may include 3GPP and non-3GPP networks. There are multiple access networks today, including Time Division Multiple Access (TDMA) in GSM/GPRS access, Universal Terrestrial Radio Access Network (UTRAN [UMTS]), cdma2000 access in 3GPP2, WLAN, cable, DSL, and WiMAX. The most recent addition to access networks is the Evolved UTRAN (E-UTRAN). More access networks are expected in the future. These will include advanced capabilities to E-UTRAN (also known as advanced LTE) as defined in 3GPP, and different variations of 4G wireless networks espoused by academia and industry.

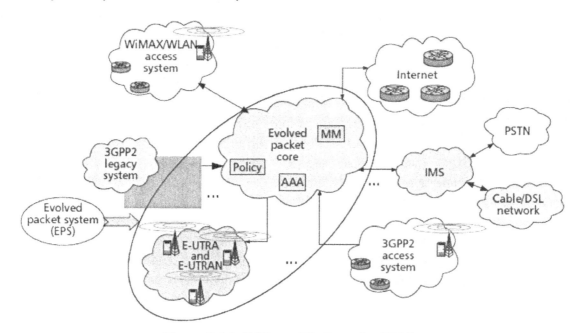

Figure 2-14: IMS as a Platform for FMC

Instead of defining all the layers as in a vertical silo network, it is beneficial to have a common service and control layer that can support multiple access networks; both wireless and fixed networks could evolve while providing backward compatibility. With the access network agnostic control and service layer, IMS provides an ideal platform for multiple access networks and for FMC. Figure 2-14 shows an example of how IMS can provide the platform for FMC. The Evolved Packet Core (EPC) acts as a gateway to different radio access networks to allow IMS to serve as the common control and session layers. Though not shown, IMS will provide open application programming interfaces (APIs) to the application layer to make the same applications available to users independently of the access network.

IMS relies on an underlying IP core for signaling and media transport. Several efforts are underway in 3GPP to realize a common IP core network that provides IP-based network control and IP transport across and within multiple access systems. The requirements and objectives of this effort, defined in the all-IP network (AIPN), include a seamless user experience for all services within and across the various access systems, the ability to transfer sessions between terminals, a high level of security, QoS, high network performance, advanced application services, the ability to select the appropriate access system based on a range of criteria, interworking with existing fixed and wireless networks, support for a wide variety of terminals, and efficient handling and routing of IP traffic.

The EPC is another development in 3GPP for an IP packet-optimized network infrastructure to support the growing demands for IP traffic in terms of both increased data rate and reduced latency. The EPC aims to make optimum use of the increased bandwidth available in access networks and to support multiple access technologies, including existing and evolving 3GPP access networks and non-3GPP access networks. Non-3GPP access networks include WLANs, 3GPP2 access networks, and WiMAX. A key objective of EPC is service continuity and terminal mobility between different access networks. Mobility must be supported at different levels, including within the evolved 3GPP and between 3GPP and non-3GPP systems. This mobility needs to provide for the seamless operation of both real-time and non-real-time services. The EPC effort does not directly impact or influence IMS procedures. However, IMS can take advantage of the core IP network to provide a stable IP address for SIP signaling.

2.5.6 Evolution of IMS to Support New Capabilities

Though IMS standardization began in 3GPP with Release 5, Release 5 provided only the basic platform for IMS. This included SIP-based signaling for registration, session initiation, interworking with the Public Switched Telephone Network (PSTN), and basic security architecture. Release 6 completed the goals of Release 5, including IMS messaging and enhancements for IMS interworking with the PSTN and the packet switched (PS) domain. Later releases added new capabilities to IMS. A summary of the key enhancements introduced to IMS in various releases is given below.

- Voice Call Continuity (VCC): As IMS-based voice services will be introduced in stages, an IMS-capable phone can only make/receive and sustain a call in selected areas where the network supports IMS. To make IMS a viable option, it is essential to include CS capability in the phone and seamlessly handoff the call from IMS to the CS domain and vice versa. VCC, introduced in Release 7, lays the foundation for anchoring a call in the IMS domain and providing continuity of a session as a user moves between areas of IMS coverage and areas without IMS coverage.

- Combining CS and IMS session (CSI): To promote IMS while IMS-enabled voice terminals are still not widely available, a capability to combine a circuit-switched voice call with an IMS data session on a different terminal was introduced in Release 7. This allows the end user to feel the experience of a multimedia session even though the voice part of the session still uses the circuit-switched domain. The other end of the session may use a similar approach or use a true IMS terminal where both voice and data are supported using IMS.

- Emergency Calls in IMS: As IMS becomes more prevalent, regulatory requirements make it necessary to support emergency calls on IMS. Release 7 introduced an Emergency CSCF (E-

CSCF) in the IMS architecture. E-CSCF allows special registration and service control without having to obtain it from the home network for making an emergency call and to query a location retrieval function to retrieve the location of the UE.

- Common IMS: Though 3GPP initiated and continues IMS standardization, many other organizations embraced IMS as a control/session layer. 3GPP2 Multimedia Domain (MMD) consists of a Packet Data Subsystem (PDS) and IMS. While PDS provides IP connectivity similar to GPRS or other 3GPP access for CDMA access networks, the IMS defined by 3GPP2 is very similar to 3GPP IMS. Similarly, Telecommunications and Internet converged Services and Protocols for Advanced Networks (TISPAN) Next Generation Networks (NGN) are based on 3GPP IMS. Though these different organizations shared a similar view of IMS, requirements derived in different organizations were in danger of getting fragmented. 3GPP decided to pool the resources and develop all the access-independent aspects of IMS in 3GPP. Known as Common IMS, 3GPP maintains one common set of IMS specifications that would be applicable to fixed, mobile, and broadband wireless networks. During 3GPP Release 8, IMS specifications from 3GPP2 and TISPAN migrated to 3GPP. From Release 9 onwards, 3GPP maintains one common set of IMS specifications.

- IMS Centralized Services (ICS): During transition from CS to IMS there will be CS networks and IP networks, some capable of supporting IMS and some not. To allow a consistent user and service provider experience as the networks migrate to IMS, it is necessary to have a common approach to service control. ICS is an initiative to provide all services and service control based on IMS mechanisms. To achieve IMS service control when accessing via CS networks, the MSC server in CS network acts as a proxy for the CS UE to the IMS core. A CS call origination appears to the IMS core as coming from an IMS UE. The service control and the call anchoring are provided in the IMS core. ICS specification started in Release 8, but reached a more stable state in Release 9.

- IMS Session Continuity (IMS SC): While VCC defines voice call continuity between the CS and IMS domains, IMS SC extends it to multimedia sessions and across both CS-to-PS and PS-to-PS domains. IMS SC also allows transferring one or more components of a multimedia session from one UE to another. For example, a person may initiate or receive a multimedia call while commuting to work and may want to transfer the video component of that session from the cell phone to a desktop device upon reaching the office. IMS SC has been a major work in Release 9 and continuing to Release 10.

2.5.7 Deployment Status

IMS Release 5/6 with basic voice capabilities became stable by 2004. Deployment and market projections at that time showed a solid growth for IMS. However, the hype behind IMS soon became disillusionment and IMS faced many challenges. The complexity of the standards, lack of a clear business case, lack of multivendor interoperability, lack of IMS handsets, etc. slowed the anticipated uptake of IMS. However, most tier 1 operators still consider IMS as a long term option and are trialing IMS. Over 50 live IMS networks were reported as of the beginning of 2010. AT&T deployed Video Share in 2007 which was based on combining circuit switched voice with IMS for data. Early deployment of IMS has been driven by fixed operators towards achieving Fixed Mobile Convergence (FMC). Ericsson leads the market with over 60 contracts followed by Huawei, Nokia Siemens Networks, and Alcatel Lucent as the major suppliers for IMS.

LTE (see section 2.7) has turned out to be a major facilitator for IMS. Demand for data over cellular networks is driving LTE deployment. Over 100 operators in nearly 50 countries had committed to LTE as of the second quarter of 2010. Since LTE does not support circuit-switched connections, there is a need to support voice over IP networks. Though other approaches (e.g., CS Fallback – where voice calls will fall back to use GSM Edge Radio Access Network [GERAN]/UTRAN circuit switched connections) are proposed as a short-term approach, operators believe that IMS-based approaches are required in the long run. A number of operators including AT&T, Verizon, Orange, Vodafone, Telefonica, and Telia Sonera are behind IMS-based voice and SMS over LTE. Many major vendors have also voiced support for this approach, known as OneVoice. The mobile industry took a significant step towards a common standard for delivering voice and SMS services over 3G LTE in the summer of 2010, after the GSM Association (GSMA) adopted the One Voice Initiative.

2.5.8 IMS Summary

IMS provides an open architecture with a common control and session layer to offer the best of the fixed, cellular, and Internet worlds. The 3GPP initiated IMS, but a number of other standards bodies soon embraced it. IMS introduces a number of new network elements. The call state control function (CSCF) is the most central node in IMS. Service control in IMS is always in the home network. This differs from the philosophy of traditional cellular networks that puts service control in the visiting network. IMS allows support for applications from both home and third-party networks. IMS protocols are based on IETF protocols; SIP and Diameter are the most dominant protocols. IMS allows interworking with CS networks and the Internet. IMS users can make and receive calls from CS networks users and also establish sessions with Internet users and applications. Seamless handoff between IMS and circuit-switched domains is defined for voice. For multimedia sessions, IMS session continuity allows seamless continuity across different PS domains and even across multiple UEs. With access-agnostic control and session layers, IMS is an ideal platform for fixed mobile convergence. A number of operators have IMS trials underway, and some have already started early deployments. Though the initial deployments were far below expectations, LTE is expected to spur IMS deployment. Deployments and a healthy growth of IMS subscribers are anticipated in the coming years.

2.6 Introduction of the Evolved Packet Core

The demand for higher data rates is mainly caused by multimedia streaming portals (e.g., YouTube), social networks (e.g., Facebook), VoIP-telephony (e.g., Skype), and web browsing. Multimedia in general and video in particular represents the major fraction of the total data traffic with an estimated 66% until 2014, followed by web and Peer to Peer (P2P) [Cis10].

The market research analyst firm Gartner announced a growth of 13.8% in the sale of mobile devices in the second quarter of 2010 in contrast to the same period in 2009, with 325.6 million devices sold worldwide. The sales volume of Smartphones reached 62 million in the second quarter of 2010, grew strongly at over 50%, and reached a 19% share of the mobile market.

A higher density overall wireless network coverage emerges from the maintenance of current 2G and 3G technologies and the roll out of new 4G technologies. The current UMTS plus new LTE and WiMAX technologies increase wireless broadband access network diversity and fade out legacy Circuit Switched (CS) access network domain, while moving to mobile All-IP Packet Switched (PS) networks. Since LTE is just one access network technology and mobile applications need to interoperate with various access network technologies, a new core network architecture is required, which aggregates the different access networks and provides transparent interfaces for applications on top of it. 3GPP standardized the Evolved Packet Core (EPC) for solving this challenge, which is described more in detail in this section.

The mobile operator's revenue from the services domain decreases, since many value-added services (VoIP, multimedia download, and news) are available for free on the Internet today. The amount of mobile data traffic (mainly voice) aligned linearly with the revenue in earlier days. There is a divergence between the user's data traffic and the operator's revenue nowadays, which will continue to grow in the near future. Data traffic today consists of voice, video, and files, which are increasing dramatically [Cis10], but the revenue of the operators has almost stagnated. The necessary reduction of costs per bit enforces a reduction of architectural components in the core network. Such a reduction in turn minimizes the complexity in the Radio Access Network (RAN) and implies a reduction of capital expenditures (CAPEX) and operating expenditures (OPEX) for network operators.

The Global mobile Suppliers Association (GSA) published an update of its Evolution to LTE report in June 2010, which confirmed 110 operators in 48 countries were investing in LTE networks. 80 operators have made firm commitments to deploy LTE networks in 33 countries. GSA anticipated that up to 22 LTE networks would be in service by the end of 2010, and at least 45 were expected to be in service by the end of 2012. The first LTE networks entered commercial service in December 2009 in Norway and Sweden [GSA10]. More recently, the GSA published an update of its Evolution to LTE report in January 2012 [GSA12]. In this update, it was reported that 285 operators are investing in LTE in 93 countries (226 operator commitments in 76 countries, 59 pre-commitment trials in 17 more countries), Meanwhile, there were 49 commercial networks in 29 countries, and GSA now forecasts that there will be 119 commercial LTE networks in 53 countries by end 2012. Four telecommunication vendors won the frequency spectrum block auction for Germany in May 2010 with about 4.4 billion Euros. O2/Telefonica, Vodafone, and Deutsche Telekom announced their LTE roll out in German cities for the end of 2010. AT&T started their commercial LTE roll out in 2011.

2.6.1 *Evolved Packet System (EPS) = LTE + EPC*

3GPP standardized the EPS, formerly known as System Architecture Evolution (SAE), which consists of Evolved UTRAN (E-UTRAN) and Evolved Packet Core (EPC), as shown in Figure 2-15.

The objectives of the Evolved Packet System [3GP11c] are to:

- Provide higher data rates, lower latency, high level of security, support of variable bandwidth, and enhanced Quality of Service (QoS);

- Support a variety of different access systems (existing and future), ensuring mobility and service continuity between these access systems – Seamless Service Continuity;

- Support access system selection based on a combination of operator policies, user preference, and access network conditions – Always-Best-Connected (ABC);

- Realize improvements in basic system performance while maintaining the negotiated QoS across the whole system; and

- Provide capabilities for co-existence with legacy systems and migration to EPS.

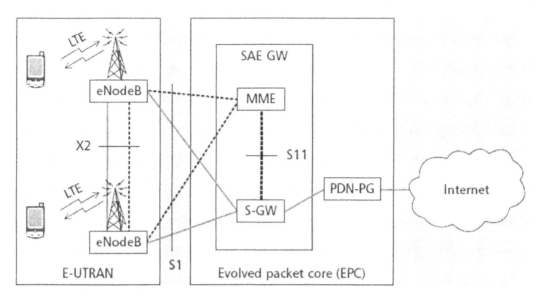

Figure 2-15: Evolved Packet System = E-UTRAN + EPC

2.6.2 Long Term Evolution (LTE)

The LTE radio access network side, E-UTRAN, is composed solely of the Evolved NodeB (eNodeB). The eNodeB controls InterCell Radio Resource Management, Radio Bearer Control, Connection Mobility Continuity, Radio Admission Control, and Dynamic Resource Allocation. E-UTRAN is connected to the EPC via the S1 interface. LTE combines OFDMA with Multiple-Input-Multiple-Output (MIMO) antenna techniques. The transmission of several independent data streams in parallel over different antennas, uncorrelated in frequency and coding, increases the overall data rate. The UE updates its channel conditions to the eNodeB using the smart Feedback Channel Concept (FCC). LTE is sometimes said to be 4G (a controversial designation!) and enables Single- Input Single-Output (SISO) download peak data rates (64QAM) of 100 Mb/s and up to 326.4 Mb/s with a 4×4 MIMO technique. The uplink peak data rate depends on the modulation level and reaches up to 86.4 Mb/s (64QAM) with a single antenna [3GP10].

2.6.3 Evolved Packet Core Architecture

This section describes the core components of the EPC, highlights their major functionalities, and outlines the main interfaces between these components. Figure 2-16 describes the EPS architecture covering the core components and Figure 2-17 depicts the main components of the EPS.

The EPC consists of three main components: Mobility Management Entity (MME), Serving Gateway (SGW), and Packet Data Network Gateway (PGW), plus several other subcomponents (see Figure 2-16).

- The key control node for LTE access is the MME, which is responsible for user tracking, SGW selection, Idle State control, and Bearer Control and which enforces user roaming restrictions.

- The SGW acts as a mobility anchor for the user while it is accessing the network through LTE or other 3GPP technologies (intra 3GPP handover), manages UE contexts, and controls the UE data path with terminating or re-establishing through paging requests depending on the current usage.

- The PDN Gateway provides connectivity from the UE to one or multiple external PDNs simultaneously. The PGW acts as a mobility anchor between 3GPP and non-3GPP technologies and performs packet filtering and charging.

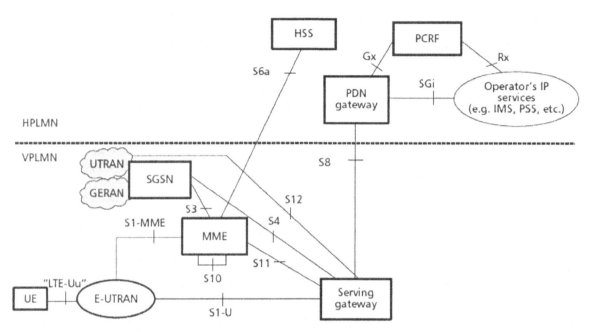

Figure 2-16: 3GPP GPRS Enhancements for E-UTRAN Access [3GP11d]

- The Home Subscriber Server (HSS) is the main database of EPC subscriber information. Static user profiles as well as dynamic information like session contexts and locations are stored in the HSS [3GP11e].

- The Policy and Charging Rules Function (PCRF) encompasses two main functions:
    - Flow Based Charging, including charging control and online credit control; and
    - Policy control (e.g., gating control, QoS control, QoS signaling, etc.) [3GP11f].

- A Serving GPRS (General Packet Radio Service) Support Node (SGSN) enables the delivery of data packets from and to mobile stations within its geographical service area. The SGSN supports mobility management, authentication, and charging in addition to packet routing and forwarding.

- The Access Network Discovery and Selection Function (ANDSF) is a new EPC element in Release 8. The ANDSF performs data management and controls functionality to assist the UE in the selection of the optimal access network in a heterogeneous scenario via the S14 interface [Cor10b].

- Authentication, Authorization, and Accounting (AAA) in the EPC are performed by the HSS, MME, and 3GPP AAA Server. AAA secures the user subscription, session key management, and security tunnel control (see Chapter 3 for more on AAA protocols).

- The evolved Packet Data Gateway (ePDG) attaches un-trusted non-3GPP access networks like WLANs to the EPC. The ePDG performs important security functions, tunnel authentication and authorization, and IPSec encapsulation/de-capsulation of packets. Alternatively, trusted WLANs can bypass the ePDG and connect directly to the PDN gateway. These two alternatives (trusted and un-trusted non-3GPP access networks connecting to the EPC) are shown in Figure 2-17.

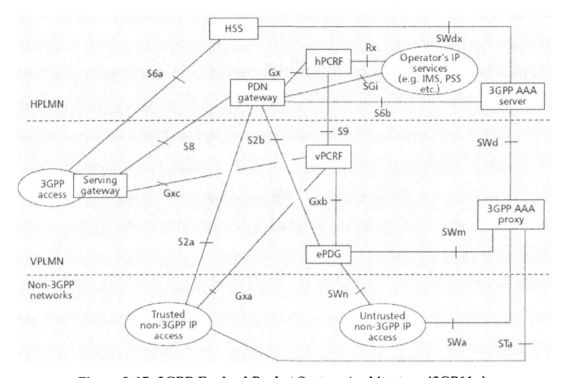

Figure 2-17: 3GPP Evolved Packet System Architecture [3GP11g]

- Home eNodeBs or Femtocells are small cellular base stations located in a private environment to improve indoor coverage and network capacity [Wan09].

- The Application Function (AF) is an abstraction of the service provider, which communicates with the PCRF to enable AAA at the application layer.

2.6.4 Protocols

Several protocols are used in the EPS within the different interfaces. Internet Protocol (IP), Mobile IP (MIP) and variations, Proxy Mobile IP overIPv6 (PMIPv6), and GTP are used in the Network Layer. The Transport Layer uses Stream Control Transport Protocol (SCTP) and TCP/UDP. The Application Layer supports OMA Device Management (DM), Diameter, and S1-AP.

2.6.5 *Applications and Services over the Evolved Packet Core*

Applications today assume that the underlying access network is transparent and that they need to perform the two tasks of network connection establishment and service execution independently, without the ability for any adaptation. This implies that a vertical handover (access network change) terminates a service before the new connection to the network is established, followed by a new service request using the new connection.

The architecture of EPC – especially the AF, PCRF and Gateways – allows service continuity and always-best-connected (ABC) access selection. Connectivity is separated from data transport; thus vertical handovers still require new connection establishment, but are able to adjust established service connections. Therefore rules from the PCRF may be applied to services in the AF to continue service delivery. An example would be a video streaming application, which is able to change its codecs and thereby alter the data rate according the network capacity of the connected client [Cor10c].

The EPC enriches its functionalities through the ANDSF, which aids in the access network (AN) selection process in the case of a handover. The handover process is supported by the ANDSF, which provides knowledge about neighboring ANs.

An IP address is assigned to the UE automatically when connecting to the network, within the process of default EPS bearer activation. The default EPS bearer comes without any QoS guarantees and transports IP data traffic using the best-effort paradigm. In contrast to the default, the dedicated EPS bearer allows different prioritization of real-time traffic or emergency calls. The concept of Traffic Flow Templates (TFT) identifies flows through source and destination IP addresses and ports. Traffic differentiation is done on the UE and in the network. The PCC enforces prioritized packet transport on the network through policy enforcement and lists [3GP11f].

3G networks separate CS voice from PS data transfer; thus voice communication still profits from CS QoS guarantees, whereas the PS domain uses the best-effort transport paradigm. 4G/LTE is completely PS. A final solution for providing voice calls in the first phase of the LTE roll out has not yet been agreed upon. Several solutions exists and the most promising is Single Radio Voice Call Continuity (SR-VCC) using IMS Centralized Services (ICS) for ICS-based call control and handover from LTE to 2G/3G via a dedicated IMS Application Server (SCC AS). Other possibilities for voice in LTE, listed for the sake of completeness, are not discussed in the scope of this chapter: Circuit Switched Fallback (CSFB), Voice over LTE via Generic Access (VoLGA), and CS over EPS [ATT09].

2.6.6 *Implementations of Evolved Packet Core*

OpenEPC is a prototype implementation of the 3GPP Release 8 Evolved Packet Core (EPC) that will allow academic and industrial researchers and engineers around the world to obtain a practical look and feel of the capabilities of the Evolved Packet Core. OpenEPC can be integrated with various access network technologies and different application domains and thus provides an excellent foundation for individual research activities and/or the establishment of Next Generation Mobile Network testbeds such as the Fraunhofer FOKUS FUSECO Playground for FUture SEamless COmmunication [Cor10].

2.7   Alternative Network Architectures – Mesh Networks

The wireless mesh network (WMN) is a relatively new concept revolutionizing the way future broadband Internet access could be provided to customers, called mesh clients (MCs). Basically, the idea is to place wireless routers (known as mesh routers, or MRs) on top of tall buildings where they form a static ad hoc network among themselves. Each MR serves neighborhood MCs. A few MRs known as Internet gateways (IGWs) are also connected to the Internet and provide Internet access for the rest of the MRs.

When the MRs and IGWs are owned and operated by different individuals the configuration is known as a community-based network. When the MRs are controlled by a single service provider, the arrangement is called fully managed. An intermediate solution known as a semi-managed WMN is possible when MRs are controlled by a few entities. An example of such a generic WMN is shown in Figure 2-18.

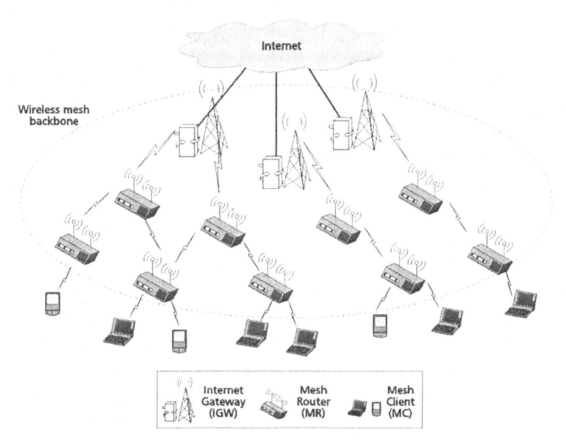

Figure 2-18: Mesh Network Architecture

An important issue in implementing a mesh network is to make sure that the MCs have sufficient wireless access coverage and bandwidth. So, to deploy an adequate number of MRs, it is important to know the number of users and their total bandwidth requirement. It is also useful to know if MRs can be placed anywhere in a given geographic area or if there are limits on their locations. Moreover, it is useful to know the busy period of each MC and the operating schedule of each MR, which are indications of the instantaneous traffic load and the number of MRs needed. MRs can be located where AC power is readily available and hence the power constraints present in a typical mobile ad hoc network (MANET) [Cor06] (see section 2.8) do not apply.

Another important issue involves WMN positioning. The topology of WMNs can be formed on an ad hoc basis, and the connectivity pattern depends on the relative distance between two adjacent MRs. But, unlike a MANET, MRs are not mobile so the topology remains static unless, due to business decisions, some MRs are taken out of commission or turned on later. Thus, the locations of the MRs and the IGWs are among the critical factors that determine WMN performance [Gup00]. If the MRs and the IGWs are randomly situated, as in a MANET, a WMN can encounter the several problems: (1) unbalanced load distribution, (2) uncontrolled interference, and (3) unreliable architecture.

The WMN positioning depends on the physical configuration of the IGWs and MRs, including their locations and the number of interfaces they have. Their configuration is subject to constraints involving geography, the maximum number of channels in the network, and traffic demand. Of particular concern is where the IGWs should be placed so as to minimize their number while satisfying the MR Internet throughput demand. One simple example of a WMN with 4 MRs is shown in Figure 2-19 and the corresponding observed bandwidth using simulation is shown in Figure 2-20 when 20 Mb/s traffic is generated at each MR and packets in each MR are processed in a FIFO manner. Important IGW selection algorithms include cluster-based IGW selection [Bej04], OPEN/CLOSE heuristic IGW selection [Pra06], and tree-based IGW selection [He07].

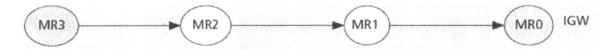

Figure 2-19. Example WMN with 4 MRs Connected in a Linear Array

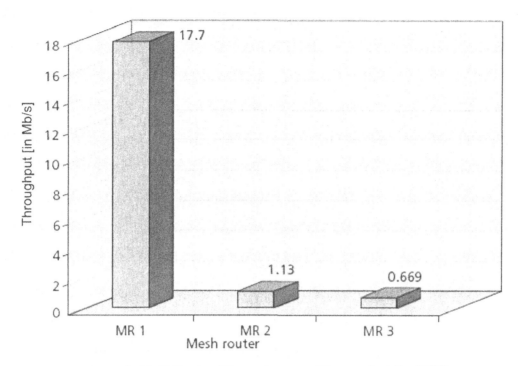

Figure 2-20. Effective Throughput as Observed at the IGW

The efficient placement of MRs is a challenging issue because of many practical constraints and contradictory requirements, including cost, link capacity, wireless interference, and varying traffic demands. To minimize costs while meeting traffic demand, any given region must be covered with a minimum number of MRs and interfaces. Thus, goals for MR positioning include providing enough network capacity; avoiding congestion incurred by balancing traffic; maximizing network capacity with a certain number of MRs and their configured interfaces; and ensuring that the network is fault tolerant. An elementary exploration of MR placement is presented in [Wan07]. The authors discuss MR placement on the condition that MRs can only be placed in pre-decided candidate positions while considering the coverage, connectivity, and traffic demand constraints. They propose a two-phase heuristic algorithm to find the close-to-optimal MR placement.

Note that traffic in a WMN is generated by MCs at the MRs and moves between MRs and IGWs and not between the MRs themselves. In fact, MRs needing access to the Internet either download the packets from an IGW or forward them to reach an IGW. Thus, besides serving MCs in its own neighborhood, each MR ought to forward packets toward an IGW from MRs further away, and the volume of traffic near each IGW is the highest. The volume of data increases closer to each IGW with the maximum data rate dictated by the cumulative bandwidth of the IGWs. Thus, responsibility for routing must be undertaken by each MR. Note that because packets must be queued and served in each MR, packets generated by MRs near an IGW tend to achieve higher bandwidth than MRs that are further away. Therefore, utmost care must be taken to process packets based on the number of MRs already traversed [Nan07a].

Even after the architecture of a WMN has been designed, many issues remain regarding the utilization of resources. Decisions are needed regarding the number channels to be allocated to MRs and the number of radio interfaces for each MR. This can affect the simultaneous utilization of multiple channels at one or more MRs, reduce interference between them, and increase the availability of bandwidth at different MRs. Accordingly it is desirable to have more and more wireless radios as one gets closer to the IGW.

Such a design can take care of the increasing concentration of traffic near the IGWs. And note that interference between nearby channels should be avoided. Merely having an adequate number of radios is not enough because it takes a finite time to change the status of an interface circuit from off to on and vice versa, and to modify the radio frequency.

Multiple channels are available in current Media Access Control (MAC) protocols. For example, 12 orthogonal channels are available with IEEE 802.11a and three with IEEE 802.11b. An MR could tune its radio interface to a channel not being used by neighboring transmitting MRs to avoid interference and thus improve throughput. For higher throughput, an MR equipped with a single radio interface can dynamically switch the frequency among multiple channels. The efficiency depends on the delay required for an interface to switch from one channel to another, which is not negligible.

Another important factor is the channel-switching synchronization for a pair of transmitting MRs, whose time information could be from either an external source, e.g., the Global Positioning System (GPS), or an internal source, e.g., beacon signals from neighboring MRs. To effectively exploit multiple channels, each MR should be equipped with multiple interfaces that simultaneously work on different channels. Each interface of an MR can be fixed at a channel that is not used in the interference region. If the channels are statically bounded to multiple interfaces, the number of channels utilized by each MR is constrained by the number of interfaces, which may force the removal of some links because a pair of MRs within the transmission region may not have a common channel. Hence, while maximizing throughput, the assignment of channels to interfaces needs to ensure the network's connectivity. Appropriate assignment of multiple channels to interfaces can eliminate much interference and enhance network performance.

Another question asks what the optimal path is from a given MR to an IGW. One can find the shortest paths from each MR to all IGWs and select the IGW closest to the MR. But the shortest path might not be the best choice because MRs are known for incorporating multi-rate links. The shortest design may be fine if the traffic is originating only from that particular MR. However, with the MRs free to serve their MCs, the volume of traffic passing through each MR can fluctuate significantly, and each MR must forward the packets towards the IGW. Therefore, availability of bandwidth along a given path not only depends on its own traffic, but also on the traffic generated by other MRs and the paths selected to reach the IGWs. In selecting an optimal path for a given flow it is thus important to have a global picture of the volume of flows through the network. Information is needed about the multiple paths from each MR to one or more MRs. Also, such dynamic selection of the paths and channels is important because the WMN topology may also change with time as some of the MRs are turned off and on.

Another important issue that arises in the design of a large-scale WMN is the fairness of the MCs regardless of their distance to the IGW. It is well known that the IEEE 802.11 network suffers from degradation of throughput in a multihop path because of interference, hidden terminal problems, and so on. Thus, careful attention should be paid to ensure that traffic from nearby MCs does not dominate the buffers of MRs near the IGW, and sufficient buffer space is guaranteed to be available to distant flows. This requires efficient queuing-management and traffic-splitting schemes [Nan07b].

As discussed earlier, besides serving their own MCs, all MRs must cooperatively forward packets towards the IGWs. However, some MRs might act selfishly so as to fully utilize the available bandwidth and provide higher Internet throughput for their MCs. This is a serious problem and such "free-riding" behavior is unacceptable in a community-based WMN network. Owing to the hierarchical architecture of a WMN, MRs that are further away from the IGW are inherently dependent on the MRs near the IGW to forward their packets. A selfish MR near the IGW can thus cause serious performance degradation. Packets dropped by the selfish MR have already consumed significant network resources as they are forwarded by other intermediate MRs.

Thus, such free-riders ought to be penalized by some kind of detection mechanism. Detecting selfishness in a multi-channel WMN becomes a challenging problem. Some traditional reputation-based approaches that depend for detection on promiscuous listening might be applicable due to the assignment of non-overlapping channels between adjacent MRs. However, due to the static nature of WMNs, a credit-based approach that assigns virtual currency to cooperative MRs largely fails. MRs in the periphery of the network fail to earn sufficient credits and are handicapped. Traffic-monitoring techniques have been devised to detect and identify such misbehavior [San06], and then isolate such MRs by not serving their associated packets.

The 802.11s standard was developed by the IEEE 802.11 working group for wireless mesh networks. Depending on wireless distribution, the architecture of 802.11s supports broadcast, multicast, and unicast; its main objective is to perform routing at the link layer. The interconnected wireless devices in 802.11s mesh networks are called mesh points (MPs). MPs may be static or mobile but they have packet-relaying capability and can connect to one another automatically to form a mesh topology. Some MPs in the network have wired connections to the Internet and act as IGWs. Unlike a traditional 802.11 WLAN, a mesh network provides Internet connectivity over multiple wireless hops. A MP is able to discover any existing WMNs within the transmission region and can get associated with them. A MP is also able to initiate a new network if no network is detected. The whole discovery process is auto-configured and does not need any user intervention. MPs in the network can exchange packets over multi-hop wireless links, selecting the path dynamically. When an MP needs to send a packet, it performs layer 2 MAC-based routing, which is different from layer 3 routing using IP addresses. Furthermore, the routing metrics of the link layer depend on the quality of the wireless links because the layer observes channel conditions and finds the route(s) satisfying the QoS requirements. Therefore, 802.11s is capable of guaranteed performance with respect to such parameters as throughput, packet loss, delay, jitter, and so on.

Many actual WMNs and many variants have been implemented at numerous institutions [Wik11] and are primarily based on 802.11-based routers. This current trend is due to many reasons such as the availability of inexpensive routers, a well-developed routing protocol in the ISM frequency band, and most MCs and end devices having an 802.11-compatible WiFi wireless interface. These are experimental systems, however, and when WMNs are actually deployed for multimedia traffic, adoption of WiMAX as the basic unit for supporting wide-area coverage is anticipated. Even though 802.11 and WiMAX use different frequency bands and 802.11 runs on the connectionless Carrier Sense Multiple Access/Collision Avoidance (CSMA/CA) protocol, whereas WiMAX runs a connection-oriented MAC, future WMNs may consist of a combination of these two types of complementary routers: WiMAX for long-distance coverage and 802.11 routers for local connectivity. This may be possible even though they use quite different quality of service (QoS) mechanisms.

Security is an important issue that has been overlooked in a WMN and becomes critical whenever handover occurs for any MC. If the network is fully managed by a single service provider, the handover and authentication may be relatively simple as all MRs follow the same protocol and are under the same management. But if the WMN is partially managed or employs a community-based approach, then handover necessitates differentiation of service providers and associated authentication steps. Handover in such system may therefore cause delays, disruption in service, and loss of packets. In addition, if any encryption is used, the mechanism has to be restarted after the handover and could be a very involved process. These will depend on the mobility of the MCs, the pattern of mobility and how MRs under different service providers have been physically placed in a WMN. Moreover, since a WMN is a kind of open architecture, many possible attacks could be easily injected and need to be monitored very carefully.

2.8   Alternative Network Architectures – Mobile Ad Hoc Networks

The mobile ad hoc network (MANET) is one of the newly emerging wireless network architectures. It is composed of wireless user devices that deliver data packets via multihop packet forwarding. Like wireless mesh networks (WMNs), a MANET differs from conventional wireless networks (e.g., cellular and WLAN networks), which are infrastructure-based and have the wireless link just at the last hop. Both WMN and MANET are multi-hop wireless networks, but WMN nodes are fixed while MANET nodes can move around. As shown in Figure 2-21, source node S transmits data packets through a multihop route to destination node D. En-route, MANET nodes act as intermediate relay nodes to forward data packets for other users.

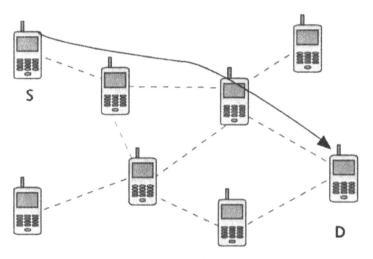

Figure 2-21: Routing in a Mobile ad hoc Network

Routing, which discovers the multihop packet forwarding paths, is a major issue in MANETs, and the user's mobility poses challenges in the routing protocol design. For example, the routing protocol needs to discover new routes while keeping track of the validity of old ones. Based on routing discovery and updating mechanisms, the MANET routing protocol can be categorized as either proactive or reactive.

Ad hoc on-demand distance vector routing (AODV) [Per03] and dynamic source routing (DSR) [Joh01] are two popular reactive routing protocols. These protocols, also known as on-demand routing protocols, discover multihop relay routes only when needed (on-demand).

In the AODV and DSR routing protocols, the basic mechanism of finding valid routes is control message flooding. When an AODV source node S would like to send a data packet to a destination node D, S first checks its routing table to see if a valid route to D exists. If a valid route exists, S just sends the data packet to D using that route; otherwise, S will initiate the route-discovery process. In this route discovery process, S floods the route request (RREQ) message to all nodes in the network. S first broadcasts the RREQ message to its neighboring nodes. When intermediate network nodes receive a RREQ message, they re-broadcast the message to their neighboring nodes. When the destination node D receives the RREQ, which indicates S is looking for a route from S to D, D replies with a route reply (RREP) message toward S. During the RREP transmission from D to S, AODV intermediate nodes store the routing information for a valid route between S and D. After S receives the RREP message, the valid route toward to D is recorded, and S then starts sending data packets to D.

Similar to the AODV operations, DSR initiates the route-discovery process if there is no valid route entry for sending data packets. The flooding of route request (RREQ) and replying of route reply (RREP) is similar to that of AODV. The main difference between AODV and DSR is that DSR is a source-routing protocol that carries routing information in packet headers. During the route-discovery phase, DSR nodes store the hop-by-hop routing information in RREQ and RREP routing messages. During the data-transmission phase, the routing information is carried in the data packets' header.

Proactive routing protocols operate similarly to the Internet routing protocol. They maintain valid routes by periodically updating routing information. Destination sequenced distance vector (DSDV) [Per94] and optimized link state routing (OLSR) [Cla03] are two popular proactive MANET routing protocols. DSDV is based on a distance-vector routing mechanism in which network nodes distribute their entire routing tables to neighboring nodes to compute valid routes. OLSR is based on a link-state routing mechanism in which the nodes distribute their neighboring connectivity to the whole network to compute valid routes.

In DSDV, each node periodically forwards the routing table to its neighbors with a sequence number that indicates the freshness of the routing information. When a wireless node forwards its routing table, the sequence number is increased and appended to the routing table. The received-neighbor routing tables are stored with the local routing information with their sequence number attached. This number is used to guarantee loop-free routing. In addition, DSDV is designed to overcome fast topology changes in a MANET by sending an immediate route advertisement when there are significant changes in the routing table. This feature enables fast routing updates for topology changes in MANETs.

In conventional link-state routing protocols, all nodes periodically broadcast the status of their neighboring links. In OLSR, the link-state routing is optimized for a MANET by reducing the broadcast of link-state information. Unlike wireline networks, wireless transmission is broadcast in nature. A transmitted message is received by all nodes in the neighborhood of the sender. As a result, not all MANET nodes need to re-broadcast messages to achieve network-wide flooding. Only a subset of OLSR nodes, called multipoint relays, are selected to broadcast link state information. The multipoint relay nodes are selected so that the link-state information flooding covers the whole network while reducing redundant transmissions.

There are also hybrid routing protocols that combine the mutual advantages of proactive and reactive routing. For example, zone routing protocol (ZRP) [Pea99] applies proactive routing within the routing zone closer to the ZRP wireless node, while applying reactive routing outside the routing zone. Moreover, some MANET routing protocols combine with location information in routing operations [Ko00]. These types of location-aided, or geographical, routing operate effectively when the wireless network topology is highly correlated with the geographic topology.

MANET technology has great potential for a wide variety of applications. The technology has been adopted by the U.S. Army for military applications. It can also be applied to vehicular wireless networks. The vehicular ad hoc network (VANET) is an extension of MANET suitable for high-speed and predictable mobility scenarios. Wireless mesh networks (section 2.7) could also be viewed as an extension of MANET. In wireless mesh networks, some infrastructure nodes are added as dedicated relays to assist multihop forwarding; hence, user device forwarding is reduced to achieve better network scalability. Wireless mesh networks and MANET technology have been applied to municipal and community wireless-access applications.

2.9 Wireless Service Technologies and Architectures

A broad range of services have been provided by a variety of wireless access technologies over the years. The key mobile communication service has been the voice call, and cellular systems have traditionally been circuit-switched and optimized for voice. Mobile data services have, however, grown significantly. As much as 15% to 30% of mobile business, in a variety of global markets, may be attributed to non-voice services. The evolution of packet/IP-based networks enables efficient development, control, integration, and delivery of IP multimedia services. At the same time, a converged service framework increasingly allows the creation and delivery of all services in a converged and common environment, allowing orchestration of services and distribution of common functions.

In their broad sense, wireless services include a wide range of services offered over wireless links or networks. Strictly speaking, services have increasingly become seamless, user-/intention-centric, and less dependent on a particular network and its access or even core. However, wireless service-enabling environments such as satellite, wide-area mobile, fixed wireless, local-area, personal-area, or femtocells, among others, do pose particular attributes, offer particular possibilities, and typically warrant relevant standardization. Furthermore, the evolution of the core network has been about enabling or facilitating a common multimedia service framework and a common session control, with such important functions as policy, differentiation, and quality of service.

The standardization, specifically service enabler definition and specifications developed by the Open Mobile Alliance (OMA), leverages the existing universal interfaces, such as those defined by the Internet Engineering task Force (IETF). Furthermore, it defines service enablers, open interfaces, common service and user profile frameworks, and dynamic delivery mechanisms, independent of underlying networks or types of implementations. Other standards bodies and industry forums invariably cover some form of service definition, but invariably work together to leverage and adopt common standards, which may be a complete enabler, a building block, or an interface. For example, the GSM Association (GSMA, www.gsmworld.com) inter-operator service initiatives may use or demand standardized enablers or building blocks, possibly within a phased evolution or set of releases. Alternatively, the Femto Forum may develop a service environment definition appropriate for its use case scenarios.

The end user, whether person, being, machine, or thing (sensor, tag, grid), must be served for its intended application, while aiming for a rich experience at high efficiency (cost-effectiveness). This experience must increasingly become ubiquitous, across different user spaces. The demands of an increasingly mobile world go hand in hand with new possibilities enabled by innovations. Fundamentally, the services may be viewed as enabling connectivity, communication, and content, upon which a broad range of applications may be developed.

2.9.1 Mobile Services

Mobile data services have been available for years, particularly with messaging, but also browsing, downloads, and location-based services in second-generation networks. The growth of mobile networks, user terminal functionality, and user experience have fostered rich communications and mobile email, high-speed access and business services, browsing, and content services, all of which are growing significantly. This is an integral part of serving an increasingly mobile society that is evolving to demand further personalized and community services and applications for such new paradigms as work, health, entertainment, information, advertising, and user-generated content on the go.

This evolution is not just about speed but also the user experience and value. The evolution will succeed in its response to the need for a user-centric, and not a platform-centric, service space that is intention-based. A service object may be dynamic and virtual. In space, it may include one or more enablers choreographed to serve the intention; in time, it can adapt to the changing parameters, needs, or communication space. But without capacity growth and cost efficiency, the exponential growth in wireless traffic cannot be accommodated.

The term "service" has been used to refer to a particular user application, which may be enabled by a particular technology platform or viewed as a customer solution. It has also had other uses as network functions, Extensible Markup Language (XML) messages, and others. In this section, service enablers are referred to as particular technology standards or platforms that typically work end-to-end to enable a particular service (or application). A service enabler may be able to provide more than one user application, particularly as services become more application-based, context-aware, personalized, or user-centric. Alternatively, multiple enablers may be leveraged to create a service. For example, a rich messaging service may be created by the use of multimedia messaging, presence, location, and possibly group (XML list) enablers.

The main global standardization body that specifies mobile service enablers is the Open Mobile Alliance (OMA), which works in close coordination with other standards bodies and industry forums. The standards, which make use of IETF protocols, try to be bearer-independent though, where needed, they may accommodate bearer-specific objects and elements. They include such service enablers as messaging and rich communication, browsing, content delivery and management, broadcast and multicast, location, presence, device and rights management, mobile commerce, and mobile advertising, among others.

Proprietary services have been broadly implemented for various business needs including time-to-market. There is a great deal of effort, however, to establish inter-working across systems and service providers.

2.9.1.1 Messaging and rich communications services

Short message service (SMS) has been a leading and superbly successful mobile data service. Typically, it includes text with a limited number of characters, mostly to communicate short messages between mobile terminals. However, variety does exist in its applications across systems, in its broadcast capability, in concatenated schemes, and in the broad use of SMS messaging within networks for other services. SMS supports both mobile originated (MO) and mobile terminated (MT) operations. It has a client-server architecture in which an SMS Center manages the operation of store and forward, re-attempts, and other administrative tasks. Because messaging, like conversational services, is expected to be seamless across users, inter-operator solutions exist, possibly through gateways.

Other messaging paradigms have evolved, becoming richer, allowing near-real-time user interaction, and leveraging other enhancers, such as user mobility information. These paradigms include mobile multimedia messaging and mobile instant-messaging enablers, in addition to presence, push-to-talk over cellular, and others. The general trend is towards a seamless user experience (e.g., converged messaging).

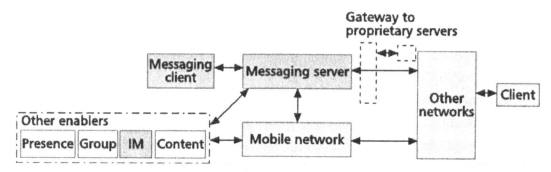

Figure 2-22: A Simplified View of a Messaging Enabler

Push-to-talk schemes are typically half-duplex communication formats in which one user has the floor at a time. They allow communication with one user or a group of users simply and nearly instantaneously. The standardized push-to-talk over cellular (POC-OMA) enabler can provide a modular architecture with a number of defined interfaces, leveraging other services or functions. Figure 2-22 shows a functional diagram of a messaging enabler, while Figure 2-23 shows the interactions between entities.

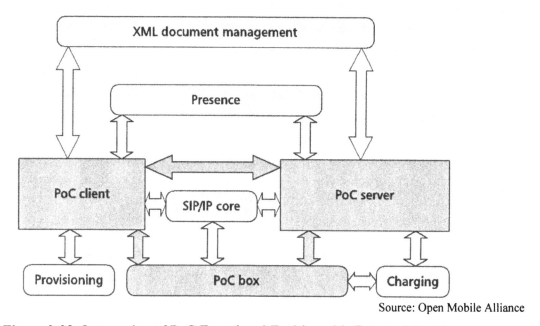

Source: Open Mobile Alliance

Figure 2-23: Interaction of PoC Functional Entities with External Entities

The GSMA has defined the Rich Communication Suite (RCS). This is based on a dynamic address book enhanced with service capabilities information, enhanced (chat) messaging, file transfer, and content sharing, defined over IMS. In its phased evolution, RCS increasingly uses standardized building blocks, basically developed by OMA. Figure 2-24 illustrates the RCS paradigm.

The core feature of RCS 1.0 (released at the end of 2008) is the "Enhanced Address Book", that appears to the users as an evolution of the usual phonebook they are used to managing. It is a contact book application that allows RCS users to know service capability information for their contacts, launch communications (Voice call, Video call, File Transfer, Messaging), and publish and be notified on Presence information, among other features.

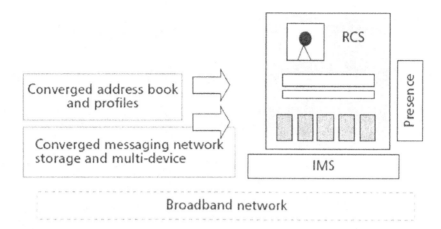

Figure 2-24: RCS as a User-Centric (through Address Book) Paradigm to Communicate, Share, and Update/Organize

RCS 2.0 (released in June 2009) extends RCS 1.0 to broadband fixed (PC) clients, and hence allows the possibility of a user having multiple clients. In addition, the Network Address Book (the NAB identified in RCS 1.0) is mandatory in RCS 2.0.

RCS 3.0 (released in February 2010) includes such features as extension of social presence, content sharing outside voice calls, more client access flexibility, and enhanced user experience with network based services. RCS 4.0 aims for a richer and more flexible platform in terms of a converged address book, profiles, integration with social networking, network-based storage, and multi-device and multi-technology environment, among other features.

Inter-operability and inter-working is particularly essential for communication services, voice, messaging or sharing alike. This has been central to the success of all messaging paradigms and RCS is no exception. Being an environment with a suite of communication mechanisms, its inter-working with legacy and non-RCS environments is also important.

The proliferation of smartphones has led to an explosive growth in wireless data traffic in general, and a range of messaging mechanisms and features across them. The ultimate success of a rich communication, networking, and sharing platform requires a user-centric environment encompassing or inter-working across user devices and messaging paradigms.

2.9.1.2 Location based services

Location-based services are intuitively necessary in the context of mobility. They include emergency, navigation, and tracking services, in addition to enhanced other services such as location-based commerce, charging, advertising, and content delivery. The location function can use different technologies, ranging from cell/sector information, to combined use of the cellular network (e.g., location of receiving base stations) and signal (e.g., round-trip-delay or triangulation) information, to GPS or assisted GPS, among others. Standardization work has allowed use of different position-determining mechanisms while defining how location is triggered and delivered. Figure 2-25 shows a high-level block diagram for a location enabler, for emergency and commercial services. Privacy loop is an important component getting the user's authorization, except when assumed for emergency services.

Other service enablers, as noted earlier, have been introduced, defined, or are emerging:

- Content management and its dynamic and personalized delivery;

- Broadcast and multicast service enabler definitions, leveraging external distribution systems;

- Mobile commerce, both radio-based (RFID) and optical based (e.g., 2D bar codes);

- User-generated content and social networking; and

- Device management for the life cycle of the user terminal, with such over-the-air features as activation, provisioning, diagnostics, firmware updates, software distribution, and others.

The reader is encouraged to explore sources in this context, including the industry forums developing standards (notably www.openmobilealliance.com) or applications definitions, profiles, certifications, and interoperability testing and guidelines.

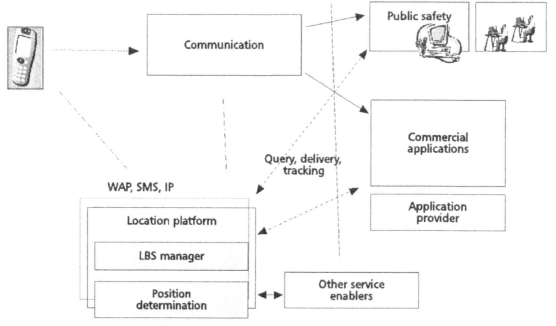

Source: Bell Canada

Figure 2-25: A Simplified Block Diagram of Location-Based Services

2.9.1.3 Content and Interaction

Services basically involve connectivity, communication, and content. They are increasingly IP-based; however, circuit-switched based services such as voice and SMS will coexist with the all-IP service environment for some time. Social networking and Internet-based services comprise much of the content. Wireless traffic has grown so explosively, reportedly doubling every year or even faster, that it has posed challenges for global service providers to meet the capacity demands. The type of content is any object of communication, including voice. However, it is such applications as video streaming, calls, or sharing, x-casting, gaming, and downloads that pose the greatest speed and capacity demands.

Wireless user terminals have grown magically in their features and capabilities, look and feel, and display and interactions. Smartphones have changed paradigms by turning mobile devices into true application platforms for end users. Interaction to launch a mode, setting, or application is multi-modal including touch, typing, and voice. The user experience, however, is not limited to the device capabilities, but also is enabled by the network behind the device. In particular, an increasingly essential performance requirement is low latency, implying a real-time or near-real-time experience and fast end-to-end interaction. The latest technologies (e.g., LTE) allow user-plane latency as low as 10-20 ms, and this is expected to go down further.

Gaining capacity is enabled not only by new wireless broadband networks, spectrum management, and new innovations (such as advanced antenna technologies, interference cancellation, self-organizing load balancing, and coverage/capacity optimization), but also at the application level. Video optimization is imperative, including such mechanisms as smart coding (adaptive, pacing, ...), caching, or others.

Content services, as expected from web/cloud services, are virtual and increasingly converged. As such, for example, information service, messaging, download, mobile commerce, and mobile advertising may be part of the same environment. Imagine acting on an advertisement banner on the side of a display, or a message at end of a call or download, to gather information or engage a search or transaction, among other possibilities. Using the user terminal's camera to point to a 2D barcode, as another example, can similarly launch any information, advertisement, transaction, and/or other services. Both mobile commerce and mobile advertisement are growing in standards, innovations, new products, and adoption. Like any other services, meeting user satisfaction and expectations (such as privacy and security), in addition to the user experience, are essential.

Machine to machine (M2M) devices and applications have been the subject of work in standards bodies and GSMA among others. Given the proliferation of these devices, there is belief that they need special considerations in terms of service architecture, application platforms, and management. Although the general expectation for M2M is connectivity and communication of little or low-bandwidth data, it must be noted that M2M will in no way be limited to this and can very well apply to high bandwidth scenarios.

2.10 Service Framework: Creation, Access, Delivery

A user-centric and efficient service architecture requires a common environment. In the past ten years in particular, with growing Internet, messaging, and mobile multimedia services, there has been work on how to enable a rich service creation environment, how to deliver dynamically, and how better to re-use and orchestrate services and common functions. Much work and innovation have been done to introduce common service creation, access, and delivery environments, using open interfaces, to define common (or so-called horizontal) enablers, and to define common or single profiles, data management, network management, and application interfaces.

Technologies and architectures for the enabling of services have grown remarkably, and yet have not reached their full potential. The evolution is about:

- Providing rich services and a rich experience without adversely affecting cost efficiency;
- Addressing the need for open and secure interfaces to the application space; and
- Meeting the need for a converged and common service framework to orchestrate services and functions to serve the user.

There is a need for open, secure, and modular integration of services, and for granularity and abstraction to be able to compose and port services, functions, and support systems. In other words, a common service framework will allow orchestrated and efficient service creation, discovery, access, and delivery.

There has been a flurry of activity within the industry to define open interfaces and service architectures and to outline a common service framework. Note, however, that there is no single approach to this, either in space (best practice, implementation-specific, co-existence with IP core, coexistence with the so-called Web 2.0, and simplified connectivity for mass social networking) or in time (phased evolution).

2.10.1 Common Service Framework

The flurry of activity to find the best practices for common service framework and delivery was greatly energized, and virtually governed (if not altered), by explosive growth in web services, social networking, or generally, the "cloud". Furthermore, as mentioned earlier, the incredible innovations in user terminals, leave much less boundary between the application environment and the device itself.

The concept of the Service Delivery Platform (SDP) has been around for years. Similarly, Service-Oriented Architecture (SOA) has been employed for such areas as Operational and Business Support Systems and network management. The main architectural considerations include the use of open, universal, protocols, particularly as defined by the World Wide Web Consortium (W3C) and the IETF, and used by third parties, enterprise networks, or for open peer-to-peer applications. In addition the framework has to allow exposure, discovery, binding, integration, re-use and distribution of functions, and choreography of services or service elements, as needed.

The aspects of this paradigm are worth exploring. On the one hand, service providers develop an evolving service architecture framework which is increasingly converged and flexible. Note that at lower layers the trends are towards seamless mobility across multi-cell (heterogeneous) communication space. The entire network is becoming all-IP with its inherent common or single entry capabilities. IMS has finally found the right timing and context for widespread adoption, which provides a common session environment and an IP multimedia service framework. Above this and within the service architecture, such protocols as Parlay X/Web/Services may provide the open interfaces for access and distribution of functions (recently, such efforts have shifted and are being pursued under the GSMA OneAPI initiative; see section 2.10.2.1). There is a variety in these interfaces and protocols themselves, and considerations such as whether the application is lightweight or more complex have a role in their definition and utilization. Furthermore, as indicated, smartphones use a variety of operating systems and application platforms with different levels of openness, though there is a trend towards more openness and toward an open application innovation and utilization environment. Cloud services on these may simply be using HTTP, and different players are working together to standardize the next-generation such as adaptive HTTP streaming.

2.10.2 Open Service Framework

Typically, converged service frameworks have a number of common attributes, no matter the difference in architecture, protocols, implementation, or best practice. Service enablers, though different in their nature, are developed to "expose" themselves, to register on a logical framework, or to be discovered. There are open interfaces, defined to external bodies such as application providers, business customers, virtual operators, or others, allowing binding using SIP, web services, or other protocols. In addition, a common framework should be able to distribute, port, or re-use functions, and to package/compose them given the abstraction inherent in the exposure.

Openness is about the use of universal (Internet) protocols that are network-independent. It refers to a combination of IT and telecom capabilities, allowing access to networks and services by internal and external entities through the use of known application programming interfaces (APIs). Furthermore, openness allows distribution and sharing of functions. The service framework, however, has measures of security and rules of engagement (or binding) based on policies.

Different service enablers that may register onto a framework may have their own specifications and protocols. Therefore, a level of abstraction may be specified to "expose," discover, or use these functions or enablers in an open-access environment. The following are examples of actions for which an open, secure, efficient, and common service framework is needed:

- To distribute content broadly and simply;

- To choreograph and re-use functional elements such as business support systems, policy, charging, and so on;

- To re-use horizontal enablers such as location and presence; and

- To integrate multiple service elements (e.g., location, streaming, presence, m-commerce, messaging) to create a composed intention-based service (e.g., a user asking about a movie).

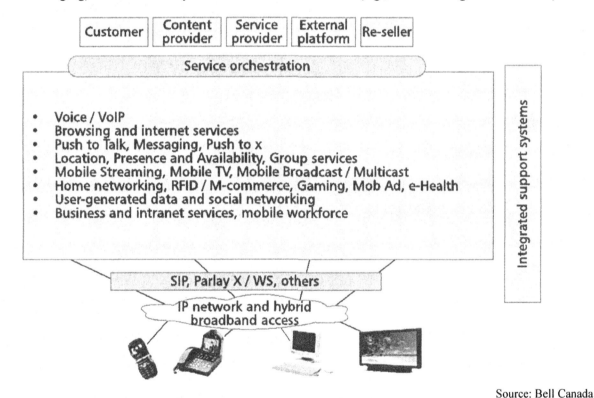

Source: Bell Canada

Figure 2-26. An End-to-End Service Enabling Framework

This framework distributes functions, integrates multiple services to form a composite serving the user's intention, and allows access by internal and (trusted or authorized) external entities. An example of such a framework is shown in Figure 2-26.

Technologies such as Parlay/OSA (www.parlay.org), with interfaces such as Parlay X/Web Services and concepts such as service-oriented architecture (SOA), service creation environment (SCE), or service delivery platform (SDP) relate to this space, and are recommended to readers for further exploration. For example, an SOA solution may be used in an operational environment to evolve from an environment with a large number of different operational and business support systems to a shared and "open" one where portability, re-usability, and integration is possible in a flexible and cost-effective manner. Similarly, new services may be *created* efficiently and with richness, or dynamically accessed, composed and *delivered,* leveraging multiple enablers and functions, internal or external, which would otherwise be in silos or hard to discover, leverage, re-use or integrate.

In addition, IP multimedia subsystem (IMS) standards allow use of a Parlay gateway when the application server is not SIP-based. Furthermore, IMS is also concerned with a common session control, common client environment, and the concept of brokering. An end-to-end system integration includes the service architecture and the IP core. This is no surprise, as the IP core should ultimately and increasingly be viewed as an IP multimedia service framework (higher layer), while increasingly enabling cooperative sub-networks and seamless mobility (lower layer). The reader is further directed to 3GPP and 3GPP2 standards and such concepts as IMS and system architecture evolution, in addition to joint work with the Parlay group.

2.10.2.1 Unleashing innovation – open AP and service interface standardization and initiatives

In the context of unleashing innovation, there has been a great deal of work towards open application programming interfaces. This will allow third parties, for example, to develop content and applications, to be able to access network enablers, and to be accessed and utilized.

GSMA OneAPI activity (an inter-operator initiative being trialed first in Canada) is an example of this, in which such services as messaging, location, and charging are involved with a common platform, representing service providers on the one side and content providers on the other. These developers can create one application and use it across multiple environments. The goal is to enable broad innovation and not limited differentiation. The phases of GSMA OneAPI are identified and standardized by the Open Mobile Alliance (OMA). In addition, the industry forum Wholesale Application Community (WAC) is aiming at this very objective of enabling an open application environment across devices and environments.

It must be noted that the context of open APIs is broad across devices and networks, and thereby it stimulates such innovations as common device management. Furthermore, there are standards, standardization work, and innovations on common user profile and next-generation service interfaces (such as the Next Generation Service Interfaces [NGSI] work at OMA). The latter can potentially allow context management using user data management, among other things.

2.10.3 A Changing Space

The services space has been changing rapidly in the past few years. Traditionally, services had been under the control of network operators. Even with the evolution of service delivery models, and the opening up to third parties through APIs (such as with OSA/Parlay and OneAPI), the assumption is that the operator is in the driver's seat, and third parties develop their applications under the restrictions and the constraints imposed by the operators' service development platform(s). With the traditional circuit-switched network, this made a lot of sense. However, with the transition to the all-IP network, other feasible models have emerged. High-level wireless operating systems running on mobile devices (high-level in contrast to the low-level real-time operating systems that lie beneath theses high-level systems) like iOS and Android provide convenient platforms for which third parties develop software. Furthermore, the mobile network no longer is the exclusive owner of resources needed by service enablers and applications. For example, mobile devices equipped with GPS own the valuable piece of information about the device location.

Thus, we can have the situation where most of the software developers target specific high-level wireless operating systems and/or mobile devices, only using the mobile network operators "bit transport" services (or in other words, their IP connectivity services). The resulting services and applications are sometimes described as "over-the-top" because they bypass the mobile operator's own services, except for bit transport. The users of over-the-top services could pay the application developers directly and only pay the mobile operators for bit transport. With the prices charged for bit transport continually decreasing, this is unattractive for mobile network operators. Different mobile network operators are thinking of different strategies for generating additional value that customers would pay for.

The rise of the "app store" adds another dimension to the picture. It appears that control of popular app stores provides another way to generate revenue. Various app stores have been set up by mobile operators, device manufacturers, and software companies.

2.11 Fundamentals of Traffic Engineering

As wireless link technologies keep improving, and higher and higher data rates are being supported (most recently with WiMAX, LTE, etc.), the network must also keep up with the increasing requirements for supporting the rapidly growing traffic. In this section, we focus on two aspects:

a) Teletraffic modeling, especially of wireless systems, to deal with the challenging problem of wireless resource provision over the air. The number of cells and channels should be limited because of costs, but what are the implications for service quality? Operators who do not plan well may need to suffer long periods of time where customers are unhappy about the quality, until the operators can upgrade their network.

b) An introduction to some of the wired network technologies that can be used in the backhaul. Although this is away from the wireless link itself, it is an important part of the network architecture.

Other topics of relevance that we are unable to cover here at this time include load balancing techniques and protocols for redundancy (making the network more robust to individual router failures, link failures or other types of failures).

2.11.1 Teletraffic Analysis

Wireless network operators want to provide high-quality service in a cost-effective manner. Service quality can be defined at many levels, including the quality of the voice encoding, error correction, and so on. At the level of wireless channel availability, the quality of service can be defined by availability metrics, two of the most important of which are:

- Blocking probability: the probability that a new call (either incoming to a subscriber or outgoing from the subscriber) is *blocked,* i.e., it cannot be allocated the necessary network resources to proceed.

- Dropping probability: the probability that an existing call is *dropped* during a handoff, i.e., the necessary resources cannot be allocated in the new cell for the handoff to complete successfully.

A call that is blocked is referred to as a blocked call and a call that is dropped is referred to as a dropped call (an event also called forced termination). The blocking and dropping probabilities are easily expressed inversely as blocking rate and dropping rate, respectively.

Wireless network operators face a tradeoff between providing high-quality service and providing it cost-effectively. On the one hand, the best quality service can be interpreted as meaning 100% availability, i.e., network resources (e.g., wireless channels) are always available for customers to use. With 100% availability, there are no blocked calls and no dropped calls. Thus, high-quality service requires the provisioning of an over-abundance of network resources. On the other hand, cost considerations would tend to limit the network resources to be provisioned. The tradeoff can be visualized in Figure 2-27.

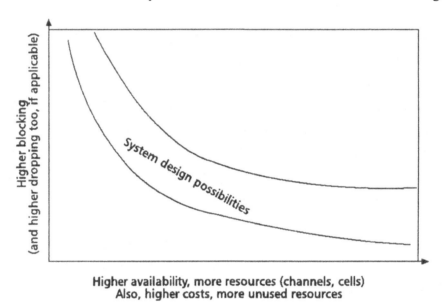

Higher availability, more resources (channels, cells)
Also, higher costs, more unused resources

Figure 2-27: Tradeoff between Quality and Cost

Teletraffic analysis deals with network availability such as call-blocking rate, call-dropping rate, channel-occupancy rate, and so on. Such analysis is of great interest to wireless service providers because they can use it to set the parameters of their network, given the limits of radio resources and network resources available. For example, they may want to minimize the call-dropping rate. Sometimes this may come at the expense of the call-blocking rate (which is the rate that new calls are blocked), since subscribers are more likely to be annoyed by the dropping of existing calls due to the blocking of handoffs than by occasional failure to begin a new call.

With the universal trend towards convergence, integrated wireless networks (where subscribers may move back and forth between cellular and WiFi or other wireless technologies) are becoming increasingly popular. Thus, the networks are becoming more complicated and it is becoming more difficult to do network planning (setting parameters such as the number of channels per base station). Therefore, teletraffic analysis is becoming increasingly essential. Moreover, to differentiate between cellular and WLAN cells, multi-layer, hierarchical modeling methods are typically used, which can be computationally very complex.

2.11.2   Teletraffic Analysis for Wireless Systems

There is an extensive body of literature on teletraffic analysis for wireless cellular systems. Because of handoffs between cells, the standard models used for decades in telephony had to be modified for use in wireless networks. Furthermore, with different traffic types, networks with different types of cells (e.g., mixed microcells and macrocells), the introduction of features like guard channels, and so on, the problem can be quite challenging. In the great majority of the papers and books, Markov models are used because of the ease in applying standard queuing theory analysis and results to the problem. Even if the actual traffic characteristics may not be exactly Poissonian, it is hoped that they are close enough so that the analytical or simulation results obtained by using a Poisson process model are still meaningful.

The Poisson process assumption has been challenged by some researchers (especially the handoff traffic, rather than new call traffic [Chl95]), as a result of which newer models have been proposed that are supposedly more accurate. One example is the proposed use of a Hyper-Erlang distribution for the inter-handoff time [Fan99], rather than an exponential distribution. Closed-form analytical expressions (like the Erlang B and Erlang C formulas) are often not derivable (even when using Poisson assumptions) because modeling of wireless networks is far more complex than modeling standard telephony networks. Thus, in general, computer simulations are used to obtain teletraffic performance of wireless systems.

One of the main thrusts of research is how to differentiate the handling of handoff traffic and new call traffic. Handoff calls are often given higher priority than new calls, since subscribers would find it more annoying if existing calls are dropped when handoff fails, than for new calls to be blocked [Sid96]. Several methods for such prioritization have been proposed. One method uses guard channels [Oh92, Oli99]. And, instead of a fixed number of guard channels, performance may be improved (at the expense of complexity) by using a variable number of guard channels [Cho00]. Another method to reduce dropped-call probability is known as channel borrowing [Kat96]. Channels are borrowed from neighboring cells to accommodate handoff calls.

The complexity of the standard queuing system model grows exponentially as the number of states increases, i.e., as the number of channels and cells increases. Approaches to simplifying the teletraffic analysis problem to something more computable have generally focused on simplifications to the model itself, or on ways to simplify the computations for a given model.

Of the attempts to simplify the model itself, making many assumptions related to homogeneity and loose coupling, some researchers have attempted to obtain useful information about a whole network by using a one-cell model. The idea is that if the coupling between cells is "loose enough," it suffices to pick a single cell in the network and examine the teletraffic dynamics of that cell alone to have a good idea what is happening in the network. The cell can be arbitrarily chosen, based on homogeneity arguments. The incoming/outgoing traffic from/to neighboring cells may be approximated as Poisson processes [Eki01]. It may be shown that such an approach allows simplification to a well-known M/M/m/m Markov model [Eki01, Ort97], whose closed-form solution is easily available from standard queuing theory. However, the accuracy of single-cell models is very questionable for sophisticated wireless networks that employ channel borrowing, dynamic guard channels, etc. Furthermore, it is not useful for multi-layer networks.

One class of simplification approaches is the use of sparse matrix algorithms. Because of loose coupling, the matrix built from the global balance equations is a sparse matrix. Sparse matrix algorithms have been investigated for use in solving packet network models [Kau81]. More recently, sparse matrix algorithms have also been used to study hierarchical wireless networks [Zho98]. Nevertheless, the savings in computation time are limited.

2.11.3 Provisioning Higher Capacity in the Core Network

The success of mobile communications and Internet services has led to significantly increased traffic worldwide. This is what is experienced today. Mobile traffic is foreseen to grow more than tenfold in the next five years. Web traffic by mobile PCs is foreseen to be dominant in the same period, driven by the attractiveness of mobile broadband. Web/mobile PC traffic will grow very strongly and most probably become the dominant application type, including P2P file sharing, music and film downloads, etc. The fast growing applications for data traffic in the large youth segment are web-based services such as YouTube and Facebook.

3GPP has implemented traffic engineering features and capabilities into 3G networks by introducing selected queuing mechanisms and priorities into the UTRAN. Meanwhile, 2G (GSM and Enhanced Date rates for GSM Evolution [EDGE]) networks remain essentially "best effort" with limited embedded traffic engineering capabilities in GERAN. The story is different, however, if a Policy Manager/Policy Enforcement platform in the network is working at the TCP/UDP level. In that case, traffic engineering can be implemented even in 2G networks.

Traffic per subscriber is in a way related to the screen size available – for a 17″ laptop it is relevant to download DVD-resolution movies, but for a handheld device with 2.5″ screen much less resolution will suffice – and represent a faster download.

The main question starting to be discussed is: what will happen in a longer timeframe, say 10-20 years? Many analysts are saying that there will be a traffic explosion. A reasonable projection that captures that phenomenon is the following.

- Today about 4 billion people generate voice traffic of about 20 million Bytes/month/user.

- Today about 20 million people generate data traffic of about 4 billion Bytes/month/user.

- Tomorrow most probably 4 billion people will generate a global mix of data traffic, including any type of service, of about 20 billion Bytes/month/user!

- <u>Result</u>: Traffic will probably increase by a factor 1000 compared to only voice in a time frame of a couple of decades!

This could appear futuristic. A question that could be asked is: where do we find a capacity increase of a factor of 1000? A possible answer, taking into consideration the mobile broadband evolution, could be the following:

- High Speed Packet Access (HSPA) and LTE technologies can bring a factor of 10 increase with respect to Wideband CDMA (WCDMA).

- Spectral efficiency improvements from 1 b/s/Hz to 2 b/s/Hz give a factor of 2 increase.

- More allocated spectrum gives a factor of 5 increase.

- A smaller cell radius (about 1/3 of that currently in use) gives a factor of 10 increase.

- In all, these multiply together ($10 \times 2 \times 5 \times 10$) to give an increase by a factor of 1000.

This simple thinking seems to assess that it is possible to cope with such a huge traffic explosion!

On the one hand it is possible to give reasons for the huge traffic growth; on the other hand, all agree that the revenue associated with the plethora of data services – a mix of voice, video, multimedia, Internet, and television – will definitely be significantly smaller than the revenue associated with voice traffic since the beginning of telephony, as shown in Figure 2-28.

As a result, network operators have a great challenge: grow their networks to sustain such a huge traffic growth while bringing down the costs of the network infrastructure, both in terms of investments (CAPEX) and operational costs (OPEX).

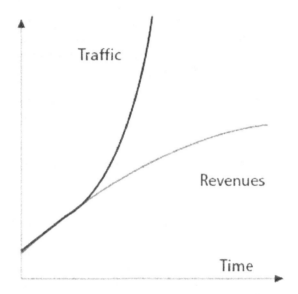

Figure 2-28: Forecast of Traffic Growth vs. Revenue Growth

This is the famous scissor curve that in reality is mitigated by a few factors like:

- The traffic does not continue growing exponentially but after a certain extent starts to saturate.
- The revenue curve is related to a pure re-selling of connectivity. In reality if a Policy Manager is applied, the most bandwidth-hungry applications could be shaped.
- Resources can be prioritized to applications that are generating the most revenue.

The final result is therefore not as negative as might first be thought.

Such a big challenge will significantly impact the transport network infrastructure: much more than in the past. For that reason the next section discusses the role of traffic engineering in this area.

2.11.3.1 Traffic engineering in next generation transport networks

A dramatic increase in traffic versus economic aspects related to lower revenues generated by new services poses great technical challenges for network operators and system manufacturers. In fact, they must be able to increase capacity and improve bandwidth utilization of their networks. This means finding the technological way to significantly increase the capacity in all network segments, from access (either fixed or mobile of any type), to the core network, up to the backbone **but** in an economical way. In practice, this means making efficient utilization of the bandwidth in any segment.

The metrics in data networks to measure performance today are mostly anchored to the cost per bit or the cost per packet, but sometimes this "box based" approach does not really lead to an efficient network.

The recovery of costs is not limited to equipment and technology improvements. Rather, most of the savings are likely to come when the architectures are changed or simplified, the vertical legacy networks' capabilities are collapsed onto the new platforms, and when the performance of the new services convinces the end users to migrate from the old-fashioned services.

From this perspective it is vital to build and operate a network ensuring the following requisites:

- Flexibility. This means the ability of the network to adapt itself according to the traffic demand evolution, and to evolve with time in a modular way, according to the principle of incremental investments. In other words, since it is absolutely not possible to plan with the same level of confidence as the traditional telephone network, it is necessary to *plan for the unexpected*. In practice this means realizing network systems and equipment with the highest level of flexibility and strongly reducing those constraints or characteristics that would lead to larger investments that would be not cost-viable for the expected revenue.

- Automation. To lower the operational cost it is necessary to increase the level of automation in the network and, eventually, to provide the network with the possibility of automatically reacting to traffic changes with time.

- Time to Service. This means lowering as much as possible the time between the request for a service from the network provider and its practical set up. Today, it could take weeks for a network operator to accommodate the request for a new service for a customer (e.g., a VPN to be upgraded in bandwidth up to a certain extent). In principle, when this time will be reduced to a minimum, it will be possible to talk about *"bandwidth-on-demand"*.

- Network simplicity and easy management. This means simplifying the life of network operators by simplifying and automating procedures, and reducing as much as possible the need for manual intervention on the field.

- Fast reaction to changes in traffic demand. Traffic demand will be more and more unstable with time. This means that not only will traffic volumes increase in an unpredictable way, but even the distribution of traffic among the network nodes could change with time. For instance, Internet traffic and, particularly, P2P can change the traffic flows and distribution even on a daily or weekly basis. The trend will therefore be to make the network able to react to traffic changes accordingly, while maintaining all the QoS requirements, differentiated per type of service, and all the recovery mechanisms. That is, the network must be resilient in any condition at any time.

Today the approach to QoS will be convergent, supporting features on both fixed and mobile networks. The Policy Manager for fixed (the European Telecommunications Standards Institute [ETSI] Resource and Admission Control Subsystem [RACS] system) and mobile (3GPP PCRF) could be implemented on the same platform. What remains peculiar are the Policy Enforcement capabilities that sometimes are implemented on GGSN/EPC nodes or in distributed probes for mobile and on distributed probes for fixed.

The migration from 2G to 3G has changed completely the performance of data connections by an order of magnitude. The introduction of HSPA has given another order of magnitude and the same will happen with LTE in the next few years and with LTE advanced into the next five years. Thus the cost per bit in mobile networks is expected to decrease by four orders of magnitude.

2.11.3.2 The role of the control plane in next generation networks

The general trend is that next generation infrastructure must be IP-based, easy-to-manage, flexible and adaptable to changes in traffic demand, multi-service, and, last-but-not least *telecom grade*. The latter requirement is extremely important because it enables providing the final user with the perception of good quality-of-experience (QoE), which means that the final user must not notice any limitations in performance, under any conditions. To fulfill this requirement, while at the same time assuring low cost in terms of investments for realizing the infrastructure and of operational costs for managing, operating and maintaining the network, it is necessary to provide different levels of quality-of-service (QoS) and resilience (robustness against any type of fault in the network), in an efficient way.

The tools that are needed in the network are essentially four:

- Queuing and priorities at the boundary of scarce resource channels;

- Policy Managers in order to properly assign network resources to users or selected applications;

- Policy enforcements at the boundaries of the network; and

- Control Plane capabilities in order to propagate the resource allocation to all the network nodes with protection included.

Mobile resource allocation signaling is largely standardized and is used to allocate appropriate PDP contexts or RABs (Radio Access Bearer) to each user.

In the fixed arena, the only way to harmonize among those tight requirements (up to telecom grade level for mission critical services) and the connectionless nature of the IP network world is to adopt a network control and management able to handle different types of "virtual connections", or "quasi-circuit". The basic technology to do that is represented by the GMPLS (General Multi-Protocol Label Switching) control plane and related protocol stacks. That is the real enabler of a next generation transport network. In addition to that, it is necessary to provide the network with the intelligence that allows it to optimize its resources, namely path-computation elements able to perform optimal grooming of paths at the different layers, and routing the path throughout the network, according to the different constraints and policies applied by the network operator.

Actually, the GMPLS control plane allows the control of tunnels at any layer: from IP/MPLS tunnels, namely the label switched paths (LSPs), down to synchronous digital hierarchy (SDH) or modern optical transport network (OTN) layers, down to the optical domain (the wavelength paths or lambdas). A hierarchical structure of paths spanning the different layers across the network could help in organizing the control and management of the network.

With GMPLS technology it is possible to control data flows at any layer, while reserving the bandwidth of the tunnels, thereby respecting the promised levels of QoS. In other words, through GMPLS a sort of circuit-switching paradigm is re-established within a packet-based network employing IP-based protocols.

In addition, the possibility of providing the network with a "cross-layer visibility" is a powerful tool to provide the network intelligence all the information to actually realize multi-layer grooming and routing.

In practice this means that the network intelligence could be able to optimally groom paths at higher levels onto paths at lower levels, then route those paths according to specific criteria. In other words, the traffic tunnels are bundled together and routed in order to save the network resources in terms of switch/router ports, transport line, optical links etc. In this case the traffic engineering is performed in a multi-layer fashion, spanning the different layers so as to minimize the resources to be used.

If the network dimensions in terms of number of nodes and numbers of paths at any layer are too big, the control and management of the network has to be organized accordingly, in order to scale up to the needed dimensions.

The ways to centralize or distribute the network control, to organize the architecture of the control (e.g., hierarchical), and to realize grooming and routing at single or multiple layers, are still topics of research in both industry and academia. The challenge of the control of the optical layer of the network further complicates this topic and makes it more challenging.

In conclusion, what could make the difference among the different telecom vendors' solutions could be the intelligence provided to the network. Traffic engineering strategies are extremely important for realizing such intelligence, because will allow network operators to lower CAPEX and OPEX, while evolving their networks according to traffic demand evolution and service requirements.

2.12 References

[3GP10] 3rd Generation Partnership Project, *Motivation for 3GPP Release 8 – The LTE Release*, http://www.3gpp.org/lte, 2010.

[3GP11a] 3rd Generation Partnership Project, Technical Specification Group, Core Network & Terminals, *General Packet Radio Service (GPRS); GPRS Tunneling Protocol (GTP) Across the Gn and Gp interface*, Specification 29.060, Release 10, 2011.

[3GP11b] 3rd Generation Partnership Project, Technical Specification Group, Service & Systems Aspects, *IP Multimedia Subsystem (IMS); Stage 2*, Specification 23.228, Release 11, 2011.

[3GP11c] 3rd Generation Partnership Project, Technical Specification Group, Service & Systems Aspects, *Service requirements for the Evolved Packet System (EPS)*, Specification 22.278, Release 11, 2011.

[3GP11d] 3rd Generation Partnership Project, Technical Specification Group, Service & Systems Aspects, *General Packet Radio Service (GPRS) enhancements for Evolved Universal Terrestrial Radio Access Network (E-UTRAN) access*, Specification 23.401, Release 10, June 2011.

[3GP11e] 3rd Generation Partnership Project, Technical Specification Group, Core Network & Terminals, *Subscriber data management; Stage 2*, Specification 23.016, Release 10, April 2011.

[3GP11f] 3rd Generation Partnership Project, Technical Specification Group, Service & Systems Aspects, *Policy and charging control architecture*, Specification 23.203, Release 11, June 2011.

[3GP11g] 3rd Generation Partnership Project, Technical Specification Group, Service & Systems Aspects, *Architecture enhancements for non-3GPP accesses*, Release 10, June 2011.

[ATT09] AT&T, Orange, Telefonica, TeliaSonera, Verizon, Vodafone, Alcatel-Lucent, Ericsson, Nokia Siemens Networks, Nokia, Samsung, and Sony Ericsson, *One Voice; Voice over IMS profile*, V1.0.0, 2009.

[Bej04] Y. Bejerano, *Efficient Integration of Multihop Wireless and Wired Networks with QoS Constraints*, IEEE/ACM Transactions on Networking, vol. 12, no. 6, 2004.

[Bor01] C. Bormann, et al., *Robust Header Compression (ROHC): Framework and four profiles: RTP, UDP, ESP, and uncompressed*, Internet Engineering Task Force RFC 3095, July 2001.

[Chl95] E. Chlebus and W. Ludwin, *Is Handoff Traffic Really Poissonian?*, IEEE International Conference on Universal Personal Communications (ICUPC), 1995.

[Cis10] Cisco Systems, *Cisco Visual Networking Index: Global Mobile Data Traffic Forecast Update, 2009-2014*, 2010.

[Cho00] S. Choi and K. Sohraby, *Analysis of a Mobile Cellular System with Hand-off Priority and Hysteresis Control*, IEEE International Conference on Computer Communications (INFOCOM), 2000.

[Cla03] T. Clausen and P. Jacquet, *Optimized Link State Routing Protocol (OLSR)*, Internet Engineering Task Force RFC 3626, October 2003.

[Col99] R. Coltun, D. Ferguson, and J. Moy, *OSPF for IPv6*, Internet Engineering Task Force RFC 2740, December 1999.

[Com06] D. Comer, *Internetworking with TCP/IP Volume 1: Principles, Protocols, and Architecture*, 5th edition, Prentice Hall, 2006.

[Cor06] C. Cordeiro and D.P. Agrawal, *Ad hoc & Sensor Networks: Theory and Applications*, World Scientific Publishing, 2006.

[Cor10a] M. Corici, F. Gouveia, T. Magedanz, and D. Vingarzan, *OpenEPC: A Technical Infrastructure for Early Prototyping of Next Generation Mobile Network Testbeds*, Proceedings of the 6th International Conference on Testbeds and Research Infrastructures for the Development of Networks and Communities, 2010.

[Cor10b] M. Corici, J. Fiedler, T. Magedanz, and D. Vingarzan, *Access Network Discovery and Selection in the Future Broadband Wireless Environment*, 3rd International ICST Conference on MOBILe Wireless MiddleWARE, Operating Systems, and Applications, 2010.

[Cor10c] M. Corici, T. Magedanz, D. Vingarzan, and P. Weik, *Prototyping Mobile Broadband Applications with the Open Evolved Packet Core*, 14th International Conference on Intelligence in Next Generation Networks (ICIN) – Weaving Applications into the Network Fabric, 2010.

[Dee98] S. Deering and R. Hinden, *Internet Protocol, Version 6 (IPv6) Specification*, Internet Engineering Task Force RFC 2460, December 1998.

[Dev05] V. Devarapalli, R. Wakikawa, A. Petrescu, and P. Thubert, *Network Mobility (NEMO) Basic Support Protocol*, Internet Engineering Task Force RFC 3963, January 2005.

[Eki01] E. Ekici and C. Ersoy, *Multi-tier Cellular Network Dimensioning*, Wireless Networks, vol. 7, no. 4, 2001.

[Fan99] Y. Fang and I. Chlamtac, *Teletraffic Analysis and Mobility Modeling for PCS Networks*, IEEE Transactions on Communications, vol. 47, no. 7, 1999.

[GSA10] The Global mobile Suppliers Association, http://www.gsacom.com.

[GSA12] The Global mobile Suppliers Association, http://www.gsacom.com.

[Gup00] P. Gupta and P. R. Kumar, *The Capacity of Wireless Networks*, IEEE Transactions on Information Theory, vol. 46, no. 2, 2000.

[He07] B. He, B. Xie, and D. P. Agrawal, *Optimizing the Internet Gateway Deployment in a Wireless Mesh Network*, Proceedings of the Fourth IEEE International Conference on Mobile Ad-hoc and Sensor Systems (MASS), 2007.

[IEE05] Institute of Electrical and Electronics Engineers, 802.11 Working Group, *Draft Supplement to Part 11: Wireless Medium Access Control (MAC) and physical layer (PHY) specifications: Medium Access Control (MAC) Enhancements for Quality of Service (QoS)*, IEEE 802.11e/D13.0, 2005.

[ISO02] International Organization for Standardization/International Electrotechnical Commission, *Information technology; Telecommunications and information exchange between systems; Intermediate system to intermediate system intra-domain routing information exchange protocol for use in conjunction with the protocol for providing the connectionless-mode network service (ISO 8473)*, Standard 10589, Second Edition, 2002.

[Joh01] D.B. Johnson, D.A. Maltz, and J. Broch, *DSR: The Dynamic Source Routing Protocol for Multi-Hop Wireless Ad Hoc Networks*, Ad Hoc Networking, vol. 1, C.E. Perkins, Ed., Addison-Wesley, 2001.

[Joh04] D. Johnson, C. Perkins, and J. Arkko, *Mobility Support in IPv6*, Internet Engineering Task Force RFC 3775, June 2004.

[Kat96] I. Katzela and M. Naghshineh, *Channel Assignment Schemes for Cellular Mobile Tele-communication Systems: A Comprehensive Survey*, IEEE Personal Communications, vol. 3. no. 3, 1996.

[Kau81] L. Kaufman, B. Gopinath, and E. Wunderlich, *Analysis of Packet Network Congestion Control Using Sparse Matrix Algorithms*, IEEE Transactions on Communications, vol. 29, no. 4, 1981.

[Ko00] Y.B. Ko and N.H. Vaidya, *Location-Aided Routing (LAR) in Mobile Ad Hoc Networks*, Wireless Networks, vol. 6, 2000.

[Mal98] G. Malkin, *RIP Version 2*, Internet Engineering Task Force RFC 2453, November 1998.

[Moy98] J. Moy, *OSPF Version 2*, Internet Engineering Task Force RFC 2328, April 1998.

[Nan07a] D. Nandiraju, N. Nandiraju, and D.P. Agrawal, *Service Differentiation in IEEE 802.11 Mesh Networks: A Dual Queue Strategy*, MILCOM, 2007.

[Nan07b] N.S. Nandiraju, D. Nandiraju, L. Santhana, and D.P. Agrawal, *A Cache Based Traffic Regulator for Improving Performance in IEEE 802.11s-based Mesh Networks*, IEEE Radio and Wireless Symposium (RWS 2007), 2007.

[Oh92] S. Oh and D. Tcha, *Prioritized Channel Assignment in a Cellular Radio Network*, IEEE Transactions on Communications, vol. 40, no. 7, 1992.

[Oli99] M. Oliver and J. Borras, *Performance Evaluation of Variable Reservation Policies for Hand-off Prioritization in Mobile Networks*, IEEE International Conference on Computer Communications (INFOCOM), 1999.

[Ort97] L. Ortigoza-Guerrero and A. Aghvami, *On the Optimum Spectrum Partitioning in a Microcell/Macrocell Cellular Layout with Overflow*, IEEE Global Communications Conference (GLOBECOM), 1997.

[Pea99] M. R. Pearlman and Z. J. Haas, *Determining the optimal configuration for the zone routing protocol*, IEEE Journal of Selected Areas in Communications, vol. 17, 1999.

[Pel08] G. Pelletier and K. Sandlund, *RObust Header Compression Version 2 (ROHCv2): Profiles for RTP, UDP, IP, ESP and UDP-Lite*, Internet Engineering Task Force RFC 5225, April 2008.

[Per94] C.E. Perkins and P. Bhagwat, *Highly dynamic Destination-Sequenced Distance-Vector routing (DSDV) for mobile computers*, Proceedings of ACM SIGCOMM, 1994.

[Per02] C. Perkins (ed.), *IP Mobility Support for IPv4*, Internet Engineering Task Force RFC 3344, August 2002.

[Per03] C. Perkins, E. Belding-Royer, and S. Das, *Ad hoc On-demand Distance Vector (AODV) Routing*, Internet Engineering Task Force RFC 3561, 2003.

[Pra06] R. Prasad and H. Wu, *Minimum-Cost Gateway Deployment in Cellular WiFi Networks*, Proceedings of the Consumer Communications & Networking Conference, 2006.

[Ros02] J. Rosenberg, et al., *SIP: Session Initiation Protocol*, Internet Engineering Task Force RFC 3261, July 2002.

[San06] L. Santhanam, N. Nandiraju, Y. Yoo, and D.P. Agrawal, *Distributed Self-policing Architecture for Fostering Node Cooperation in Wireless Mesh Networks*, Proceedings of the Personal Wireless Communications Conference, 2006.

[San10] K. Sandlund, G. Pelletier, and L-E. Jonsson, *The RObust Header Compression (ROHC) Framework*, Internet Engineering Task Force RFC 5795, March 2010.

[Sch03] H. Schulzrinne, S. Casner, R. Frederick, and V. Jacobson, *RTP: A Transport Protocol for Real-Time Applications*, Internet Engineering Task Force RFC 3550, July 2003.

[Sch08] H. Schulzrinne, *Session Initiation Protocol (SIP)*, http://www.cs.columbia.edu/sip, 2008.

[Sch10] H. Schulzrinne, *RTP News*, http://www.cs.columbia.edu/~hgs/rtp, 2010.

[Sid96] M. Sidi and D. Starobinski, *New Call Blocking Versus Handoff in Cellular Networks*, IEEE International Conference on Computer Communications (INFOCOM), 1996.

[Ste07] R. Stewart, Ed., *Stream Transmission Control Protocol*, Internet Engineering Task Force RFC 4960, 2007.

[Var03] V.K. Varma, K.D. Wong, K.-C. Chaing, and F. Paint, *Integration of Wireless LAN and 3G Wireless Technologies,* IEEE Communications Magazine, November 2003.

[Wan07] J. Wang, B. Xie, K. Cai, and D.P. Agrawal, *Efficient Mesh Router Placement in Wireless Mesh Networks*, Proceedings of The Fourth IEEE International Conference on Mobile Ad-hoc and Sensor Systems (MASS), 2007.

[Wan09] L. Wang, Y. Zhang, and Z. Wei, *Mobility Management Schemes at Radio Network Layer for LTE Femtocells*, Vehicular Technology Conference, 2009.

[Wik11] Wikipedia, *Mesh Networking*, http://en.wikipedia.org/wiki/Mesh_networking, 2011.

[Won03] K.D. Wong and V.K. Varma, *Supporting Real Time IP Multimedia Services in UMTS,* IEEE Communications Magazine, Vol. 4 No. 11, November 2003.

[Won05] K.D. Wong, *The IP Multimedia Sub-system*, Wireless Internet Telecommunications, Chapter 12, Artech House, 2005.

[Zho98] Y. Zhou and B. Jabbari, *Performance Modeling and Analysis of Hierarchical Wireless Communications Networks with Overflow and Take-back Traffic*, IEEE International Symposium on Personal, Indoor and Mobile Radio Communications, 1998.

2.13 Suggested Further Reading

I. F. Akyildiz, X. Wang, and W. Wan, *Wireless Mesh Networks: A survey*, Computer Networks and ISDN Systems, vol. 47, no. 4, 2005.

E. Belding-Royer and C. K. Toh, *A Review of Current Routing Protocols for Ad hoc Mobile Wireless Networks*, IEEE Personal Communications, vol. 6, 1999.

M. Blanchet, *Migrating to IPv6: A Practical Guide to Implementing IPv6 in Mobile and Fixed Networks,*John Wiley and Sons, 2006.

J. Broch, D.A. Maltz, D.B. Johnson, Y.C. Hu, and J. Jetcheva, *A Performance Comparison of Multi-hop Wireless Ad hoc Network Routing Protocols*, Proceedings of the 4th annual ACM/IEEE International Conference on Mobile Computing and Networking, 1998.

G. Camarillo and M-A. Garcia-Martin, *The 3G IP Multimedia Subsystem (IMS): Merging the Internet and the Cellular Worlds*, 2nd Edition, John Wiley and Sons, 2006.

D. Collins, *Carrier Grade Voice Over IP*, 2nd Edition, McGraw-Hill, 2002.

J. Davidson, J. Peters, M. Bhatia, S. Kalidindi, and S. Mukherjee, *Voice over IP Fundamentals*, 2nd Edition, Cisco Press, 2006.

K. Drage, et al., *Internet Protocol (IP) Multimedia Call Control protocol Based on Session Initiation Protocol (SIP) and Session Description Protocol (SDP); Stage 3*, 3GPP Technical Specification 24.229, Release 7, 2006.

K. Drage, et al., *Presence Service Based on Session Initiation Protocol (SIP); Functional Flows, Information Flows, and Protocol Details*, 3GPP Technical Specification 24.841, (Release 6), 2004.

IETF MANET Working Group, http://www.ietf.org/html.charters/manet-charter.html

Y-B. Lin and I. Chlamtac, *Wireless and Mobile Network Architectures*, 1st Edition, John Wiley and Sons, 2000.

M.A. Miller, *Voice over IP Technologies: Building the Converged Network*, Hungry Minds, 2002.

D. Minoli and E. Minoli, *Delivering Voice over IP Networks*, 2nd Edition, John Wiley and Sons, 2002.

M. Mouly and M-B. Pautet, *The GSM System for Mobile Communications*, Telecom Publishing, 1992.

N. Nandiraju, D. Nandiraju, L. Santhanam, B. He, J. Wang and D.P. Agrawal, *Wireless Mesh Network: Current Challenges and Future Directions of Web-in-the-sky*, IEEE Wireless Communications, vol. 14, no. 4, 2007.

S. Olson, G. Camarillo, and A.B. Roach, *Support for IPv6 in Session Description Protocol (SDP)*, Internet Engineering Task Force RFC 3266, June 2002.

C. Perkins, *Mobile IP: Design Principles and Practices*, Prentice Hall, 1998.

C. Perkins, *RTP: Audio and Video for the Internet*, Addison-Wesley Professional, 2003.

M. Poiselka, A. Niemi, H. Khartabil, and G. Mayer, *The IMS: IP Multimedia Concepts and Services*, 2nd Edition, John Wiley and Sons, 2006.

A. Raniwala, K. Gopalan, and T. Chiueh, *Centralized Channel Assignment and Routing Algorithms for Multi-channel Wireless Mesh Networks*, Mobile Computing and Communications Review, vol. 8, no. 2, 2004.

T.S. Rappaport, *Wireless Communications: Principles and Practice*, 2nd Edition, Prentice Hall, 2002.

A.B. Roach, *Session Initiation Protocol (SIP)-Specific Event Notification*, Internet Engineering Task Force RFC 3265, June 2002.

J. Rosenberg and H. Schulzrinne, *An Offer/Answer Model with Session Description Protocol (SDP)*, Internet Engineering Task Force RFC 3264, June 2002.

J. Rosenberg and H. Schulzrinne, *Reliability of Provisional Responses in Session Initiation Protocol (SIP)*, Internet Engineering Task Force RFC 3262, June 2002.

J. Rosenberg and H. Schulzrinne, *Session Initiation Protocol (SIP): Locating SIP Servers*, Internet Engineering Task Force RFC 3263, June 2002.

H. Sinnreich and A.B. Johnston, *Internet Communications Using SIP: Delivering VoIP and Multimedia Services with Session Initiation Protocol (Networking Council)*, 2nd Edition, John Wiley and Sons, 2006.

H. Soliman, *Mobile IPv6: Mobility in a Wireless Internet*, Addison Wesley Professional, 2004.

J. Solomon, *Mobile IP: the Internet Unplugged*, Prentice Hall, 1997.

J. Varga, et al., *Signalling Flows for the IP Multimedia Call Control Based on Session Initiation Protocol (SIP) and Session Description Protocol (SDP); Stage 3*, 3GPP Technical Specification 24.228, Release 5, 2006.

B. Xie, Y. Yu, A. Kumar, and D.P. Agrawal, *Load-balanced Mesh Router Migration for Wireless Mesh Networks*, Journal of Parallel and Distributed Computing, 2008.

K.D. Wong, *Wireless Internet Telecommunications*, Artech House, 2005.

Chapter 3

Network Management and Security

3.1 Introduction

The purpose of Network Management is to maintain the operational state of the network during changes to underlying customers and technologies. The operational processes of a telecommunications company include the provisioning of customer services, billing for customers' usage, and the management of network faults. Network management is involved in telecommunications operational processes where there a need to use information from a Network Element (e.g. Node B, Radio Network Controller RNC, Serving GPRS Support Node [SGSN], Gateway GPRS Support Node [GGSN]).

The computer systems responsible for maintaining network and resource information are referred to as Operational Support Systems (OSS). These responsibilities include product activation, network inventory, fault detection, and network performance. Activities that deal directly with the network are performed by a Network Operations Centre (NOC). The NOC has the primary responsibility for network management in a telecommunications company

There is also the need for Customer Management systems as a part of telecommunications business processes. These systems maintain customer and product information and are known collectively as Business Support Systems (BSS). These systems maintain a company's relationship with the customer. The interactions with the customer through service requests are received by Front of House (FoH). These systems and customer care groups do not interact directly with network elements.

The following reference frameworks define the operational procedures for network management.

- FCAPS (Fault, Configuration, Accounting/Administration, Performance, and Security) is the International Organization for Standardization (ISO) Telecommunications Management Network model and framework for Network Management Functions [ITU97]. Each category defines a set of network management functions.

- The Information Technology Infrastructure Library (ITIL), maintained by the Office of Government Commerce (OGC) of the United Kingdom, defines processes for managing a Service Lifecycle.

- The enhanced Telecom Operations Map (eTOM), maintained by the TeleManagement Forum (TMF), defines building blocks of Fulfillment, Assurance, and Billing (FAB) commonly used in organizational structures.

The latter two are industry models that outline all the operational tasks, from the strategy and design phases to the operation of a product.

The key function of network management is the exchange of information between network elements and network management systems. This requires the use of common protocols in the network; the following protocols provide a communication channel for management information:

1. Common Management Information Protocol (CMIP) is the telecommunications network standard for information exchange defined by the ISO.

2. Simple Network Management Protocol (SNMP), defined as an Internet Engineering Task Force (IETF) standard, has become the de facto standard as network elements become increasingly driven by Transmission Control Protocol/Internet Protocol (TCP/IP).

The security of information is necessary in network management for the purpose of authentication, billing, and data integrity. The need for security is defined by the following principles:

1. Confidentiality: only allow authorized users and systems to access network information.

2. Integrity: network and user information cannot be modified by an unauthorized third party.

3. Availability: users can be granted access and sufficient bandwidth to wireless networks.

4. Non-repudiation: billing and usage data can be provided to users challenging their accounts.

The intent of this chapter is to highlight the important network management models, protocols and security capabilities relevant to wireless communications.

3.2   Network Management Concepts

This section covers general IT Network Architecture concepts and terminology. Figure 3-1 is an introduction to common network management terms and concepts.

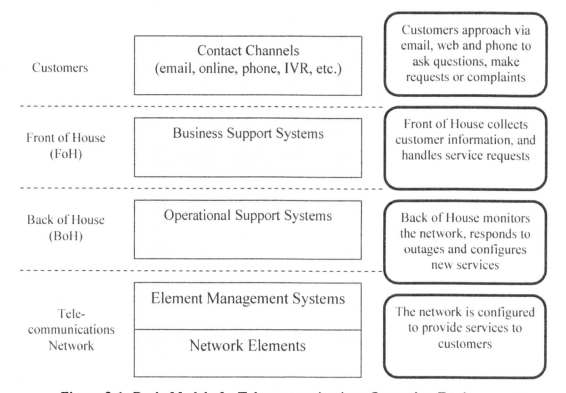

Figure 3-1: Basic Model of a Telecommunications Operating Environment

- Front of House (FoH) – a contact center responsible for collecting customer information and handling customer service requests;

- Business Support Systems (BSS) – IT systems for supporting Front of House in addressing customer service requests, such as product catalogues and customer relationship management and billing systems;

- Back of House (BoH) – a technical group responsible for operational processes that deal directly with the network, often centrally located in a Network Operations Centre (NOC);

- Operational Support Systems (OSS) – IT systems for supporting Back of House support groups, including network inventory, service provisioning, network configuration, and fault detection;

- Network Elements (NEs) – the building blocks of a telecommunications network for the transportation of data; for example, optical termination units, core routers and switches; and

- Element Management Systems (EMS) – interface across NEs to execute common management tasks.

The key challenge for IT application providers is providing Commercial off the Shelf (COTS) software that meets the operational requirements of telecommunications companies.

- In reality a telecommunications network is made up of multiple vendors based on competitive factors, acquisitions, and technology evolution.

- Each vendor will bring a different Element Management System into the NOC.

- In a real environment, an operator in the NOC needs to work across multiple network elements.

- It is not uncommon for base station equipment to be from one, SGSNs from another, and Home Location Registers (HLRs) from a third vendor.

- Each will almost certainly have different configuration and diagnostic commands.

There have been many attempts to standardize at least the basic operations across core elements. The goal is to allow network equipment from one manufacturer to be replaced by equipment from another vendor while continuing to work seamlessly with IT management systems.

3.2.1 Fault, Configuration, Accounting, Performance, Security

To introduce the basic concepts of Network Management we refer to the ISO Telecommunications Management Network model and framework for network management. FCAPS summarizes the responsibility of a network operations environment and the activities associated with problem management. A NOC has ownership of the following five activities. For the most part, telcos will attempt to automate these activities using technology wherever possible.

3.2.1.1 Fault management

The availability of a service impacts the company's ability to generate revenue and affects customer churn to other service providers. Service outages can be due to hardware failures, network changes, or external factors (e.g., weather, physical damage, vandalism). The primary objective of a NOC is to respond to network faults and maintain the availability of the network. The activities associated with Fault Management are:

- Detect – the continuous monitoring of the network for a disruption to customer service;

- Isolate – identify the key network infrastructure responsible for the service outage;

- Communicate – notify customers and stakeholders of the impact and estimated restoration times;

- Restore – mitigate the customer impact through a temporary work-around and/or a permanent fix.

The primary method of fault detection and isolation is Network Alarming, in which a network element detects a fault condition and sends an alarm (CMIP, SNMP) to an Alarm Management system Performance monitoring also plays an important role in fault detection through Active Probes (simulating customer transactions) and Passive Probes (monitoring actual customer traffic).

For example:

A service outage occurs with a loss of mobile coverage to a rural country town. The fault is detected by the NOC through a loss of connectivity to the Base Station which generates a "Link Down" critical alarm into the Alarm Management System.

The NOC sends a technician to the site, known as a truck-roll, to isolate the problem to Base Station hardware or the transmission link.

At this stage the NOC already suspects a loss of coverage to the site and can communicate the impact to customers. The severity of the incident will depend on factors such as location (urban, rural, priority site) and checking whether this was listed as a Capacity cell (traffic can be shared onto neighboring cells) or a Coverage cell (no neighboring cells exist). The technician confirms the Base Station hardware is operating correctly and that the transmission link (optical fiber) is down.

A temporary work-around is investigated to restore service and resolve the customer impact. The availability of other fibers on the transmission link is checked to determine whether there was a partial cable cut and traffic can be re-routed onto another sheath. If the site is high-priority, a portable cell tower may be deployed to the location. The permanent fix would involve splicing the cable cut or re-trenching of cable to the Base Station.

3.2.1.2 Configuration

The network configuration defines the current state of the network. It affects behavior such as routing of traffic and the customer's interaction with services. The most common uses of configuration are:

- Customer services – activation of a new service;

- Traffic routing – route network traffic for capacity planning purposes or as part of dedicated customer services;

- Security settings;

- Element provisioning – introduction of new elements into the network and loading a default configuration; and

- Configuration updates – addition of software/firmware to the network for new features, security concerns, etc.

A telecommunications company needs the ability to provision new customer services. This involves checking the current configuration and adding/removing/changing settings. Front of House (FoH) receives requests from customers that need to be logged in a Customer Relationship Management (CRM) system for action. The new service then needs to be configured into the network by updating the network configuration, either automatically or with some manual provisioning steps.

Some challenges include maintaining an interface between FoH provisioning systems and each of the network platforms that store the service configuration. Also, each new telecommunications product that is introduced must be integrated into those provisioning systems. Without an interface it is difficult to maintain accuracy between billing systems (what the customer pays for) and the network configuration (what the customer uses).

For example:
    A customer calls FoH to request that a voicemail service be enabled on his/her mobile service. The operator updates the CRM system, which reflects the service as being enabled in the customer profile and the billing system. The configuration settings now need to be updated in the network to enable call forwarding to voicemail the next time a call is not answered by the customer. The CRM system pushes the settings to the voicemail platform through a mediation layer.

3.2.1.3 Accounting
Customers receive a regular bill concerning their services from their telecommunications provider. Accurate monitoring of customer usage is necessary for this.

The accounting function tracks service usage and enables usage restrictions, throughput shaping, and customer billing. Customer billing may include features such as usage discounts, service bundling, zero-rating of traffic, and value-based charging.

Customer Relationship Management (CRM) and Billing Systems play an important role in maintaining customer satisfaction. A telecommunications company is required to maintain billing records for legal reasons as well as handling customer complaints.

Usage monitoring relies on Authentication, Authorization and Accounting (AAA) protocols covered later in this chapter, including RADIUS and Diameter.

3.2.1.4 Performance
Performance information can be used real-time for fault detection by tracking deviations from normal operation. It is also used long-term for the improvement product performance and tracking of customer Service Level Agreements (SLAs).

- Network performance – performance of the physical network and transport layers of the wireless network (e.g., network coverage, carrier to noise ratios, bit error rates, number of retransmissions).

- Service performance – performance of services delivered over the network infrastructure, and a measure of customer experience (e.g., Mean Opinion Score of a voice call, time to send a Short Message Service [SMS] message).

Performance monitoring relies on the use of active and passive monitoring.

- Active Monitoring – simulating customer transactions from locations in the network to test the service end to end.

- Passive Monitoring – monitoring actual customer traffic to detect a deviation in call volumes from normal operation.

Performance measurements include availability, throughput, and latency.

3.2.1.5 Security

Security plays a support role to other network management functions such that it requires its own treatment. Security is relevant in the following areas:

- Configuration – restricted access to network elements to reduce the amount of change, e.g., tracking configuration changes to assets; and

- Billing – data security through authentication and encryption to reduce fraudulent transactions and maintain accurate records.

3.3   Operations Process Models

The Telecommunications Management Network (TMN), introduced by the International Telecommunication Union (ITU), defined a model to provide a common management layer across network elements and telecommunication infrastructures. Although the original TMN standards have been superseded, they introduced the concepts of Network Management and OSS across multiple vendors, technologies, and management protocols. The following concepts were introduced by the TMN:

- Operations System – This represents one of the Operational Support Systems (OSSs) that perform mediation and telecommunications management functions.

- Workstation – Operators interact with OSSs and Network Elements directly through Command Line Interface (CLI) or via an Element Management System (EMS).

- Data Communication Network (DCN) – This is the communications network for interacting with Network Elements.

- Network Elements (NEs) – Routers, switches, application servers, etc. contain the management information that is being managed and updated via the OSS Management Information Base (MIB).

The TeleManagement Forum (TMF) is a standards body that defines a Business Process Model and IT Architecture for Telecommunications providers. This process model will be discussed at length in a later section.

3.3.1 The Information Technology Infrastructure Library

The Information Technology Infrastructure Library (ITIL) provides standards-based definitions of processes and descriptions for Incidents, Problems, and Known Errors involved with Network Management. There are five volumes in the ITIL version 3 standard.

- ITIL Service Strategy
- ITIL Service Design
- ITIL Service Transition
- ITIL Service Operation
- ITIL Continual Service Improvement

The United Kingdom Office of Government Commerce describes the ITIL as being a public domain framework, providing a best practices framework for "service management". The nature of a service itself is defined in very broad terms as "a means of facilitating outcomes customers want to achieve without the ownership of specific costs" [OGC07a]. In a telecommunications context, therefore, the scope can be construed to mean the lifecycle of a particular telecommunications service from inception to deletion.

The ITIL is in the public domain insofar as the published best practices are not proprietary. Crucially however, the ITIL is not prescriptive at a technology level. That is, it does not specify particular technologies, protocols, etc. to be used in the execution of its aims. Rather, it specifies a high level framework but stops short of further specificity.

In its present form (v3), the ITIL is divided into several Core Guidance Topics comprising the overall Service Lifecycle. Their arrangement is shown in Figure 3-2. Particularly important for our purposes are the Service Operation (maintain a level of service for end users) and Service Transition (delivery of a service into operational use) aspects of this lifecycle. Within these particular Core Guidance Topics reside process definitions which set the context for the protocols and technologies discussed later in this chapter.

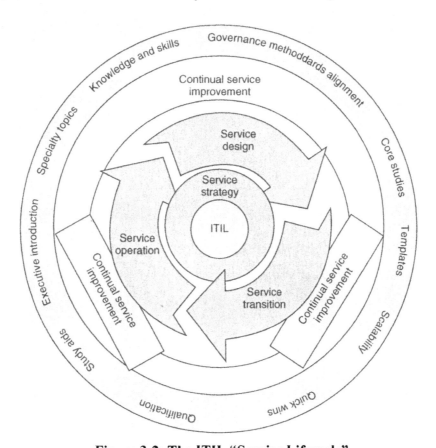

Figure 3-2: The ITIL "Service Lifecycle"

3.3.1.1 Service Operation – the relationship of Incidents, Problems, and Known Errors

Consider a hypothetical telecommunications company, complete with customers and their service expectations. The Service Operation aspect of the ITIL proposes a series of processes and concepts for managing the performance of such services within the customers' expectations.

The first concept is that of the *Incident*, which the ITIL defines as any event which disrupts, or could disrupt, a service [OGC07a]. Such events are quite common in the life of telecommunications companies, and hence an *Incident Management* process is required to ensure the goal of "restoring normal service operation as quickly as possible and minimizing the adverse impact on business operations" [OGC07a]. It is important to note here that the Incident itself is the occurrence of a service discontinuity or disruption, specific to one or more end users. It is not necessary to understand the root cause of the Incident in order to fully describe it. Indeed, the ITIL conveniently delineates the matter through the definition of a second concept: the *Problem*.

An ITIL Problem is defined to be "the unknown cause of one of more incidents" [OGC07a]. There are two particular prerequisites implicit in this definition. First, it is impossible to have a Problem without also having one or more associated Incidents. This prerequisite gives the Problem its operational context; it is the undesirable condition which gives rise to service disruptions. Second, it is important to note the unknown nature of its root cause. This may seem a little disingenuous, but once a Problem's root cause is known, it is no longer a Problem. Rather, at this point, following successful diagnosis of its root cause, the Problem becomes what is defined as a *Known Error*.

Known Errors are essentially the determined root causes of previously established Problems. They may also, optionally, have a work-around associated with them (as may Problems themselves) to alleviate their disruptive effects in the short term until a more permanent solution can be arrived at and implemented. This permanent solution usually connotes a *Change*, which is a concept we come to in the next section.

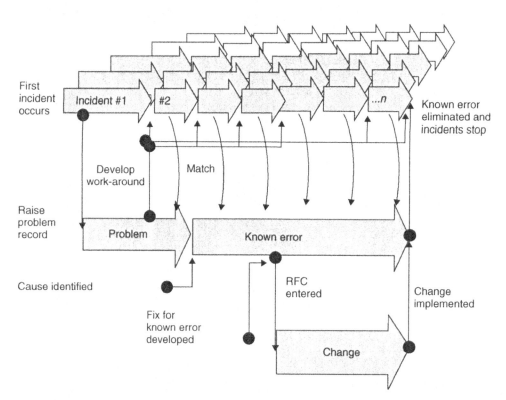

Figure 3-3: The Relationship of Incidents, Problems, Known Errors, and Changes

Figure 3-3 shows a simple example of the interaction between these four major ITIL concepts. This occurs in four discrete steps:

1. Incidents pertaining to a common root cause start to occur, and after some time operational staff notice the commonality and respond by raising a Problem record. From that point onwards the Incident Management process continues to recognize and record new incoming Incidents and match them to the Problem record.

2. The *Problem Management* process determines a work-around, which can be passed back to the customers reporting correlated Incidents. At some point later, the root cause is authoritatively diagnosed, a Known Error record is created, and the Problem record is associated with it.

3. Further along in the process, a more permanent solution is identified to resolve the Problem in question. This involves the introduction of a Change via a mechanism which will be discussed in some detail shortly.

4. The introduction of this Change redresses the Known Error, and therefore the preceding Problem. Any preceding Incidents are similarly resolved at this time, and no further Incidents with that root cause recur in the future.

3.3.1.2 Service Transition – change management

Whereas Service Operation pertains to services which are in their steady state, Service Transition focuses on the procedures the ITIL specifies for modifications to telecommunications services. This aspect of the ITIL describes the procedures which should be employed when managing planned activities. The particular sub-aspect *Change Management* has a very specific bearing on the management of wireless networks. The ITIL's definition of a *Service Change* is quite broad: "The addition, modification or removal of an authorized, planned or supported service or service component and its associated documentation" [OGC07b].

This definition can be reasonably construed to encompass the modification of any element, however insignificant, making up the provision of a service to a customer. The ambiguity is arguably deliberate, however, as it allows the organization in question to determine the scope to some extent.

Change Management under the ITIL is then the process through which the implementation of such Changes is controlled. A key consideration is the continuity or stability of the business, as change activity itself has been shown to be a significant driver of Incidents. Changes also need to be executed in an expedient matter, originating as they often do out of a need to resolve Known Errors, and also out of a need to build new capabilities into existing infrastructure. The Change Management process therefore adds value by facilitating such modifications, while managing and controlling the inherent risks. Understanding that change activity varies greatly in terms of urgency and risk, the most recent version of the ITIL proposes three variants of the Change Management process.

- The Normal Change Model is the baseline, depicted in Figure 3-4, from which the other models derive.

- The Standard Change Model is invoked in the case of frequently-repeated Changes. In such situations, the risk is generally well understood and very low, and hence the assessment and authorization stages present in the Normal Change Model are deemed superfluous and are omitted.

- The Emergency Change Model is invoked in the case of urgent Change activity that needs to be applied immediately in order to mitigate an actual or imminent major service disruption. Incident or Problem activity is a frequent antecedent, in the manner described in Figure 3-3.

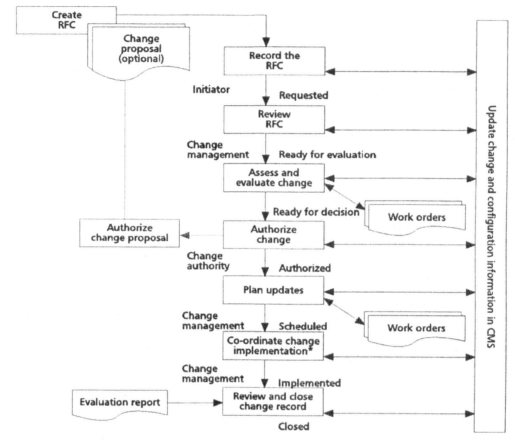

Figure 3-4: ITIL Change Management Process for "Normal" Changes

The Normal Change Model as described in Figure 3-4 commences with the creation of a Request for Change (RFC). This document will generally be required to contain a description of the change process, along with various other metadata, including but not limited to an estimate of any possible disruptions, the reasons motivating the Change, and contingencies in the event of failure [OGC07b].

The assessment stage refers to an independent verification of the risks associated with the requested change. This is sometimes lumped in with the authorization stage, whose primary purpose is to weigh the relative risks and benefits of the request and then recommend either its approval or its rejection. Duly authorized or approved changes are then scheduled for implementation, and then executed by personnel. Finally, they are formally closed. This workflow is commonly managed by the company in question by a dedicated software system.

3.3.2 The Enhanced Telecom Operations Map

Developed by the TMF as an industry standard, the enhanced Telecom Operations Map (eTOM) provides an accepted business process framework for service providers. It introduces standardized names and descriptions (e.g., Incident Management, Problem Management, etc.). Telecommunication providers are encouraged to align their business processes with the eTOM process framework.

The eTOM is a part of the New Generation Operational Support Systems (NGOSS) program, which consists of:

- eTOM, which defines a set of business processes for telecommunications. This includes all aspects of running a large telecommunications company, but also, and importantly, processes relating to network management; monitoring resource performance, analyzing service quality, and resolving customer-impacting incidents.

- SID (Shared Information/Data Model), which specifies a data model for telecommunications services and determines which IT applications need to share this data (e.g., customer billing records, service configurations).

- TAM (Telecom Application Map), which links processes and data models to actual IT applications that can be purchased.

The eTOM model [TMF07] follows a hierarchical decomposition of processes into levels. The first level (L0) defines three process areas and horizontal groupings based on the level of interaction with the customer. The three process groupings, shown in Figure 3-5, are:

- *Strategy, Infrastructure and Product*, which includes processes that develop strategies and commitments to them within the enterprise;

- *Operations*, which includes all operational processes that support the customer and network management; and

- *Enterprise Management*, which includes basic business processes required to run and manage any large business: Financial Management, Human Resources, etc.

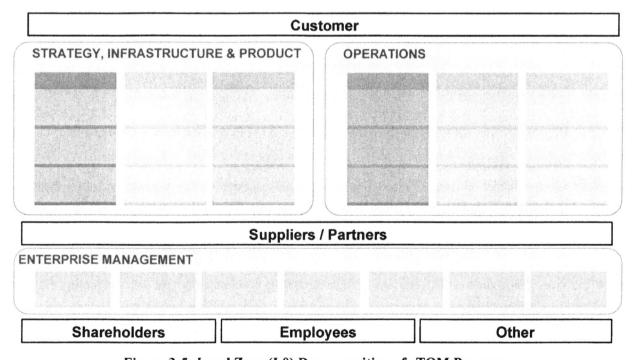

Figure 3-5: Level Zero (L0) Decomposition of eTOM Processes

3.3.2.1 Operations processes

Operations processes are at the core of network management. These processes are technology and product agnostic and can readily be applied to the domain of wireless communication. The first level of these processes (L1) is divided vertically, as shown in Figure 3-6, to define the day-to-day operations of maintaining a customer service; service configuration, fault handling, and billing management.

- *Fulfillment* addresses customer order handling, service configuration, activation, and resource provisioning.

- *Assurance* addresses problem handling, quality of service, fault troubleshooting, and performance monitoring.

- *Billing* addresses bill creation, customer data records, collection, and rating.

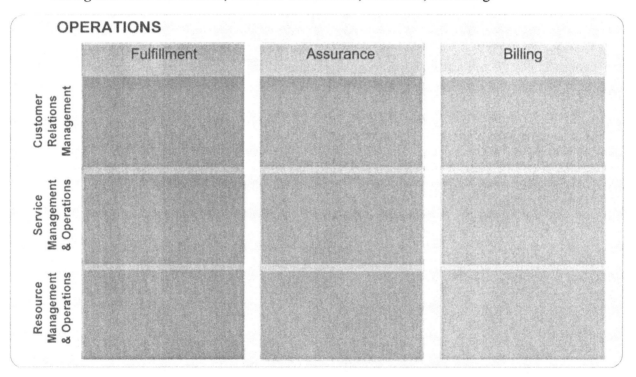

Figure 3-6: Level One (L1) Decomposition of eTOM Processes within *Operations*

3.3.2.2 Network assurance

The most relevant section to network management is Assurance. Each section is also divided horizontally (as with L0 shown in Figure 3-5) to define the level of interaction with the customer. These groupings specific to Assurance, shown in Figure 3-7, are:

- Customer Relationship Management - create trouble tickets, track and manage customer SLAs;

- Service Management and Operations - diagnose service problems, monitor service quality; and

- Resource Management and Operations - monitor resource performance, control quality of service.

The next level of detail (L2) identifies the first specific processes that can be addressed by Operational and Business Support Systems (OSS/BSS). These processes can be used to make sure that current OSS/BSS have adequate coverage when introducing a new Product/Service and implementing the necessary Network Management to maintain that Product/Service. The eTOM vocabulary states that a Product (e.g., Mobile TV) is made up of one or more Services (e.g., Video Streaming, Customer Subscriptions) and enabled by many Resources (e.g., IT infrastructure, Streaming Servers, 3G Network).

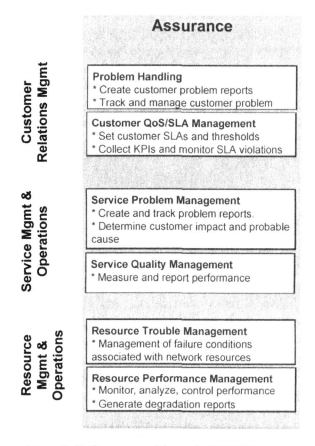

Figure 3-7: Level two (L2) Decomposition of eTOM Processes within *Assurance*

At level 3 (L3), Assurance is broken down into the sub-processes linked within their corresponding (L2) processes by flow diagrams. There is also an inter-relation between (L2) processes in a way that defines a necessary course of action that traverses Customer, Service, and Network boundaries; for example, to address a Customer Problem with a Service we "Create Customer Trouble Report", "Survey and Analyze Service Problem", "Localize Resource Trouble", and so on.

3.4   Network Management Protocols

Network Management is dependent on being able to read and update the information in the network. Interconnectivity between multiple Network Elements is achieved through a standard interface that views all information as managed resources. The managed information in the network is stored in a management information base (MIB). The MIB can change based on the Network Element and vendor. A management protocol defines a process for getting and setting attributes in the MIB.

Management networks need to support multiple Network Element management protocols. The de facto standard in telecommunications networks is SNMP (Simple Network Management Protocol). Other protocols include extensible Markup Language (XML), Common Object Request Broker Architecture (CORBA), and Simple Object Access Protocol (SOAP). Older network systems may operate on the Common Management Information Protocol (CMIP) or on ASCII message-based protocols.

3.4.1 The Simple Network Management Protocol (SNMP)

Having established the operational context through discussion of the ITIL and eTOM, it is now meaningful to discuss the protocols and technologies that translate best practice into reality. One of the most significant protocols is the Simple Network Management Protocol (SNMP), which was originally defined through the Internet Engineering Task Force (IETF) beginning in 1988, and which has been in operation ever since. It is, as the name suggests, a relatively simple protocol, insofar as it had four defined operations in its initial formulation [see RFC 1157].

It is also worth noting at the outset that the protocol forms a part of the Internet Protocol (IP) suite, given that its messages are encapsulated within User Datagram Protocol (UDP) datagrams which are themselves encapsulated within IP packets. This fact limits the applicability of SNMP to IP elements, although with the expansion of IP reachability within Radio Access Networks, etc. the effective scope of the protocol is expanding significantly.

Central to the architecture inherent to SNMP is the notion of two distinct elements:

- The *SNMP Agents*, of which there are several, and possibly quite a large number. The Agents reside on the devices or nodes under management, and generally do three things. They service requests for information (in the form of "Objects") originating from the SNMP Manager pertaining to the node. They also respond to requests from the Manager to set a particular condition. And they periodically send unsolicited messages to the Manager in reaction to a particular predefined condition or threshold being reached.

- The *SNMP Manager*, of which there is generally only a single instance and which receives information pertaining to the state of the SNMP Agents.

Version 1 of the SNMP protocol [RFC 1157] specifies four basic operations, which are retained in present implementations. The first three pertain to the retrieval and setting of information on a network element under management. Note that the information in question is arranged in an object hierarchy, which will be described in some detail shortly.

- *get*, which is an instruction initiated by the Manager towards a specific network element (identified by its IP address). The agent is commanded to return the value of a specific Object retained on the Agent. This instruction, like the one following, is commonly issued by the Manager in order to determine a performance statistic or something similar.

- *get-next*, which is a partially stateful version of the previous instruction, aimed at traversing the tree-structure of the information retained on the Agent. When iterated, this allows the Manager to obtain all of the values in the Agent's information base (which will be described shortly).

- *set*, which is an instruction initiated by the Manager towards a specific network element, which specifies that a particular Object on the Agent be set to a particular value. This instruction is commonly issued in order for the Manager to configure the node under management.

The remaining operation originates from the Agent and is the only one initiated by the Agent:

- *trap*, which is a message sent from the Agent to the Manager advising that a particular pre-defined condition has been reached. These tend to be element-specific; however, there are several standardized traps which will be discussed later.

More recent versions of SNMP use much the same model but offer some improvements. For SNMP v2 and v3, the basic differences with SNMP v1 are:

- SNMP v1 offers a simple request/response protocol and the operations listed above. The only security measure implemented is community strings, which have been shown not to be particularly effective.

- SNMP v2 offers additional protocol operations and multiple requested values (GetBulk and Inform), but still lacks strong security attributes. The definition of SNMP v2 can be found in RFC 1901.

- SNMP v3 adds security and remote administrative capabilities.

The SNMP v3 header has both security fields and non-security related fields. SNMP v3 provides a more secure level of authentication and privacy, and uses Data Encryption Standard (DES) encryption to encrypt the data packets.

The definition of SNMP v3 can be found in:

RFC 3411, *Architecture for Describing SNMP Management Frameworks*

RFC 3412, *Message Processing and Dispatching*

RFC 3413, *Various SNMP Applications*

RFC 3414, *User-based Security Model (USM), providing for both Authenticated and Private (encrypted) SNMP messages*

RFC 3415, *View-based Access Control Model (VACM), providing the ability to limit access to different MIB objects on a per-user basis*

The differences among the various SNMP versions are fully described in RFC 3584.

3.4.2 The Structure of Management Information and the Management Information Base

Having discussed the basic interactions between the Agent and Manager elements, we now turn our attention to the structure of the information stored within these elements [Zel99].

First there is the Structure of Management Information (SMI) [RFC 1155], which defines how information, arranged as "Objects," is to be named and defined. Each piece of information, or Object, has a specific place in a hierarchical tree structure. This particular identity is called the OBJECT IDENTIFIER, a series of integers which authoritatively names the Object. Consider the example shown in Figure 3-8.

Under the SMI scheme, the OBJECT IDENTIFIER for the *enterprise* Object in Figure 3-8 can be represented as:

{iso (1) org (3) dod (6) internet (1) private (4) enterprise (1)}

or by shorthand as

1.3.6.1.4.1.

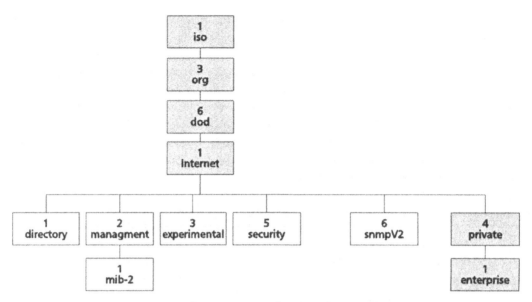

Figure 3-8: Example Showing the Naming Convention for the MIB

Now that we know how to address or name a particular Object in the hierarchy, we turn our attention to the types of data which can appear in such a hierarchy. RFC 1155 refers to these data generically as "Managed Objects" and defines a syntactical representation of these called the OBJECT-TYPE. Any such definition comprises five fields:

- *OBJECT*, which is a textual name for the managed object, along with its OBJECT IDENTIFIER denoting its place in the hierarchy.

- *SYNTAX*, which is an integer denoting the *type* of the data wrapped by the managed object. Generally, only the primitive types available in Abstract Syntax Notation One (ASN.1) are valid.

- *Definition*, which is a textual description, usually in ASCII, of the OBJECT TYPE's function.

- *Access*, which is an enumerated type specifying the read/write status of the object.

- *Status*, which is also an enumerated type specifying whether implementation of this object is required in order to claim full compliance of a MIB group.

Therefore, an OBJECT TYPE, named sampleObject, residing as the first leaf under the enterprise Object drawn in Figure 3-8, might be defined in the following ASN.1 notation:

sampleObject OBJECT-TYPE

SYNTAX DisplayString (SIZE (0 … 255))

ACCESS read-only

STATUS mandatory

DESCRIPTION

"This is a sample OBJECT TYPE. Nothing more."

::= { enterprise 1 }

Readers are encouraged to familiarize themselves with SNMP v2 and v3. These are described in RFCs 1441-1452 and 3411-3418, respectively.

3.5   Security Requirements

The objective of communication security is the preservation of three principles: Confidentiality, Integrity, and Availability [ISO05].

- Confidentiality: the communication data are only disclosed to authorized subjects.

- Integrity: the data in the communication retain their veracity and are not able to be modified by unauthorized subjects.

- Availability: authorized subjects are granted timely access and sufficient bandwidth to access the data.

The security mechanism that is needed is determined from the threat type and the affected principle. No one mechanism is able to address all such principles, so combinations of mechanisms are common. Wireless infrastructure has many common threats [NIS08], the major categories of which are listed in Table 3-1.

Threat Category	Description	Principle Affected
Denial of service	Attacker prevents or prohibits the normal use or management of networks or network devices.	Availability
Eavesdropping	Attacker passively monitors network communications for data, including authentication credentials.	Confidentiality
Man-in-the-middle	Attacker actively intercepts the path of communications between two legitimate parties, thereby obtaining authentication credentials and data. Attacker can then masquerade as a legitimate party.	Confidentiality/ Integrity
Masquerading	Attacker impersonates an authorized user and gains certain unauthorized privileges.	Integrity
Message modification	Attacker alters a legitimate message by deleting, adding to, changing, or reordering it.	Integrity
Message replay	Attacker passively monitors transmissions and retransmits messages, acting as if the attacker were a legitimate user.	Integrity
Traffic analysis	Attacker passively monitors transmissions to identify communication patterns and participants.	Confidentiality

Table 3-1: Common Threats to Wireless Infrastructure

Other concepts that are important to know include:

- Identification – This occurs when a user (or client) presents itself to a system to claim its identity.

- Authentication – The mechanism by which the system verifies the identity of a user (or client).

- Authorization – The mechanism by which the system verifies that the user (or client) is permitted to execute a particular action.

- Accountability – The ability of a system to determine the actions and behavior of a user (or client).

- Privacy – The level of confidentiality and privacy protection given to a user by a system.

- Non-Repudiation – The user (or client) cannot deny that he or she is the individual who made a particular transaction.

3.5.1 Network Access Control

Physical access control poses a special problem, particularly given the near-ubiquity of the air interface of some wireless systems. Given that this ubiquity is often one of the key design objectives, *logical* access control is the mechanism used to achieve the required security.

Network access control is based on the procedure described before: to authenticate, authorize, and account for a user or client. There are many approaches to managing these issues, and it is with this understanding that we describe some of the enabling technologies.

3.5.1.1 RADIUS

RADIUS (Remote Access Dial In User Server) is a standard protocol adopted by the IETF in RFC 2865 and RFC 2866 for carrying authentication, authorization, accountability, and configuration information between a Network Access Server and a centralized Authentication Server [RFC 2865]. RADIUS's basic operation is shown in Figure 3-9.

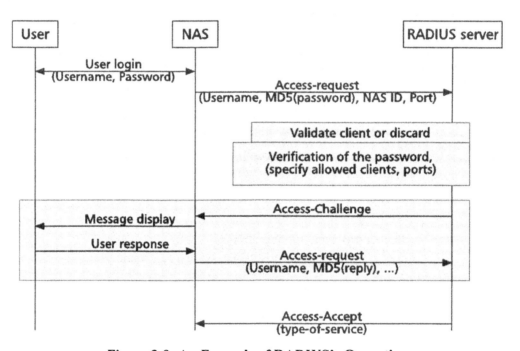

Figure 3-9: An Example of RADIUS's Operation

In this example, the end user presents his credentials to the Network Access Server (NAS). The NAS generates an Access-Request message to the RADIUS server. Once the RADIUS server receives the request it generates a challenge and sends it back to the user. The user then performs a mathematical operation using the challenge (e.g., a hash) and sends it back to the RADIUS server. The RADIUS server then checks the credentials and generates either an Access-Accept or and Access-Reject message to the NAS, which itself confirms or denies the connection to the end user. In wireless networks, the NAS may be implemented within a specific access point such as the PDSN (Packet Data Serving Node), GGSN, etc.

It is important to note that this is just the basic operation of the RADIUS protocol; all detailed options are described in RFC 2865. RADIUS packets are carried over UDP (User Datagram Protocol) and have many attributes (e.g., username, password, etc.); the complete list is available in RFC 2865. There is also a capability for some attributes to be vendor-defined.

3.5.1.2 Diameter

Diameter is an Authentication, Authorization, and Accounting protocol based on RADIUS for such applications as network access or IP mobility. Diameter is defined in RFC 3588 and is seen as a successor to RADIUS.

The primary ways in which Diameter improves upon RADIUS are:

- Failover – RADIUS does not define failover mechanisms. Diameter supports application-layer acknowledgment and defines failover algorithms.

- Confidentiality – RADIUS defines the application-layer authentication and integrity scheme but not confidentiality (just optional support for Internet Protocol Security [IPSEC]). By contrast, IPSEC support is mandatory for Diameter and Transport Layer Security (TLS) is optional.

- Reliable transport – In order to provide well defined transport behavior, Diameter runs over the Transmission Control Protocol (TCP) or Stream Control Transmission Protocol (SCTP) protocols.

- Server-initiated messages – Support for server-initiated messages is optional in RADIUS, making it difficult to implement unsolicited disconnection or re-authentication/re-authorization across heterogeneous deployments. This support is mandatory in Diameter.

- Auditability – RADIUS does not define data-object security mechanisms and because of that, untrusted proxies may modify attributes or packet headers without being detected. The implementation of security objects is not mandatory in Diameter, but it is supported.

- Capability negotiation – RADIUS does not support error messages, capability negotiation, or mandatory/non-mandatory flags for attributes (Attribute Value Pairs [AVPs]). Diameter does provide this support.

- Peer discovery and configuration – Diameter enables dynamic discovery of peers using the Domain Name System (DNS). Also, the derivation of the dynamic session keys is enabled via transmission-level security.

- Roaming support – Diameter supports user roaming. By contrast, the IETF Roaming Operations (ROAMOPS) Working Group investigated whether RADIUS could be used to support roaming and concluded that it was ill-suited to handle the interdomain exchange of user and accounting information.

3.5.1.3 Extensible Authentication Protocol

In 1998 the Extensible Authentication Protocol (EAP) was defined by the IETF in RFC 2284. Subsequently, in June 2004 the technology was revised in RFC 3748. RFC 2284 focuses primarily on the message types and packet format. The security considerations and interactions with other protocols were addressed in the latter RFC [NIS07].

EAP may be used on dedicated links as well as switched circuits, and on wired as well as wireless links. To date, EAP has been implemented with hosts and routers that connect via switched circuits or dial-up lines using Point to Point Protocol (PPP) [RFC 1661]. It has also been implemented with switches and access points using IEEE 802. EAP encapsulation on IEEE 802 wired media is described in IEEE 802.1X, and encapsulation on IEEE wireless LANs in IEEE-802.11i [RFC 3748].

In EAP, the party that wants to be authenticated is called the *supplicant* and the party that demands proof of authentication is called the *authenticator*. Four types of messages are defined in EAP: *request*, *response, success,* and *failure*. The authenticator sends a request message to the supplicant asking for a response message to authenticate. If the authentication is successful, a success message is sent to the supplicant; if not, a failure message is sent.

EAP supports many authentication methods, including passwords, digital certificates, tokens, etc. It is possible to combine these methods and use asymmetric methods for mutual authentication. The pass-through feature enables authenticators to forward the messages directly to the authentication server, offering the ability to vendors to generate new specific methods. These EAP methods perform the authentication transactions and generate the key material used to protect subsequent communications. RFC 3748 defines three authentication methods:

- MD5-Exchange (mandatory),
- One-Time-Password (OTP) (optional), and
- Generic Token Card (GTC) (optional).

However, none of these methods provide key material and so are not suitable for WLANs. This is because WLANs use TLS-based EAP. RFC 3748 recommends that key establishment and generation should be based on mature and well established techniques. TLS has thus emerged as the dominant protocol for EAP methods that support WLANs. Common TLS-based methods are:

- EAP-TLS,
- EAP Tunneled TLS (EAP-TTLS),
- Protected EAP (PEAP), and
- EAP Flexible Authentication via Secure Tunneling (EAP-FAST).

A description of each of these methods is beyond the scope of this chapter, and the reader is referred to RFC 3748 for details.

3.5.1.4 IEEE 802.1x

The wide success of Local Area Networks, both wired and wireless, and their growth in size have made requirements for security essential and the designation of a specific security standard within IEEE 802 desirable. In 2001, IEEE approved standard 802.1x, which specifies the port-based network access control for wired and wireless networks. IEEE 802.1x uses EAP as an auxiliary protocol to transmit the authentication data. When EAP is transmitted on a LAN (e.g., IEEE 802.3, IEEE 802.5, IEEE 802.11, etc.), it is encapsulated by the Extensible Authentication Protocol over LANs (EAPOL).

The following definitions are important in order to understand 802.1x.

- The supplicant is the client that wants to authenticate to the network.

- The authenticator is the Wireless Access Point (802.11) or Ethernet Switch (802.3) that facilitates authentication for the supplicant.

- The authentication server is the element that determines whether a supplicant is authenticated or not; it is generally a RADIUS server.

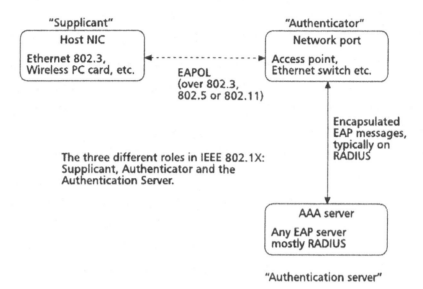

Figure 3-10: An Example of 802.1x Operation

Figure 3-10 shows the authentication of a user on an IEEE 802.1x network. The process starts when a non-authenticated supplicant tries to connect to the authenticator (the access point). The access point will only allow the client to generate EAP traffic (EAPOL) until it has been authenticated. The client sends the EAP-Start message, and the access point sends the EAP-Request Identity message to the client. The client answer is then sent to the RADIUS server to be verified. If verification is successful, the authenticator will then allow the client to generate any kind of traffic.

3.5.2 *Wireless LAN Security*

In order to be able to communicate data on a wireless network, a wireless station (STA) must establish an association with the access point (AP). Only after that association is established is an exchange of data allowed. In the infrastructure mode, the association process proceeds as follows [Arb01]:

- Unauthenticated and unassociated,

- Authenticated and unassociated, and

- Authenticated and associated.

The original standard IEEE 802.11 defines two mechanisms to authenticate the access to the wireless LAN: open system authentication and shared key authentication, neither of them secure[1]. The shared key authentication scheme, based on a unilateral challenge-response mechanism, is usually referred to as Wired Equivalent Privacy (WEP), which is described below. Open system authentication is actually a null

---

1   WEP (Wired Equivalent Privacy) encrypts the response by XORing the challenge with a pseudo-random key stream generated using a WEP key. The attacker can XOR the challenge and the response to expose the key stream, which can subsequently be used to authenticate [NIS07].

authentication mechanism, as it does not provide true identity verification. The client is authenticated to the AP having provided the following information:

- Service Set Identifier (SSID): The SSID is the name assigned to the wireless LAN used by the client to identify the network. The SSID is generally transmitted as plain text, although a security control can generally be set to not broadcast it.

- Media Access Control (MAC) address: Many wireless LAN implementations allow the administrator to specify the MACs allowed to access the network using access lists. This too is an imperfect precaution, as the MAC is also sent as plain text and is thus susceptible to spoofing.

Shared key authentication is based on a common secret shared between the access point and the client, a cryptographic key known as a WEP key. It uses a simple challenge-response scheme whereby the STA initiates an Authentication Request with the AP, and the AP generates a random 128-bit challenge. Once the client receives the management frame from the responder, it copies the contents of the challenge text into a new management frame body. Using the WEP key and the Initialization Vector (IV), the STA encrypts this new management frame and sends it back to the AP for verification. If the verification at the AP is successful, the AP and STA switch roles and the process begins again to ensure mutual authentication. The procedure is shown in the Figure 3-11 [Arb01]. An additional limitation of shared-key authentication is that it only authenticates the client, not the actual user.

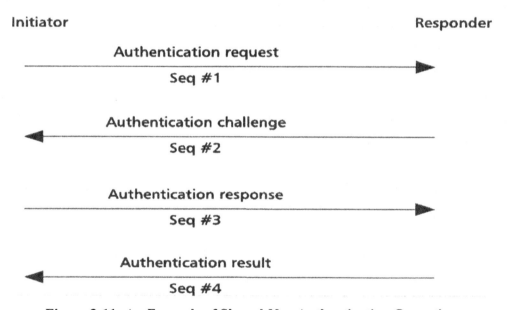

Figure 3-11: An Example of Shared Key Authentication Operation

3.5.2.1 Wired Equivalent Privacy

The Wired Equivalent Privacy (WEP) protocol was designed to provide confidentiality for wireless network traffic. It is based on the RC4[2] cipher algorithm. The standard defines 40-bit and 104-bit keys; however, many vendors have implementations with longer keys (128-bit, 256-bit, etc.). WEP also uses a 24-bit Initialization Vector (IV) as a seed value to initialize the cryptographic system. For example, a 40-

---

2   RC4 does not meet Federal Information Processing Standards (FIPS) requirements for cryptographic algorithms. Accordingly, some US government agencies and others requiring FIPS-validated solutions cannot use WLAN security solutions based on RC4, including both Temporal Key Integrity Protocol (TKIP) and WEP.

bit WEP key and an IV are used to generate a 64-bit key for the system. RC4 is a stream cipher algorithm that generates a pseudo-random sequence, which is unified with the message using an XOR operation. Of course, a different sequence is needed for each message; that is where the IV becomes significant. The IV's short length was one of the main weaknesses of the protocol, as evidenced by the many open source tools available today which are capable of attacking it.

For integrity checking, WEP uses a 32-bit Cyclic Redundancy Check (CRC-32). This is computed for each payload before encryption takes place. The CRC-32 is then encrypted and becomes the Integrity Check Value (ICV) on the frame to be transmitted. The CRC-32 is vulnerable to bit-flipping attacks, allowing an attacker to know which bits change when message bits are modified. The bit-flipping flaw persists after a stream-cipher step such as RC4. For this reason, WEP protocol integrity control is vulnerable to bit flipping attacks.

3.5.2.2 Temporal Key Integrity Protocol

The Temporal Key Integrity Protocol (TKIP) was designed to strengthen the WEP protocol without causing significant performance degradations. It mainly uses the same algorithms as WEP and because of that no hardware replacement is often required for an equipment upgrade from WEP to TKIP. WiFi Protected Access (WPA) certification, from the WiFi Alliance, requires the implementation of TKIP. TKIP provides the following security features which improve on WEP [NIS07]:

- Confidentiality protection, using the RC4 algorithm;

- Integrity protection, using the Message Integrity Code (MIC) based on the Michaels algorithm;

- Replay prevention, using a frame sequencing technique; and

- Use of a new encryption key for each frame.

TKIP uses three keys during the encapsulation process, two of them for integrity control (each half duplex channel) and the other for encryption. Table 3-2 lists the primary characteristics of TKIP and the security principle achieved.

TKIP Characteristic	Security Principle
Two 64-bit message integrity keys are used for the MIC. The MIC is computed over the user data, source and destination addresses, and priority bits.	Integrity
A monotonically increasing TKIP Sequence Counter (TSC) is assigned to each frame. This provides protection against replay attacks.	Integrity
A key-mixing process produces a new key for every frame. It uses the Temporal key (explained later) and the TSC to generate a dynamic key.	Confidentiality
The original user frame, the MIC, and the source address are encrypted using RC4 using the per-frame key.	Confidentiality

Table 3-2: TKIP Characteristics

3.5.2.3 Counter Mode with Cipher Block Chaining MAC Protocol

The Counter Mode with Cipher Block Chaining MAC Protocol (CCMP) was developed with many of the same objectives as TKIP; however, CCMP development was done with fewer hardware restrictions. It is based on a generic authenticated encryption block cipher mode of the Advanced Encryption Standard

(AES) [NIS01] Counter with Cipher Block Chaining Message Authentication Code (Counter with CBC-MAC, or CCM) mode [RFC 3610]. CCM is a mode of operation defined for any block cipher with a 128-bit block size. In few words, AES is to CCMP as RC4 is to TKIP [NIS07, Leh06].

CCMP implementation is mandatory for WPA2 WiFi Alliance certification. In most cases, the cryptographic characteristics of CCMP require a hardware change for the upgrade from WEP.

CCMP generates an integrity control (MIC) using Cipher Block Chaining (CBC-MAC). The key used for encryption and integrity is the same, but with different initialization vectors. The integrity protection is done for the payload and header, while the encryption (confidentiality protection) is done only for the payload. This allows for easier detection of a wrong packet.

CCMP processing expands the original MAC Protocol Data Unit (MPDU) size by 16 octets, 8 octets for the CCMP Header field and 8 octets for the MIC field (see Figure 3-12). The CCMP Header field is constructed from the Packet Number (PN), ExtIV, and Key ID subfields. PN is a 48-bit PN represented as an array of 6 octets. PN5 is the most significant octet of the PN, and PN0 is the least significant. Note that CCMP does not use the WEP ICV [IEE04]. CCMP key space has size 2^{128} and uses a 48-bit PN to construct a cryptographic nonce, which is a number or bit string used only once in security engineering, to prevent replay attacks. The construction of the nonce allows the key to be used for both integrity and confidentiality without compromising either.

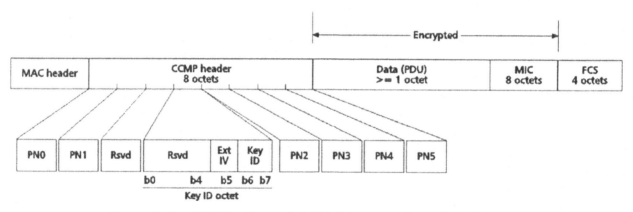

Figure 3-12: CCMP's Expanded MAC Protocol Data Unit (MPDU)

The encapsulation process is as follows [IEE04].

- Increment the PN to obtain a new PN for each MPDU. Retransmitted MPDUs are not modified.

- Use the fields in the MPDU header to construct the additional authentication data (AAD) for CCM. The CCM algorithm provides integrity protection for the fields included in the AAD. MPDU header fields that may change when retransmitted are muted by being masked to 0 when calculating the AAD.

- Construct the CCM nonce block from the PN, A2, and the Priority field of the MPDU, where A2 is MPDU Address 2. The Priority field has a reserved value set to 0.

- Place the new PN and the key identifier into the 8-octet CCMP header.

- Use the temporal key, AAD, nonce, and MPDU data to form the cipher text and MIC. This step is known as CCM originator processing.

3.5.3 Robust Security Networks (IEEE 802.11i/WPA2)

Amendment IEEE 802.11i to 802.11 introduces the concept of Robust Security Networks (RSNs) and Robust Security Network Associations (RSNAs). These new concepts imply enhanced security features beyond the simple shared key challenge-response authentication just described [NIS07]. In RSNAs, IEEE 802.1x provides authentication and controlled port services, while key management is done using IEEE 802.1x and IEEE 802.11 working together. All STAs in an RSNA have a corresponding IEEE 802.1X entity that handles these services [IEE04].

In addition to WEP and IEE 802.11 authentication, an RSNA defines [IEE04]:

- Enhanced authentication mechanisms for STAs;

- Key management algorithms;

- Cryptographic key establishment; and

- An enhanced data encapsulation mechanism, called CTR (counter mode) with CCMP and, optionally, TKIP.

IEEE 802.11i offers two general classes of security capabilities for IEEE 802.11 WLANs. The first class, pre-RSN security, includes the security mechanisms explained before: open system or shared key authentication for validating the identity of a wireless station, and WEP for the confidentiality protection of traffic. The second class of security capabilities includes a number of security mechanisms to create RSNs [NIS07]. Figure 3-13 illustrates the security mechanisms for RSN and pre-RSN networks.

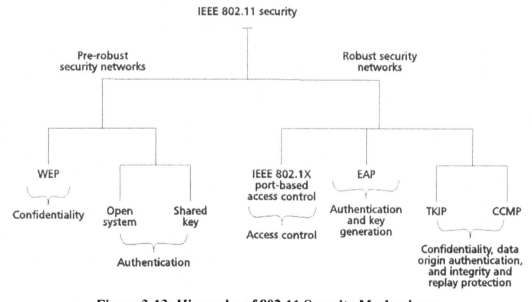

Figure 3-13: Hierarchy of 802.11 Security Mechanisms

3.5.3.1 Key hierarchies

Among the most important cryptographic characteristics are the keys used during the process. Pre-RSNs use manual WEP keys; only one key is used and no key management exists. For RSNs, many keys are needed for integrity and confidentiality protection. IEEE 802.11i defines two key hierarchies for RSNs: the Pairwise Key Hierarchy (PKH) for unicast traffic and the Group Key Hierarchy (GKH) for multicast/broadcast traffic [NIS07]. In PKH, the process begins with one of the two root keys. From those keys, all key material is produced. These pairwise keys are used by both TKIP and CCMP. The Pairwise Master Key (PMK) is taken directly from the root key. This key has at least 256 bits, depending on the generation method (Phase Shift Keying [PSK] or Adaptive Antenna Array [AAA]).

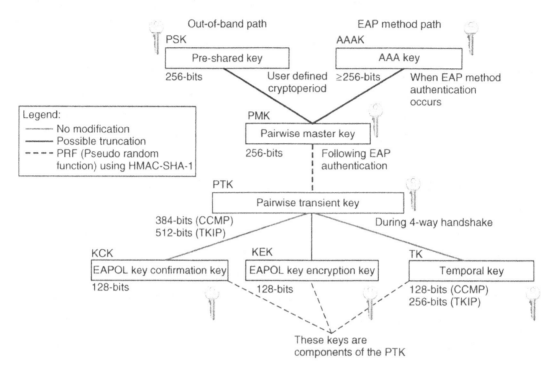

Figure 3-14: Pairwise Key Hierarchy

As illustrated in Figure 3-14, in PKH, after the Pairwise transient key (PTK) is generated from the PMK using a pseudo-random function and keyed-Hash Message Authentication Code (HMAC-SHA1), the PTK is partitioned into the EAPOL key confirmation key (KCK), the EAPOL key encryption key (KEK), and the Temporal key (TK). These keys are used to protect unicast communication between the authenticator's and supplicant's respective clients. PTKs are used between a single supplicant and a single authenticator [IEE04].

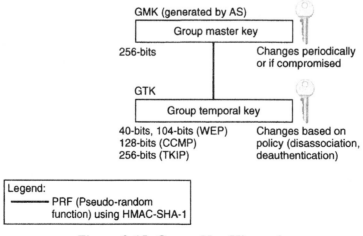

Figure 3-15: Group Key Hierarchy

The other key hierarchy is the Group Key Hierarchy. As shown in Figure 3-15, it consists of a single key: the Group temporal key (GTK). The GTK is generated by the authenticator and transmitted to all the stations.

3.5.4 3GSM Security

3GSM (3rd Generation Global System for Mobile Communications) security principles are similar to other networks, such as WLANs. However, user expectations for fast communications and ease of use, as well as the propensity for handsets to be easily lost or stolen, present a number of unique challenges. Under the first-generation Advanced Mobile Phone System (AMPS), only the device's serial number was used to validate the handset. In short order, many devices were developed to compromise the system, for example by intercepting serial numbers over the air and thus cloning handsets.

The second generation GSM system was developed with security in mind. The responsibility for user security is largely in the hands of the Home Environment (HE) operator. The HE operator can control the use of the system by the provision of Subscriber Identity Modules (SIMs), which contain user identities and authentication keys.

The design of third generation security was based on the following principles [3GP01]:

- 3G security will build on the security of second generation systems. Security elements within GSM (e.g., SIMs) and other second generation systems that have proved to be needed and robust will be adopted for 3G security.

- 3G security will improve on the security of second generation systems: 3G security will address and correct real and perceived weaknesses in second generation systems.

- 3G security will offer new security features and will secure new services offered by 3G.

3.5.4.1 GSM legacy

One of the most important elements of GSM security is the SIM. This is a removable security card which is terminal-independent and requires almost no user intervention (aside from the input of a PIN at device-boot time). The SIM is managed by the HE and contains all the identification data and cryptographic keys needed for the subscriber to make or receive a call. The authentication process is shown in Figure 3-16.

The Authentication Center (AuC) associated with the Home Location Register (HLR) contains the identification material for the user (i.e., the shared secret key). The Mobile Station (MS) attempts authentication to the network via the Mobile Switching Center/Visitor Location Register (MSC/VLR), which in turn passes the request to the HLR/AuC. It responds with a challenge (128-bit RAND), a key for encryption (Kc), and an expected value of the user expected response (XRES) to the MSC/VLR. The network sends the RAND to the MS, the MS transfers the parameter to the SIM, and the SIM generates the RES (32 bits) and Kc (64 bits) using the RAND and the shared secret key Ki. After that, the MS sends the response (RES) to the network and the network compares the RES to the XRES for verification. The functions that generate the RES and the Kc are called A3 and A8 respectively.

During call establishment, an encrypted mode of transmission is established where the MSC/VLR transports the current Kc to the base station and then instructs the MS to select the same Kc generated in the SIM during authentication. The encryption algorithm is called A5 and it is a stream cipher. Currently three different A5 algorithms have been standardized, called A5/1, A5/2, and A5/3. The encryption process is done before modulation but after the interleaving process using a stream cipher (function A5) and the Kc. Note that this encryption is only for the air interface. Encryption of the channel is switched on or off by the base station, which also selects the algorithm in use.

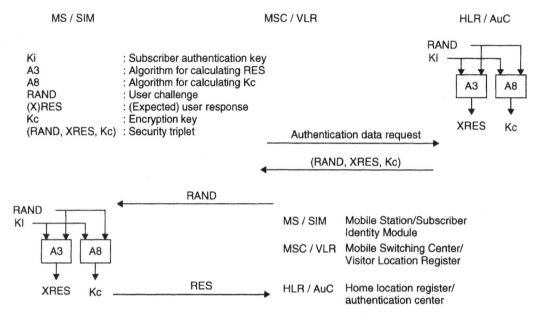

Figure 3-16: GSM Authentication Mechanism

The primary security issues affecting GSM are:

- The network is not authenticated by the MS, so impersonation attacks are possible using rogue network hardware.

- Only the air part of the communication is encrypted.

- Cipher keys and authentication values are transmitted unencrypted within and between networks (IMSI [International Mobile Subscriber Identity], RAND, SRES [Signed Response], Kc).

- Protection against radio channel hijacking and also channel integrity rely on this encryption.

- Some of the algorithms have been proven weak through successful attacks.

3.5.4.2 3GSM security domains

The first part of the 3GSM security model is the concept of Security Domains. A Security Domain is a collection of elements that is under the control of one security authority, and performs a security function. In particular the following domains are specified:

- Home Network (HN),

- Visited Network (VN),

- Access Network (AN), and

- Third-Party Application Service Providers (ASPs).

Note that the AN is not necessarily part of the VN, but could be a third-party entity such as an Internet café or a hotel network. In many cases, this network does not participate other than to provide carriage.

The domains are illustrated in Figure 3-17 in the context of the European Telecommunications Standards Institute (ETSI) Next Generation Network (NGN) concept. The two third-party application providers have relationships with the networks. However, the two could be the same provider. For example, a content downloader could have relationships with more than one carrier. Thus, a customer of one roaming on the other will be able to access content through either carrier. It is expected that there may be services from one carrier not available to another, for example a business locator in a particular city, which would be used by customers of both – e.g., lost tourists in a city.

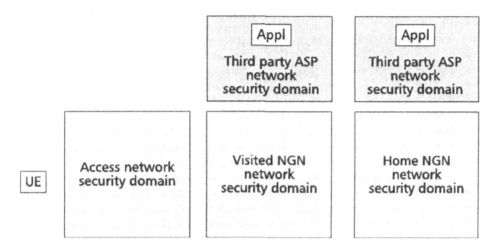

Figure 3-17: NGN Security Domains

The AN is abstracted from the 3G network. This is the case in other standards, and it is useful to distinguish these networks for a number of reasons: it makes the security associations and relationships clearer, and it may be imposed by regulation and reselling. The standards assume that some sort of Security Gateway Function (SEGF) exists between the domains. The details of the security gateway are not specified within 3G, leaving it to the domains to determine their requirements.

3.5.4.3 Security Associations

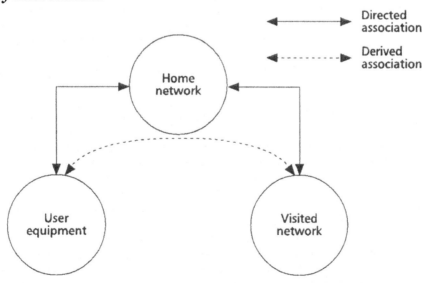

Figure 3-18: The Security Associations of UMTS/GSM

Security domains can form Security Associations (SAs) between themselves. Two direct SAs are defined and are shown in Figure 3-18: between the User equipment (UE) and the HN and between the HN and the VN.

Under 3GSM, the UE is assumed to incorporate a SIM, which provides certain functions such as authentication and key agreement, and provides secure storage and processing. Note that the HN also has a security association with third-party ASPs. There is no direct relationship between the UE and the VNs. The other relationships form the basis of the model. Trust is established by the UE being authenticated to the network. The 3G Security Architecture is illustrated in Figure 3-19, which identifies the five SAs.

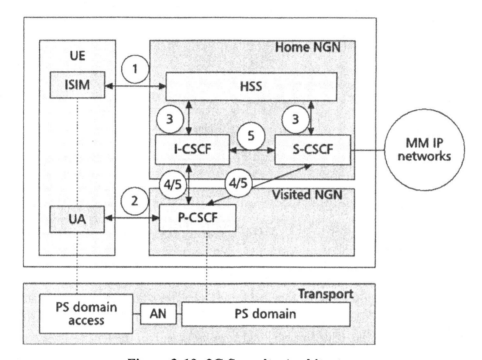

Figure 3-19: 3G Security Architecture

The standard explains this diagram using the following numerical key.

1. This is the long term SA between the IP Multimedia Services Identity Module (ISIM) and the HSS. It provides mutual authentication. Note that the authentication is usually carried out by the Serving Call Session Control Function (S-CSCF), although the HSS generates the challenges and expected responses.

2. This is the secure link between the UA and the Proxy-Call Session Control Function (P-CSCF). It supports data origin authentication.

3. This is the security association for the internal elements of the network. This is more fully described in 3GPP Technical Specification 33.120 [3GP01].

4. This is the security association for Session Initiation Protocol (SIP) nodes. If the P-CSCF is in the Home Next Generation Network (NGN), then this relationship is covered by 5.

5. This is the internal security association for SIP nodes within the network.

Figures 3-18 and 3-19 show the explicit links for the home network and for the visited network. Note that in both of these cases the transport layer is assumed to be independent.

3.5.4.4 Trust model

In the case where the UE does not have a relationship with the network, the home network provides the authentication vectors – that is, a set of challenges, the expected responses, the responses expected by the UE, and the cryptographic keys. Thus, the trust relationships use a delegated Trust Model. The user trusts the VN because the user trusts the HN which has a trust relationship with the VN. This is the model used in mobile networks and roaming. This model is expected to be extended to wireline services.

Under GSM the handset does not authenticate the network; it implicitly trusts the network equipment. This led to some security weaknesses which allowed the confidentiality of calls to be compromised by unauthorized third parties. The security weakness has been addressed by having the home and visited networks also provide responses to the handset. The handset compares these responses against the expected responses from the network. If the network fails this authentication, then the handset will not connect. An important aspect of this set up is that all signaling and control traffic is sent from the VN to the HN. This is used for billing and to validate the signaling from the UE.

3.5.4.5 Interconnect security

The security domains are controlled by one security authority and are linked to each other by security gateways. The 3GSM standards do not prescribe what protocols or mechanisms should be used to connect these gateways or to protect the traffic. Consequently, carriers need to determine requirements for the connections between domains. Among these requirements should be:

- standards-based protocols such as IP security;
- strong authentication – probably certificate-based;
- regular re-authentication;
- use of standards-based cipher; and
- integrity mechanisms.

The key lengths and algorithms should be reviewed regularly to ensure that they will continue to satisfy the security objectives. Initially, the security gateways should use IPSEC with 2048-bit RSA public key moduli for authentication, Internet Key Exchange (IKE) to provide key exchange and management, 128-bit Rijndael as the cipher between the gateways, and Message Digest algorithm 5 (MD5) providing integrity. The security gateway is expected to be a network gateway or edge router. Once the gateways are secured, the security domains trust each other. The risk of trusting other security domains is that fraudulent traffic may originate in the trusted domain. This may be due to a compromise within the domain or to malicious activity by operators of the domain. Thus, due diligence is required when establishing a trusted domain. There exist many commercial procedures to establish trust relationships. These include requirements that any trust domain should:

- comply with international risk management standards such as ISO/IEC 27002 [ISO05];
- have fraud management processes;
- have dispute resolution procedures; and
- have procedures for carrying out investigations, including joint investigations with other carriers and ISPs.

These, when used with other commercial procedures, will reduce exposure to fraud.

3.5.4.6 Authentication and Key Agreement

The fundamental protocol used within 3G for authentication and key exchange is the Universal Mobile Telecommunications System Authentication and Key Agreement (UMTS AKA). This is a zero-knowledge protocol. That is, no secret information is passed between two parties. It also provides mutual authentication of the UE and the network. There are three steps in the protocol.

- The UE requests registration and access.

- The network responds with a random challenge and a network authentication token.

- The UE validates the network's authentication token and responds with its own token.

After these exchanges, the network validates the response from the UE. If this response is validated, then the cryptographic keys are derived. If the network is a VN, then the VN passes the request to the HN, which responds by passing authentication vectors to the VN. The authentication vectors consist of the random challenge to be used, the expected response from the UE, the network authentication token, and the cryptographic keys. Note that at no point do either the UE or the HN pass on the shared secret.

3.5.4.6.1 Keys and other parameters

The authentication key K is a fixed 128-bit key held within the SIM. It is the long-term secret key shared between the SIM and the AuC. See Table 3-3 for a summary of the size of this and the following parameters.

Parameter Sizes										
K	RAND	SQN	AMF	CK	IK	MAC	MAC-S	AUTN	AUTS	AK
128	128	48	16	128	128	64	64	64	64	48

Table 3-3: Parameter Sizes

The random value RAND is generated by the serving network at the time of request.

The sequence number SQN is 48 bits and is used as a counter. This counter is kept by both the MS and the HN. It is a generated value, although there is no prescribed method to generate this value. However, there are methods for maintaining synchronicity between the HN and the MS.

The Authentication Management Field (AMF) is a 16-bit parameter; it may vary from domain to domain. The only constraint is that the format and interpretation of the AMF must be the same for all implementations in a domain. Among the examples of the AMF are: support for multiple authentication algorithms and keys; changing sequence number verification parameters; and setting threshold values to restrict the lifetime of cipher and integrity keys.

All other values are derived from these four values just defined.

- The cipher key CK is the session key. It has 128 bits.

- IK is an integrity key. It has 128 bits.

- The message authentication code MAC is a 64-bit field. It is used in the authentication of the mobile device and the network. Note also that the parameter XMAC is used by the SIM to authenticate the network.

- MAC-S is an authentication token used in resynchronization and to construct AUTS.

- Authentication Token (AUTN) is a network authentication token.

- AUTS is a token used in resynchronization.

- AK is an anonymity key used to conceal the sequence number, as the latter may expose the identity and location of the user. The concealment of the sequence number is to protect against passive attacks only. If no concealment is needed then f5 (see below) $\equiv 0$ (AK = 0).

3.5.4.6.2 Cryptographic primitives

Nine cryptographic primitives have been specified for use in 3G. These have been specified at two levels: the first with the parameters and lengths, and the second with the actual algorithms. This means that if a weakness is found with a particular algorithm, then another algorithm with the same parameters can be specified. However, if the length of authentication, for example, is no longer considered sufficient, then these parameters will need to be re-specified. These algorithms provide the functions used to:

- authenticate the UE and the network,

- derive cipher keys,

- derive anonymity keys, and

- resynchronize keys.

The confidentiality and integrity ciphers are based on the block cipher Kasumi. Kasumi encrypts blocks of 64 bits and uses a 128-bit key. The functions for authentication and key agreement use a cryptographic primitive that is a block cipher with a 128-bit key and 128-bit block size. The standards do not specify which block cipher, only the parameters. The cryptographic primitives or functions used are:

- f1 – Message authentication function used to compute MAC;

- f1* – Message authentication function used to compute MAC-S;

- f2 – Message authentication function used to compute RES and XRES;

- f3 – Key generating function used to compute CK;

- f4 – Key generating function used to compute IK;

- f5 – Key generating function used to compute AK in session set-up;

- f5* –Key generating function used to compute AK in resynchronization procedures;

- f8 – Data confidentiality algorithm for the session (uses CK); and

- f9 – Data integrity algorithm for the session (uses IK).

More specifically:

$$
\begin{aligned}
\text{MAC} &= \text{f1(K, RAND, SQN, AMF)} \\
\text{XMAC} &= \text{f1(K, RAND, SQN, AMF)} \\
\text{RES} &= \text{f2(K, RAND)} \\
\text{XRES} &= \text{f2(K, RAND)} \\
\text{CK} &= \text{f3(K, RAND)} \\
\text{IK} &= \text{f4(K, RAND)} \\
\text{AK} &= \text{f5(K, RAND)}
\end{aligned}
$$

These are used in the protocols for authentication and key exchange. Thus the functional types of these are (GF = Galois Field):

$$f1: \{GF(2)\}^{320} \rightarrow \{GF(2)\}^{64}$$

$$f3: \{GF(2)\}^{256} \rightarrow \{GF(2)\}^{128}$$

$$f4: \{GF(2)\}^{256} \rightarrow \{GF(2)\}^{128}$$

$$f5: \{GF(2)\}^{256} \rightarrow \{GF(2)\}^{48}$$

The function f2 will be fixed within the domain; however, its output varies between 4 and 16 octets (32 to 128 bits). There may be other constraints, such as buffer size in the SIM or the packet sizes in the exchange.

3.5.4.6.3 Description of protocols

There are two main protocols: the distribution of authentication data from the HE to the SN, and the authentication between the SN and MS. In the first protocol, the SN requests the authentication information from the HE. Essentially the IMSI (i.e., the permanent number) is sent from the SN to the HE. The HE responds with: {AV1, ..., AVn}, where each AVi is defined:

$$AV = RAND \parallel XRES \parallel CK \parallel IK \parallel AUTN, \text{ and}$$

$$AUTN = SQN \parallel AK \parallel AMF \parallel MAC.$$

Note that a fixed number of authentication vectors are transferred. This allows the SN to authenticate a number of times without reference to the HE. Now that the SN has the authentication vectors, the authentication protocol is straightforward.

- The SN sends the MS the request: RAND $\parallel$ AUTN.
- The MS responds with RES.
- The SN compares RES with XRES. If they differ, authentication fails.
- The MS generates XMAC and compares it with MAC. If they differ, the MS rejects user authentication and sends an error code.
- If both the SN and the MS accept the authentication, then the session is established.

Note that the user authenticates the network. This mutual authentication overcomes a major shortcoming of the GSM protocol.

The Trust Model is essentially a delegated trust model. Trust is delegated by tokens. A network, N, will accept a registration attempt from UE that claims to originate either from N or from a home network (HN) with which N has a relationship. The UE will attempt to register with the network. N acts as a proxy for the registration messages. If the registration is successful the HN will send a message to N indicating that the registration has been successful. This message is essentially confirmation that the HN will accept charges incurred by the UE. Similarly, when the UE registers, it will determine expected responses from N. If the responses received from the network correspond with the expected response, the UE will trust N. Note that in this way, both the UE and the network authenticate each other. Note also that in the roaming case, there is no direct security association between the UE and the HN.

3.5.4.6.4 Analysis of security protocols

For security analyses, a distinction is made between protocols and cryptographic primitives, such as ciphers. A protocol is a series of messages exchanged between two or more parties; the messages are processed and other messages are generated and decisions are made on the basis of that processing. Cryptographic primitives are the functions used to generate and process messages and responses. In the protocol analysis, the cryptographic primitives are assumed to be secure.

AKA is the protocol that is used for authentication and key agreement. It is an extension of the authentication and key agreement protocol used in GSM. In GSM the network is not authenticated, which allowed user confidentiality to be compromised by a rogue base station. This flaw has been addressed in 3G, and no exploit of AKA has been published. However, this does not mean that no exploit exists. For protocols, particularly key exchange protocols, methods of evaluating and validating these protocols have been developed. Either a protocol logic, such as the Burrows-Abadi-Needham (BAN) logic, or model checking is used.

Køien has completed a formal mathematical analysis of the AKA protocol [Køi02]. In this model, the cryptographic primitives are assumed to be secure. The formal analysis found no flaw in the protocol and indicated that it was sound. A caveat indicates that the model may be incomplete. However, the protocol is simple and involves a small number of steps, so the model is accurately described in the formal language and therefore the probability that a flaw exists is small. Thus, we have confidence that the UMTS AKA protocol is sound. The security is therefore dependent on the security of the cryptographic primitives.

3.5.4.6.5 Analysis of cryptographic primitives

The cryptographic primitives can be grouped into two sets:

- primitives used for confidentiality and integrity of the traffic between the UE and the network, and
- primitives used for user authentication, key agreement, and generating nonces (random values).

Those in the first set are based on the Kasumi block cipher. Those in the second set use another kernel, or block cipher; the actual cipher is not specified, just its parameters. The standards use the Rijndael cipher in the examples, and the test data assume the use of this cipher.

Note that for authentication, key agreement and nonce messages are passed between the SIM and the HN, while the VN does not process these messages. Thus, the cryptographic primitives used by the HN and the SIM do not have to be shared with the VN, and therefore do not need to be standardized.

The 3G standards define the confidentiality and integrity functions (ciphers). These are based on the Kasumi block cipher. The other cryptographic primitives assume that a kernel is used; the kernel is assumed to be a block cipher with a 128-bit key and a 128-bit input/output. The example cipher used is Rijndael, the cipher chosen for the Advanced Encryption Standard (AES), and all the test vectors assume that Rijndael is used.

Unlike protocols, there is no way of being able to prove the soundness or security of a cipher or cryptographic algorithm [Col97]. The approach used by cipher designers has been to use strong structures and design criteria to avoid known weaknesses. These ciphers are then submitted to the cryptographic community for scrutiny.

The primitive Kasumi is a block cipher based on the Misty block cipher, designed by Mitsuru Matsui and published in 1995. The details of Kasumi were published in 1999. The most successful attack on Kasumi was published in 2005 by Biham, et al. [Bih05]. This attack requires $2^{54.6}$ chosen plaintext/ciphertext pairs (about 43 petabytes) and needs $2^{76.1}$ operations of Kasumi – the equivalent of encrypting the entire 43 petabytes two million times. This will recover one session key of Kasumi – the session key being updated regularly, at least daily. This attack is not feasible and does not appear to lead to more efficient attacks. Since this cipher is used to protect the RF interface, it is sufficiently secure for the medium term, say the next 10–15 years.

The 3G standards specify the other functions and primitives using a block cipher. The block cipher is not specified, only that it has a 128-bit key and a block size of 128 bits. The functions are built upon the cipher. The cipher used as an example in the standards is Rijndael, the cipher chosen for AES. The test data published by ETSI and 3GPP assume that Rijndael is used. Vendors are expected to support the use of Rijndael in these functions. Thus, the security of these functions will depend on Rijndael. Rijndael was designed by Joan Dæmen and Vincent Rijmen and was chosen to be the Advanced Encryption Standard (AES) in 2000 by the US National Institute of Standards and Technology. The US National Security Agency (NSA) still approves of the use of Rijndael to encrypt TOP SECRET information.

The most successful attack on Rijndael was published by Courtois and Peiprzyk in 2002 [Cou02]. This attack reduces the strength of 128-bit Rijndael to 91 bits – that is, about 10^{28} operations are required. This indicates that Rijndael will be secure for at least the next 10-15 years. If Rijndael is exploited, then the cryptographic primitive used by 3G can be changed, although this will require significant updates for the SIMs. Thus, the primitives used for key generation and the generation of nonces are considered secure.

The authentication string is 64 bits. If the authentication function is considered to be strong this means that there is a 2^{-64} ($\sim 10^{-19}$) chance that a malicious user will be able to impersonate a user and steal sessions or gain access to other functions. Note that in the banking industry, PINs on ATM cards are four to six digits long, and three attempts are permitted. This means that there is a chance of at least 10^{-6} of accessing the account by guessing the PIN. Thus, the authentication using the SIM is significantly more secure than these cards. If the 64 bits of authentication were ever considered insufficient, then 3G includes a re-authentication that would be used to authenticate the SIM twice. Upon receipt of the first authentication, immediate re-authentication would be requested with another random challenge; that is to set the re-authentication period to be small on the first registration. This will allow the 128 bits of the shared authentication key to be used.

3.5.4.7 3GSM summary

The security model used by 3G is fundamentally sound, the main risk being fraud originating from a trusted domain. Carriers therefore must take steps, both technically and commercially, to ensure that their exposure to this risk is small.

The security protocols used by 3G assuming the use of a SIM have been analyzed and have been found to be sound. The cryptographic primitives are currently secure and are expected to be secure in the foreseeable future. The cryptographic primitives for authentication and key agreement are chosen by the carrier – although Rijndael will be supported by vendors. Therefore, if the cryptographic primitives for authentication and key agreement are compromised or are discovered to be weak, then the carrier is able chose other primitives.

If the confidentiality and integrity primitive, Kasumi, is compromised, then changes to the standards and equipment will be required. This will be a global problem affecting all 3G deployments; therefore, finding a solution will be an industry problem and the expense will be amortized over many carriers and users. However, it is very unlikely that in the next 10–15 years such a compromise will be published.

3.6 References

[3GP01] 3rd Generation Partnership Project, Technical Specification Group, Services and System Aspects, *3G Security; Security principles and objectives*, Specification 33.120, Release 4, 2001.

[Arb01] W. Arbaugh, N. Shankar, and Y.C. Wan, *Your 802.11 Wireless Network has no Clothes*, University of Maryland, March 2001.

[Bih05] E. Biham, O. Dunkelman, and N Keller, *A Related-Key Rectangle Attack on the Full KASUMI*, 11th International Conference on the Theory and Application of Cryptology and Information Security, 2005.

[Cle06] A. Clemm, *Network Management Fundamentals*, Cisco Press, 2006.

[Col97] B. Colbert, *On the Security of Cryptographic Algorithms*, PhD Thesis, University of New South Wales, 1997.

[Cou02] N. Courtois and J. Pieprzyk, *Cryptanalysis of Block Ciphers with Overdefined Systems of Equations*, 8th International Conference on the Theory and Application of Cryptology and Information Security, 2002.

[IEE04] Institute of Electrical and Electronics Engineers, *IEEE Standard for Information Technology – Telecommunications and information exchange between systems – Local and Metropolitan area networks – Specific Requirements. Part 11 Wireless LAN Medium Access Control (MAC) and Physical Layer (PHY) specifications Amendment 6: Medium Access Control (MAC) Security Enhancements*, IEEE 802-11i, 2004.

[ISO05] International Organization for Standardization/International Electrotechnical Commission, *Information Technology – Security Techniques – Code of practice for information security management*, Standard 27002, 2005.

[ITU96] International Telecommunication Union, Telecommunication Standardization Sector (ITU-T), *Principles for a telecommunications management network*, Publication M.3010, 1996.

[ITU97] International Telecommunication Union, Telecommunication Standardization Sector (ITU-T), *TMN management functions*, Publication M.3400, 1997.

[Køi02] G.M. Køien, *A Validation Model of the UMTS Authentication and Key Agreement Protocol*, Technical Report R&D N 59/2002, Telnor, 2002.

[Leh06] G. Lehembre, *Seguridad WiFi - WEP, WPA, WPA2*, Hakin9, 2006.

[NIS01] National Institute of Standards and Technology, Federal Information Processing Standards, *Announcing the Advanced Encryption Standard (AES)*, Publication 197, 2001.

[NIS07] National Institute of Standards and Technology, *Establishing Wireless Robust Security Networks: A Guide to IEEE 802.11i*, Special Publication 800-97, 2007.

[NIS08] National Institute of Standards and Technology, *Guide to Securing Legacy IEEE 802.11 Wireless Networks*, Special Publication 800-48, Revision 1, 2008.

[OGC07a] Office of Government Commerce, United Kingdom, *The Official Introduction to the ITIL Service Lifecycle*, 2007.

[OGC07b] Office of Government Commerce, United Kingdom, *Service Transition*, 2007.

[TMF07] TeleManagement Forum, *Enhanced Telecom Operations Model (eTOM) – The Business Process Framework*, 2007.

[Zel99] D. Zeltserman, *A Practical Guide to SNMPv3 and Network Management*, Prentice Hall. 1999.

In addition, the following *Requests for Comments* (RFCs) generated by the Internet Engineering Task Force are referenced by number in this chapter.

[RFC 1155] *Structure and identification of management information for TCP/IP-based internets*, 1990.

[RFC 1157] *Simple Network Management Protocol (SNMP)*, 1990.

[RFC 1661] *The Point-to-Point Protocol (PPP)*, 1994.

[RFC 1901] Force, *Introduction to Community-based SNMPv2*, 1996.

[RFC 2284] *PPP Extensible Authentication Protocol (EAP)*, 1998.

[RFC 2865] *Remote Authentication Dial In User Service (RADIUS)*, 2000.

[RFC 2866] *RADIUS Accounting*, 2000.

[RFC 3584] *Coexistence between Version 1, Version 2, and Version 3 of the Internet-standard Network Management Framework*, 2003.

[RFC 3588] *Diameter Base Protocol*, 2003.

[RFC 3610] *Counter with CBC-MAC (CCM)*, 2003.

[RFC 3748] *Extensible Authentication Protocol (EAP)*, 2004.

Chapter 4

Radio Engineering and Antennas

4.1 Introduction

Wireless systems include radio communication links. The radio signals are transmitted through and received by antennas. The antenna parameters indicate its suitability for the system structure, the link and application, and performance requirements. Through the link, the radio propagation carrying the information experiences losses depending on frequency, distance, and link attributes. Modeling these channels can become complex and approximations are used, particularly for mobile systems.

Radio engineering and link design involve understanding all aspects related to radio propagation, frequency considerations, antennas, gains and losses, link attributes and design, coverage, and performance requirements, among other factors. Radio systems and access technologies are based on radio engineering.

4.2 Radio Frequency Propagation

4.2.1 Background

In mobile communications, the radio channel is the key limiting factor in controlling system performance. Radio waves from the transmitter to the receiver can either have line-of-sight (LOS) propagation or the path can be obstructed by buildings, trees, undulations of the ground, and other natural or man-made objects. Reflections, diffraction, and scattering are all factors that influence the link design. The complexity of the path is a function of the surroundings, which from a topological perspective have been classified as urban, suburban, or rural. (From a capacity perspective, the traffic density is of course the primary factor.) It is possible to have LOS propagation between the transmitter and receiver in an open area (e.g., rural), but often difficult in urban and suburban settings. Thus, in general, the modeling process is based on statistics and measured data. A deterministic approach based on an assumption of LOS propagation is only the starting point.

The propagation model in general predicts the average received level (large-scale variation) for an assumed separation between the transmitter (Tx) and the receiver (Rx), as well as the variability of the signal in an area surrounding the receiver (small-scale variation). The variability is the result of multipath, where the signal level can change by as much as 30–40 dB within a fraction of a wavelength. This multipath phenomenon follows a Rayleigh distribution.

Propagation inside a building, tunnel, or other enclosed area is no different than outside, where a typical receive signal is the composite sum of many signals reflected from walls, floors, ceilings and other fixed objects. The modeling process is similar to propagation outside a building, and is based on statistics or on empirical equations derived from measured data.

For reasons of consistency, the symbols and general flow used in these next sections follow the wireless communications textbook by Rappaport [Rap02].

4.2.2 Frequency Considerations

In radio propagation, the frequency at which the transmission is done plays a fundamental role. Several considerations should be made, starting with the spectrum availability for radio transmission, the specific use that can be made of the radio spectrum depending on international and national regulations, and the effect on the signal of the frequency at which it is transmitted.

Within the electromagnetic spectrum, and according to the International Telecommunication Union (ITU) [ITU08], the radio spectrum is usually considered to be placed below the infrared band (300 GHz), providing a wide range of frequencies where wireless transmission of information can be achieved (see Figure 4-1).

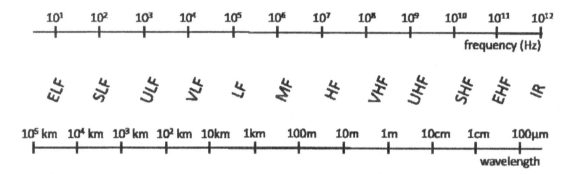

Figure 4-1: Radio Frequency Bands

Despite the intrinsic universal nature of the wireless spectrum, it is a limited resource with a limited capacity, and its usage is regulated and partitioned to provide different wireless services. Regulatory bodies at international and national levels have been in charge of determining which services access which bands and if the access is made on an exclusive usage basis (primary radio service) or shared usage basis (secondary radio service). As an example the ITU, as an agency of the United Nations, sponsors the forum that establishes radio allocation and management in the World Radiocommunication Conference (WRC). In this conference the ITU Radio Regulations [ITU08] are revisited and within these regulations we can find the allocation of the spectrum to over 40 defined radio communication services in the three Regions into which the world has been divided: Region 1 comprises Europe, Africa, the Middle East, and the north of Asia; all the Americas are in Region 2; and Region 3 includes the rest of Asia and Australia.

Thus, the use of different frequencies for transmission allows sharing the universal radio resource through a technique called Frequency Division Multiple Access (FDMA). However, the frequency at which the transmission is made conditions the wireless channel response and also the way the waves propagate.

The wireless channel does not have a uniform response in the frequency domain and this, from the transmission point of view, translates into different attenuation levels. As a general rule, attenuation increases as the transmit frequency increases, in an inverse relation with the signal wavelength, in this way limiting the range of the communication link. Thus, the underlying reading of this concept would be that low frequency transmission is more appropriate for wide-range coverage especially if not limited by capacity (or traffic density).

Also, waves transmitted at different frequencies propagate differently, leading to the following propagation modes [Rab08].

- Ground wave propagation: For frequencies below 3 MHz (MF bands and below), the propagation follows the curvature of the earth and has wide-range coverage. Its applications are usually for radio navigation and maritime mobile service.

- Ionosphere wave propagation: This propagation mode typically occurs when frequencies are in the range of 3 – 30 MHz (HF band) and the propagation occurs through refraction in the ionosphere. This propagation mode is not available 24 hours per day, since the ionosphere changes its refraction characteristics during the day. Typical applications make use of the wide coverage possible with this propagation mode.

- Direct (line-of-sight) wave propagation: This mode of propagation is typically used for frequencies over 30 MHz (VHF and above). Here the wave commonly travels directly from the transmitter to the receiver, although reflections from the ground plane and nearby objects may happen. The range of propagation is limited and, as was previously mentioned, it decreases as the frequency increases. In this range of frequencies many services are allocated, to mention some of them: radio, television, and satellite broadcasting, radiolocation services, mobile phone services, etc.

- Troposphere dispersion wave propagation: In this case propagation is due to dispersive reflections in the troposphere layers that occur because of changes in the refraction index. The transmission in this case can go beyond the line of sight and it usually has strong losses and fading.

It is clear from this discussion that the frequency is a key parameter in radio propagation. However, it is not the only one; the propagation medium and the path have great influence through the physical mechanisms that are discussed in the next sections [Rab08, Rap02].

4.2.3 Free-Space Loss

In every transmission there is a power loss in the signal strength that depends on the physical medium through which the transmission is taking place. Free-space loss assumes that there is an unobstructed line of sight between the transmitter and receiver and that there are isotropic antennas at both ends of the transmission link. Although free-space is not a realistic approach, it is typically used as a reference for the minimum loss that can be expected in any transmission.

Free-space loss depends on the transmission frequency f and on the distance d between transmitter and receiver as follows:

$$l_{\text{free}} = \left(\frac{4\pi d}{\lambda}\right)^2 = \left(\frac{4\pi df}{c}\right)^2 \tag{1}$$

Where λ is the wavelength and c is the speed of light.

This same expression can be given in dB and in practical units by taking logarithms on both sides of equation (1):

$$L_{\text{free}}(\text{dB}) = 32.45 + 20\log f(\text{MHz}) + 20\log d(\text{Km}) \tag{2}$$

Free-space communication follows the Friis equation where the received power falls off as the square of the transmitter-receiver separation:

$$P_r(d) = [P_t G_t G_r \lambda^2] / [(4\pi)^2 d^2] \qquad (3)$$

Here, if the gain $G_t = 1$, the antenna is regarded as an isotropic radiator. This implies that the received power decays with distance at a rate of 6 dB when the range is doubled, or 20 dB/decade. The product $P_t G_t$ is the effective isotropic radiated power (EIRP).

In practice, the effective radiated power (ERP), which is the radiated power from a dipole antenna having a numeric gain of 1.64 or 2.15 dB above an isotropic gain, is often used instead of EIRP. The Friis formula is valid for a received signal P_r in the far field.

The maximum permissible line-of-sight distance between transmitter and receiver is a function of the heights of the transmit and receive antennas and is limited by the curvature of the earth.

4.2.4 Reflection

Reflection is one of the propagation mechanisms that occurs when a wave impinges on an object that has a wavelength considerably longer than the propagation wave. The object could be the surface of the earth, a building, etc. Reflection is a typical effect in the direct wave propagation mode.

The analysis of this effect is complex and is characterized by the Fresnel reflection coefficient defined as the coefficient of the electric field intensity of the reflected and transmitted wave. This coefficient depends on the angle of incidence, the object's material properties, and the frequency; it is expressed in terms of its magnitude and phase $R = |R|e^{-j\beta}$. However, if we focus again on the power loss that reflection generates, the analysis can be significantly simplified.

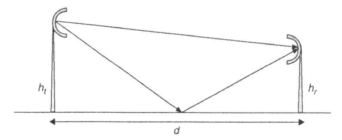

Figure 4-2: Two-Ray Propagation Model for the Study of Reflection

In a simple scenario we can assume that the transmitter and receiver are separated at a distance d that allows the earth between them to be assumed flat and that there is a single ground-reflected path. This leads to a two-ray model shown in Figure 4-2: the signal arrives at the receiver through two different paths, the direct path (line-of-sight, where the free-space loss applies) and the reflected path. The reflected wave travels farther and thus it has a different strength and also a different phase when impinging on the receiver. Both waves sum in the receiver and therefore a loss that encompasses the total signal strength loss can be obtained from the free-space loss by adding a correction factor. This factor is given by the ratio of the power through the line of sight path and the sum of the powers through both paths.

$$l_{2\text{-ray}} = \left(\frac{4\pi d}{\lambda}\right)^2 \frac{1}{\left|1 + |R|e^{-j(\beta+\Delta)}\right|^2} = \frac{\left(\frac{4\pi d}{\lambda}\right)^2}{1 + |R|^2 + 2|R|\cos(\beta+\Delta)} \qquad (4)$$

Defining h_t and h_r as the height of the transmit and receive antennas respectively, and assuming d is large compared to $h_t + h_r$, the difference of the path lengths is given approximately by $2h_t h_r/d$ and therefore the phase difference is $\Delta = 4\pi h_t h_r/d\lambda$. Also if d is large enough compared to the antenna heights, the angle of incidence can be considered zero and then $|R| \approx 1$ and $\beta \approx \pi$. Under these assumptions, equation (4) can be further simplified:

$$l_{2\text{-ray}} = \frac{\left(\dfrac{4\pi d}{\lambda}\right)^2}{2 + 2\cos\left(\pi + \dfrac{4\pi h_t h_r}{d\lambda}\right)} = \frac{\left(\dfrac{4\pi d}{\lambda}\right)^2}{4\sin\left(\dfrac{2\pi h_t h_r}{d\lambda}\right)^2} \approx \frac{\left(\dfrac{4\pi d}{\lambda}\right)^2}{4\left(\dfrac{2\pi h_t h_r}{d\lambda}\right)^2} = \frac{d^4}{\left(h_t h_r\right)^2} \tag{5}$$

And this leads to a total loss model where the power loss decreases as the *fourth* power of the distance between transmitter and receiver instead of the distance squared for free-space.

Alternatively note that if $d \gg h_t$, h_r the model can be expressed as

$$\frac{P_r}{P_t} = G_t G_r \left(\frac{h_t h_r}{d^2}\right)^2 \tag{6}$$

The above equation in dB is

$$\begin{aligned}
PL(dB) &= 10\log_{10}\left(P_t / P_r\right) \\
&= -10\log_{10}\left(G_t\right) - 10\log_{10}\left(G_r\right) - 20\log_{10}\left(h_t\right) - 20\log_{10}\left(h_r\right) + 40\log_{10}\left(d\right)
\end{aligned} \tag{7}$$

In other words, the equation, when applied to mobile systems, suggests the fourth-power relationship of the path loss with the distance from the cell-site antenna, and a 6-dB per octave variation because of antenna heights.

4.2.5 Diffraction

The diffraction mechanism occurs when between the transmitter and receiver there is an object in the propagation direction or its vicinity that modifies the electric field. The perceived effect would be that the wave propagates "around" the obstacle and this leads again to a power modification in the receiver compared to free-space transmission.

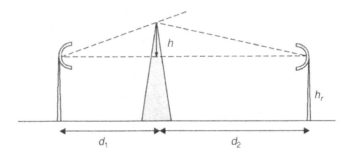

Figure 4-3: Knife-edge Propagation Model with a Single Obstacle

Figure 4-3 shows a propagation scenario where a single knife-edge obstacle produces the diffraction of the propagated wave. In this simple scenario the diffraction gain is given by the following formula:

$$g_{\text{diffraction}} = \frac{1}{2}\left(\left(\int_u^\infty \cos\left(\frac{\pi t^2}{2}\right)dt\right)^2 + \left(\int_u^\infty \sin\left(\frac{\pi t^2}{2}\right)dt\right)^2\right) \qquad (8)$$

Where u is the dimensionless Fresnel-Kirchoff diffraction parameter:

$$u = h\sqrt{\frac{2(d_1 + d_2)}{\lambda d_1 d_2}} \qquad (9)$$

In practice, graphical or numerical techniques are used to compute diffraction loss.

As mentioned above, the height of the ridge above or below the line joining the transmitter and receiver antennas is important. The radio waves occupy the volume in space known as the Fresnel zone [Bul77] with the height of the Fresnel zone measured in terms of wavelength λ. Practically, no additional loss (other than free-space loss) is encountered if the ridge is below the direct path by λ or more. The loss will increase as the clearance is reduced below λ and the waves will bend (diffraction loss) if the obstruction completely covers the direct path.

4.2.6   Refraction and K-Factor

Any wave travelling through a particular medium may modify its direction of propagation if there is a change in the medium properties that produces a change in the propagation speed. To characterize this change, the refractive index n of a particular medium is defined as the ratio between the speed of the wave in vacuum and the speed in this medium.

The change of direction between two different media with refractive indexes n_1 and n_2 respectively is given by:

$$\frac{\cos(\theta_1)}{\cos(\theta_2)} = \frac{n_2}{n_1} \qquad (10)$$

Where θ_1 is the angle of the wave impinging at the interface between the two media and θ_2 is the new angle of propagation.

This refraction effect is present in waves travelling through the troposphere since the refractive index of this medium changes with altitude, $n(h)$. This change in the direction leads to a curvilinear path in the propagation (see Figure 4-4), which may affect the transmission differently.

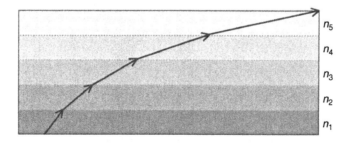

Figure 4-4: Curvilinear Propagation Path Due to Changes in the Refraction Index

It can be shown that the curvature of the wave propagation ρ depends on the refractive index as follows:

$$\rho = -\frac{dn(h)}{dh} = -\Delta n \qquad (11)$$

Depending on the position of both ends of the communication system and if the curvature of the propagation path is bigger (smaller) that the earth curvature—for example, in terrestrial links—it may increase (decrease) the range of the system.

To account for this effect, shown in Figure 4-5, we can define the effective earth curvature as the difference between the earth curvature ρ_0 and the wave curvature ρ. Considering the curvature is the inverse of the radius we can define:

$$\frac{1}{kR_0} = \frac{1}{R_0} - \frac{1}{R} \qquad (12)$$

where R_0 is the real earth radius, R is the wave radius, and kR_0 is the effective earth radius.

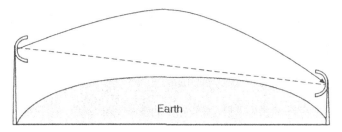

Figure 4-5: Propagation Model that Takes into Account the Curvature of the Earth and the Curvature of the Wave

The k-factor is then defined as the ratio between the effective earth radius and the real earth radius.

$$k = \frac{1}{1 - \frac{R_0}{R}} = \frac{1}{1 + \Delta n R_0} \qquad (13)$$

It should be noted that depending on the value of the k-factor, we can have any the following scenarios (see **Figure 4-6**):

- $k > 1$: the effective earth radius is bigger than the earth radius, then the coverage range is bigger.

- $0 < k < 1$: the effective earth radius is smaller than the earth radius, therefore the coverage range is smaller.

- $k < 0$: the effective earth radius is negative.

Assuming now the effective earth radius, we can compute the transmitter line-of-sight d_h from the following relation:

$$\left(kR_0 + h_t\right)^2 = d_{h_t}^2 + \left(kR_0\right)^2 \qquad (14)$$

$$d_{h_t}^2 \approx 2kR_0 h_t \qquad (15)$$

and extending this relation to the receiver side, we can compute the total line-of-sight distance d_{LOS} as follows:

$$d_{\text{LOS}} = d_{h_t} + d_{h_r} \approx \sqrt{2kR_0 h_t} + \sqrt{2kR_0 h_r} \tag{16}$$

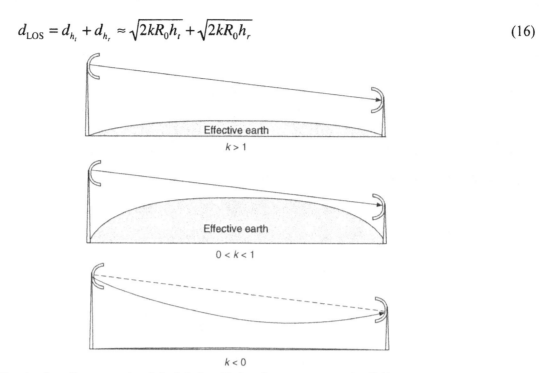

Figure 4-6: Equivalent Propagation Model that Takes into Account the Effective Curvature of the Earth Depending on the *k*-factor

4.2.7 Scattering

Scattering occurs when the medium of transmission has many objects with dimensions smaller than the wavelength. Typical such objects in mobile surroundings include foliage, street signs, and lamp posts. To predict the propagation loss over an arbitrary surface of the earth, the roughness of the ground must be taken into account. The criterion generally accepted for defining the surface roughness is the Rayleigh criterion. This states that if the path-length difference ($r_2 - r_1$) between wavefront A and B (on two paths) does not exceed one-quarter of a wavelength, then the surface is regarded as electrically smooth [Meh94]. Under this constraint the electrical phase difference from Figure 4-7 is given by:

$$\Delta\psi = (2\pi / \lambda)(2H\sin\theta) \cong \lambda / 8\theta \tag{17}$$

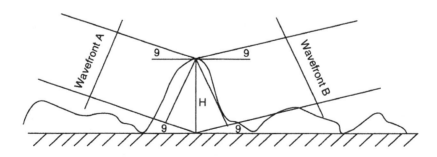

Figure 4-7: Electrical Phase Difference Beween Two Wavefronts

This equation states that for a small grazing angle, θ, the physical size of the irregularity can be relatively larger without destroying the electrical smoothness of the reflecting surface. On the other hand, for the same value of θ, at a higher frequency the irregularity must be kept small. As an example, at 900 MHz and for a grazing angle θ of 1 deg, the computed value for the Rayleigh height is 2.38 m. Unless the irregularity height exceeds 2.38 m, the surface will be regarded as smooth. A smoother surface generates higher reflected energy, which lowers the loss.

4.2.8 Atmospheric Losses

In direct wave propagation, the troposphere itself provides additional losses. The presence of gasses (oxygen, steam), rain, clouds, or fog may produce absorption of the electromagnetic energy. This effect is noticeable in frequencies above 10 GHz.

The loss due to gasses and water vapor also depends on the distance between transmitter and receiver, and a different loss scaling factor γ, in dB/km, is used depending on the origin of the loss (oxygen and/or water):

$$L_{\text{atmosphere}}(\text{dB}) = \left(\gamma_{\text{oxyg}} + \gamma_{\text{water}}\right)d \qquad (18)$$

As an example, for oxygen, in the span of frequencies from 10 GHz to 60 GHz, the factor γ_{oxyg} ranges from 0.06 dB/km to 20 dB/km. The same range of loss scaling occurs for the factor γ_{water} in the frequency range from 10 GHz to 350 GHz.

The loss for rain (or snow or hail) is evaluated in a statistical way, accounting for the rain intensity R_p (mm/h) that is passed $p\%$ of time, using the loss-scaling factor $\gamma_{rain}(R_p)$ (dB/km).

4.3 Antennas

4.3.1 Background

Without an antenna, a wireless system cannot perform its function. From the wireless system's perspective, the primary use of the antenna is to propagate an information-carrying signal as a transmitter or capture the radio energy as a receiver. The antenna functions here as a transducer. It converts a guided wave from a waveguide, coaxial cable, or transmission line into a propagating electromagnetic wave in free space (or air). The guided wave excites electrical currents that flow in the antenna structure, and these electrical currents in turn develop an electromagnetic (EM) field surrounding the antenna according to Faraday's Law and Ampere's Law. This EM field generates the wave that propagates through free space outward from the transmitting antenna toward the receiving antennas.

Antennas are typically reciprocal devices: they can receive and process EM waves while they transmit. The reverse process occurs when the antenna receives—the incoming EM wave excites electrical currents on the antenna structure. These currents are converted into guided waves that carry the information signal to the RF circuitry for processing and demodulation. Two spatial regions surround an antenna: the *near field* and *far field*. The near-field region is closest to the antenna and contains the reactive and oscillating EM field. The far-field region is farther away, where only the transverse-propagating EM field exists. The distance to the far-field region is generally approximated as $R \geq 2D^2/\lambda$, where D is the largest dimension of the antenna and λ is the wavelength. The transition between near- and far-field regions is not abrupt but is gradual, as the reactive near field sheds energy into the propagating wave. The far-field distance is generally a good approximation of the distance required to see only the outward propagating wave.

An antenna can also be considered as an impedance transformer because it converts a nominal RF system impedance (e.g., 50 ohms) into the free-space wave impedance (~377 ohms). Antennas generally operate over a narrow or limited bandwidth, hence they can also be regarded as bandpass filters in the overall communication system. Depending on the antenna's bandwidth, a pre-selection filter in the receiver may not be needed. Antennas can also be designed for wideband or broadband operation.

4.3.2 Antenna Parameters

Any antenna for a wireless system cannot be thoroughly designed or analyzed without an understanding of its basic parameters. However, the relative importance of each parameter depends on the intended application. The following sections are adapted from reference [Kra02], and depend on IEEE Standard 145-93, *IEEE Standard Definitions of Terms for Antennas*, for some definitions.

4.3.2.1 Input Impedance

The antenna input impedance is measured at its feed terminals (i.e., where the guided wave structure is connected to the antenna). It depends on the physical construction of the antenna and varies with frequency. Antenna input impedance is an important parameter because it generally must be matched to the characteristic impedance of the guided wave line and RF circuitry (e.g., 50 ohms) through an impedance transformer or balun at the feed terminals. Any large mismatch between the antenna impedance and the RF system impedance produces a large return loss and reduces the antenna gain and effective radiated power (ERP). This could degrade system performance to the point that the communication link would not connect in applications where link margins are small. Most system designers require the antenna's return loss to be < 2:1 or lower over the bandwidth. The antenna structure may include the impedance-matching circuitry and it often includes an RF connector that links it to the rest of the system. Antenna input impedance is determined either by closed-form solutions, through computer modeling and simulation, or by measurement with a network analyzer.

4.3.2.2 Size, Weight, and Power (SWAP)

For certain applications, size, weight, and power requirements are critical, particularly when the antenna is portable or mounted on a vehicle (aircraft, satellite, etc.). SWAP considerations are critical for satellite applications because space in the satellite and inside the launch vehicle is extremely limited. So is the power-generating capability onboard a satellite and added weight is very costly to launch. SWAP may be of secondary importance for fixed metropolitan Wi-Fi installations where AC power is conveniently available and weight is not a major concern. For a fixed operating frequency, the higher the gain required by the system, the larger the antenna (and hence weight). For a fixed physical size, operating at a higher frequency provides higher gain, while a lower frequency provides lower gain.

4.3.2.3 Field and Power Patterns

Virtually all antennas radiate electromagnetic waves more strongly in some directions than in others. This behavior is measured by using *field patterns*, which are three-dimensional (3D) vector quantities that are functions of θ and φ, the angles used in a standard spherical coordinate system. The field patterns describe the variation of the components of the electric (or magnetic) field vector as a function of θ and φ. For example, a normalized field pattern is defined by:

$$E_\theta(\theta,\phi)_n = E_\theta(\theta,\phi) / E_\theta(\theta,\phi)_{max} \qquad (19)$$

and

$$E_\phi(\theta,\phi)_n = E_\phi(\theta,\phi) / E_\phi(\theta,\phi)_{max} \tag{20}$$

where E_θ and E_φ are the θ and φ components of the spherical electric-field vector. There is no radial component of the electric field, E_r, in the far field of the antenna due to the transverse nature of the electromagnetic waves. Field patterns are typically displayed in 3D color plots, or by using several two-dimensional planar cuts. These 2D cuts, usually shown as polar plots, are often the principal planes (*xz* and *yz* planes). They are called the E-plane cuts, while the *xy* plane is known as the H-plane cut. *Power patterns* are visualized in 3D or in 2D planar cuts in exactly the same way as field patterns. Power patterns are defined by

$$P_n(\theta,\phi)_n = S(\theta,\phi) / S(\theta,\phi)_{max} \tag{21}$$

where S is the Poynting vector given by

$$S(\theta,\phi) = \left[E_\theta^2(\theta,\phi) + E_\phi^2(\theta,\phi) \right] / Z_0 \tag{22}$$

and Z_0 is the free-space wave impedance (~377 ohms). Field and/or power patterns for an antenna are determined either from closed-form solutions, from computer modeling and simulation, or from antenna measurements.

4.3.2.4 Beamwidth

One of the most important antenna parameters is beamwidth. An antenna that is directive and radiates its power more strongly in one direction has one major (or main) lobe, also called the *main beam*, in its field (or power) pattern and several minor lobes, or sidelobes. The beamwidth (in degrees) is a specification of the antenna's main beam; it is a measure of how strongly the antenna radiates in the main beam's direction. For example, a half-wave dipole antenna has a beamwidth of 78°, but a large parabolic reflector antenna may have a beamwidth of only 1°. Thus, the parabolic reflector is more highly *directive* than the dipole because it radiates more strongly along its main beam. The lower the beamwidth of the antenna, the more directive it is and thus it has higher gain. Beamwidth is inversely proportional to antenna gain, and thus is inversely proportional to antenna size. For a fixed physical antenna size, increasing the operating frequency increases the gain and therefore decreases the beamwidth. Beamwidth is most often expressed as the –3 dB (or half-power) beamwidth, which is defined where the field pattern

$$E_\theta(\theta,\phi)_n = 1 / \sqrt{2} \tag{23}$$

This point corresponds to the angle where the power pattern becomes

$$P_n(\theta,\phi)_n = 1 / 2 \tag{24}$$

because power is related to E_θ^2 and -3 dB = 10log(0.5). Other beamwidth specifications may also be used such as -10 dB or -20 dB beamwidth, which correspond to where $P_n(\theta, \varphi)_n = 0.1$ and $P_n(\theta, \varphi)_n = 0.01$, respectively. Half-power beamwidth is also expressed in the E- and H-planes as θ_{HP} and φ_{HP}, where θ_{HP} is defined in the *xz* or *yz* plane and φ_{HP} is defined in the *xy* plane.

4.3.2.5 Directivity, Gain and Aperture

The *directivity, D,* of an antenna is defined as the ratio of the radiation intensity to the radiation intensity averaged over a sphere. Directivity is a function of θ and φ and is mathematically expressed by

$$D(\theta,\phi) = U(\theta,\phi)/U(\theta,\phi)_{avg} = S(\theta,\phi)/S(\theta,\phi)_{avg} \qquad (25)$$

where $U(\theta, \varphi)$ is the radiation intensity. If no direction is specified, the direction of maximum radiation intensity is implied and the directivity becomes a single value. The gain, G, of an antenna is less than the directivity due to ohmic and other losses. The gain is expressed mathematically as $G = kD$, where k is an efficiency factor between 0 and 1.

An *isotropic radiator* (or isotropic point source) is a theoretical antenna that has an omnidirectional pattern and a directivity D = 1. This means that the isotropic source radiates energy uniformly in all directions over 4π steradians. All antennas have directivity D ≥ 1. Gain is usually specified in units of dBi, which means dB over isotropic. Thus, an isotropic radiator has a gain of 0 dBi. For example, a half-wave dipole has a directivity D = 1.64, which is equivalent to a gain of 2.15 dBi, and a gain pattern that is approximately $sin^3(\theta)$. The 3D gain pattern of a half-wave dipole can be visualized as a doughnut, with the dipole antenna positioned vertically at the doughnut's center. The pattern has a half-power beamwidth (HPBW) of 78° in the E-plane (elevation plane) and an omnidirectional (uniform) pattern in the H-plane (azimuth plane).

Although mathematically different, the terms gain and directivity are often used interchangeably. Most antenna engineers plot 2D and 3D directivity or gain patterns in place of field or power patterns. This is because the gain or directivity pattern is very similar in definition to the normalized power pattern, and the gain is often the single most important parameter of the antenna. In any wireless communication system, the antenna must have enough gain to close the link and perhaps provide some link margin as well. An approximate expression for directivity is given by D = 40,000/($\theta_{HP}\varphi_{HP}$), where θ_{HP} and φ_{HP} are the E- and H-plane half-power beamwidths in degrees. Directivity and gain of an antenna can be determined from either closed-form solutions, computer modeling and simulation, or measurements.

Every antenna has an *effective aperture* or A_e. Effective aperture can be viewed as that area over which the antenna effectively "captures" incoming electromagnetic waves. This aperture generally differs from the antenna's physical aperture, which is the cross-sectional area of the antenna's physical structure. Thus, the effective aperture is essentially an "electrical area" of the antenna, which represents its electromagnetic collection area. The effective aperture can be larger or smaller than the physical aperture, depending on the antenna type. The directivity (and hence gain) of an antenna is related to the effective aperture by the expression

$$D = 4\pi A_e / \lambda^2 \qquad (26)$$

Therefore, more gain requires either a larger effective aperture or area, or operation at a shorter wavelength (i.e., a higher frequency).

4.3.2.6 Polarization

The polarization of an antenna describes the orientation of the electric-field vector of the radiated electromagnetic field in the antenna's far field as the field propagates away from the antenna. Antennas radiate linear, elliptical, or circular polarization. In linear polarization (LP), the electric-field vector oscillates along a straight line as the wave propagates outward. Examples include antennas that are vertically polarized, horizontally polarized, or perhaps have a linear 45° polarization. Dipole or monopole antennas are good examples of linearly polarized antennas. An automobile radio antenna is linearly polarized. An antenna designed for vertical polarization will not receive or transmit horizontally polarized signals. In this case, the horizontally polarized signal represents the *cross-polarization, or cross-pol,* term.

For circular polarization, the electric-field vector rotates and traces out the locus of a circle as the radiated wave propagates. Circular polarization (CP) is further divided into left-hand circular polarization (LHCP) and right-hand circular polarization (RHCP), depending on the sense of rotation. With the thumb of the right hand pointed in the direction of propagation, curling the fingers of the right hand give the sense of RHCP. Thus, looking in the direction of propagation, the electric-field vector rotates clockwise for RHCP and counterclockwise for LHCP. Antennas designed for RHCP will not receive LHCP and vice-versa. LHCP represents the cross-pol for RHCP and vice-versa. A helical antenna is a prime example of a circularly polarized antenna.

With elliptical polarization (EP), the electric-field vector traces out the locus of an ellipse as the radiated wave propagates. Elliptical polarization can also be given a RH or LH sense in the same way as circular polarization. In fact, both linear and circular polarizations are a special case of elliptical polarization. A measure of an antenna's polarization is its *axial ratio (AR)*, which is the ratio of the major to minor axes in the polarization ellipse. For linear polarization, $AR = \infty$, and for circular polarization, $AR = 1$.

Polarization mismatch can be a significant concern in wireless system design, implementation, and operation. For example, a linearly polarized antenna can receive CP signals, but with 3 dB less power due to the polarization mismatch. Similarly, a CP antenna can receive LP signals, but again with the 3 dB loss. Therefore, any sound link-budget analysis must always include a loss term for polarization mismatch, depending on the system application. For maximum power transfer and signal reception, the Tx antenna must be of the same polarization and orientation as the Rx antenna. If the Tx is vertical LP, the Rx must also be LP and oriented vertically.

4.3.3 Antenna Types

Antennas can be classified in many ways—by their frequency bandwidth (narrowband, broadband, ultrawideband), by their physical design (e.g., wire antennas, reflectors, printed circuit, aperture), by their pattern characteristics (omnidirectional, high gain), by their electrical size (small, large), by their operational mode (traveling wave, surface wave, guided wave) or by their scanning characteristics (none, mechanical, electrical). A single antenna design may span several categories but perhaps the most straightforward way to classify antennas is by their physical design.

Wire antenna designs may involve loops, dipoles, folded dipoles, rhomboids, long wires, twin lines, or helices, and may include phased arrays of these individual elements. Many wire antennas are used in applications where lower gain (2–3 dBi), linear polarization and omnidirectional patterns are acceptable. Applications include portable radios, automobiles, mobile handsets, and wireless Internet modems and routers. Some wire designs such as helix antennas can have more moderate gains in the 8–15 dBi range. Wire antennas are typically narrowband antennas with bandwidths on the order of 10%–20% due to their resonant behavior. Other types of wire-like antennas such as Yagi-Udas, log periodic, and conical spirals operate over a much broader bandwidth than dipoles or other wire antennas. Log-periodic antennas are a mainstay for receiving broadcast television signals.

Aperture antennas consist of reflectors, parabolic dishes, lenses, spirals, flat panels, frequency-selective surfaces (FSS), and horns that radiate electromagnetic energy from a physical aperture. Such antennas are used in applications calling for moderate gains ($\approx$ 3 dBi to 20 dBi) or high gains (> 20 dBi) with moderate to very narrow beamwidths. Applications for aperture antennas include satellite communications, radar, antenna measurements and microwave telephone relays. An in-depth discussion of each antenna type is beyond the scope of this chapter, but any good undergraduate or graduate-level antenna textbook provides detailed descriptions. An exhaustive reference is the *Antenna Engineering Handbook* [Vol07].

Antennas typically operate over a narrow bandwidth (< 20%) due to their high Q structure, abrupt transitions, and resonant behavior, but they can be designed to operate over moderate-to-large bandwidths by ensuring, for example, smoother variations and transitions in the antenna structure. Broadband antennas are often called frequency-independent antennas, and can exhibit moderate-to-large bandwidths, anywhere from 2:1 up to 33:1 or more. Rumsey's principle states that the impedance and pattern properties of an antenna will be frequency-independent if the antenna shape is specified only in terms of angles [Kra02]. Examples of these broadband types are the biconical, planar log-spiral, and conical log-spiral antennas. A broadband antenna may also combine several different elements or regions of varying size or length so that the active region of the antenna shifts as the frequency changes. An excellent example of this type is the log-periodic antenna.

Aperture antennas can also be made broadband by including structures with smooth transitions and variations such as the quad-ridge horn antenna or impulse-radiating antenna. A new class of broadband antennas called fractal antennas are being developed and deployed based on self-similar geometrical structures [Wer00].

4.3.4   Phased Arrays

There are many wireless applications where the antenna pattern must be reconfigured or the antenna beam moved to different angular locations to communicate with different end users or to scan the surrounding area. These applications are addressed by moving (i.e., steering or scanning) the antenna beam either mechanically or electrically. Mechanically scanned antennas are typically some form of aperture antenna mounted on a pedestal that rotates in the azimuth direction, and they often include some rotation or movement in the elevation direction. Antennas for radars, satellite communication, and radio astronomy are examples of mechanically scanned designs. Electrically steered antennas are called *phased array antennas* or simply *phased arrays*. A phased array antenna generally remains fixed in space while the antenna beam is scanned (or even reconfigured) electronically. However, some applications may use a mechanically scanned phased array. In this case, the mechanical scan may cover the azimuth while the phased array scans in elevation.

A phased array antenna is composed of identical antenna elements in a regularly spaced lattice of some kind. A linear phased array places the antenna elements at equal intervals along a straight line, usually ¼ to ½ wavelength apart. A linear phased array can scan only along the direction of the array. A flat-panel (or 2D) phased array places the antenna elements at equal *x-y* space intervals on a planar surface. A flat-panel phased array has two scanning degrees of freedom.

The phased array works based on the principle of constructive or destructive interference. The signals feeding each antenna are given different amplitude weights (i.e., an *amplitude taper*) and relative phases (i.e., a *phase taper*). Signals from each antenna traveling to the same point in the far field of the array experience different phase changes due to the distance traveled; combined with the relative signal phases, they add or subtract when summed together. The amplitude taper and the interelement spacing control the sidelobe levels and the phase taper controls the position of the main beam. Thus, the beam is electronically scanned by changing the relative phase of the signal feeding each antenna element.

Phase shifters are usually digitally controlled, and it is important to understand the relationship between a one-bit change in phase control and the amount of change in scan angle. The interelement spacing between elements in the phased array is an important design parameter. A spacing greater than one wavelength introduces *grating lobes*, which are unwanted sidelobes that can reach levels almost equal to the main beam. This phenomenon degrades system performance by transmitting or receiving signals in unwanted directions.

The *array factor* for a phased array is the antenna pattern produced using a chosen amplitude and phase taper by assuming each antenna element is a point source. The amplitude, interelement spacing, and phase taper are iteratively adjusted by examining the array factor until the desired sidelobe-level performance is achieved. If the antenna elements in an array are identical, then the composite phased array antenna pattern is found simply by multiplying the array factor by the individual element antenna pattern. If the array elements are not equal, the composite array pattern is obtained by a vector sum in the far-field region of the array of the electromagnetic field radiated from each array element.

4.3.5   *Beamforming and Smart Antennas*

A traditional phased array antenna has a pattern with a single main beam and various sidelobes. The angular position of the beam can be electronically steered by varying the phase of the phase shifter at each antenna element. The level of the sidelobes is controlled by varying the amplitude of the signal to each antenna and the physical spacing between antennas. A phased array antenna can also produce multiple beams by using a beamforming network behind the antenna. This network is a complex interconnection of phase shifters, directional couplers, power dividers or combiners, or transmission lines that feed each of the array elements. Many different types of beamforming networks exist, including a power divider, a Butler matrix, Blass and Nolen matrices, a Wullenweber array, a McFarland 2D matrix, a Rotman lens, a Bootlace lens, and a dome lens [Han98]. These beamforming techniques produce multiple simultaneous fixed beams; and an in-depth discussion is beyond the scope of this chapter. Multiple independently steerable beams can be produced by having more than one set of phase shifters and amplitude-control elements connected to the array elements.

Equivalently, multiple independent beams can also be produced with a subarray. This is simply a set of phased array elements that is part of a larger phased array system. For example, a 32×32 element phased array antenna could be spatially subdivided into four 8×8 element subarrays to provide four independently steerable beams. However, the subarray elements need not necessarily be grouped together spatially. The 32×32 array could be subdivided according to the rule that every 4th element belongs to one of the subarrays. In this way, the subarrays are interdigitated with each other and should produce independent beams provided the subarrays have sufficient electrical isolation.

Smart antennas incorporate beam-steered phased arrays and use signal-processing techniques to shape the beam pattern according to certain optimum criteria [Vol07]. For example, a smart antenna can create a null in the antenna pattern in the direction of an interfering source or jammer. Smart antennas can increase the capacity of wireless communication networks through space division multiple access (SDMA). Some of the basic signal-processing algorithms used in smart antennas are the least mean squares (LMS), sample matrix inversion (SMI), recursive least squares (RLS), conjugate gradient method (CGM), constant modulus algorithm (CMA), and the least-squares constant-modulus algorithm (LS-CMA) [Vol07]. Smart antennas are also often called digital beamformed or adaptive arrays.

4.3.6 Antenna Design and Measurements

To begin an antenna design, the requirements must first be defined. These requirements typically flow from a rigorous system-level design and analysis. Parameters such as gain, return loss, beamwidth, axial ratio, polarization, and bandwidth must be well-defined and realistic. Mechanical considerations of the system may indicate the type of basic antenna structure that best suits the application. For example, an aircraft antenna might need to conform to the aircraft surface; thus, a spiral or patch antenna may work best. Onboard a satellite, a helix antenna or reflector antenna may be more suitable.

The design can proceed once the antenna requirements have been defined. An antenna can be designed in several ways depending on the basic antenna structure. For very simple antennas, a closed-form solution may be available. Often, a set of parametric curves can be used to set the various physical antenna parameters to achieve the desired gain, bandwidth, etc. Another popular design approach is to use commercial electromagnetics software to design the antenna and analyze its performance by computing the gain, bandwidth, return loss, axial ratio, and beamwidth. These commercial suites are based on four primary numerical solution techniques for Maxwell's equations—the finite element method (FEM), the finite-difference time-domain (FDTD) method, the method of moments (MoM), and physical optics (PO). The High Frequency Structure Simulator (HFSS) from Ansoft Corp. uses the FEM, and Microwave Studio from Computer Simulation Technology, Inc. (CST) uses a FDTD-like technique. The Numerical Electromagnetics Code (NEC) is a MoM software tool, and the GRASP tool, which uses PO, is produced by TICRA. GRASP is used almost exclusively for reflector antenna design and analysis, particularly for large reflectors.

The design process is usually iterative, starting with a basic design and then either fabricating and testing the antenna to gauge its performance, or performing several design iterations through computer analysis and simulation. The design is modified after each iteration, and design by computer is generally preferable because it eliminates expensive fabrication and testing. Once the final design is chosen, a prototype can be fabricated and experimentally characterized.

Antenna gain, patterns, and effective radiated power (ERP) are determined in one of two ways: by computer simulations or experimentally through measurements. Often, both techniques are used, with measurements generally viewed as the "right" answer to validate the computer analyses. Antenna measurements simulate how the actual physical device will operate in practice, provided the measurement process is sound. Therefore, they take precedence in verifying the final antenna performance. IEEE Standard 145-93 defines ERP: "in a given direction, the relative gain of a transmitting antenna with respect to the maximum directivity of a half-wave dipole multiplied by the net power accepted by the antenna from the connected transmitter." From the same standard, the effective isotropic radiated power (EIRP) is defined as: "in a given direction, the gain of a transmitting antenna multiplied by the net power accepted by the antenna from the connected transmitter." ERP and EIRP are often confused in practice. EIRP is generally used in satellite-link budget calculations for RF system design and analysis.

Antenna gain, patterns, and impedance (or return loss) are determined experimentally through antenna measurements. Impedance (return loss) involves a simple measurement with a network analyzer. Gain patterns are measured by far-field or near-field measurement procedures. Far-field measurements are done on a far-field antenna range. Such ranges are usually one of two types: an outdoor open area test site (OATS) or an indoor compact range/anechoic chamber. An OATS usually consists of two towers or other outdoor locations separated by line-of-sight over a fixed distance. One tower contains the transmitting antenna along with the measurement equipment such as a network analyzer and antenna positioner control. The other tower holds the antenna under test (AUT). It may contain a positioner to scan the AUT in azimuth and/or elevation and to change orientation for cross-pol measurements.

The indoor anechoic chamber on the other hand is usually a moderate-to-large room electromagnetically shielded from outside stray EM signals. Electromagnetic absorber material is placed on the ceiling, walls and floor of the chamber to reduce or eliminate reflections of test signals that could corrupt the measurements. The chamber uses a transmit antenna and the AUT is the receive antenna. A compact range is a smaller version of the anechoic chamber and uses a reflector to generate a plane wave that impinges on the AUT. The transmit antenna for the compact range is typically a calibrated horn antenna that is used as a feed to illuminate the reflector, usually in an offset configuration.

In a far-field measurement, the antenna measurement range and equipment are first calibrated. Typically, the range is calibrated using standard-gain horn antennas because the gain and patterns of these antennas are very well known. Then the AUT is mounted on a positioner that mechanically moves the antenna in azimuth and often in elevation. The positioner can also rotate the antenna to provide measurements of both polarizations. Both the magnitude and phase of the received signal are recorded and are often transformed into antenna gain by the measurement control software. The measurement software automates the measurement procedure.

In a planar near-field measurement, the AUT transmits while a small waveguide probe or horn antenna is mechanically scanned across a planar aperture in front of the AUT. Both magnitude (I) and phase (Q) of the electromagnetic field are recorded at each sample point, and the sampling density must meet or exceed the Nyquist criterion. Control of the longitudinal probe position is critical to achieve accurate and reliable results. The I and Q data are then transformed to the far field in software by aperture integration to obtain the far-field pattern. The main advantage of planar near-field scanning is its mathematical simplicity because the software transformation can employ an FFT algorithm. The main disadvantage is that the far-field pattern is accurate only over a limited angular range. Planar near-field measurements are best suited to high-gain antennas.

Cylindrical and spherical near-field scanning are also available, which scan the probe over a cylindrical or spherical surface surrounding the antenna. A similar but more complex aperture-integration technique is used on the sample data to compute the far-field pattern. Spherical near-field scanning works best for low-gain or more omnidirectional antennas. For any near-field scanning technique, the equipment is also initially calibrated, and final data must be modified to compensate for the field pattern of the probe.

4.4   Radio Engineering and Wireless Link Design

4.4.1 Radio Signal Power

The radio signal power represents the integrity of the information carried and the ability of the receiver to extract it. There are a number of attributes in play.

The transmitted signal itself is limited in power to avoid interference, in addition to safety, cost, and other considerations. The antenna represents a gain over a theoretical isotropic radiation case which would propagate power in all directions. Otherwise, the transmitted signal experiences losses as it propagates as discussed earlier. At the receiver, the signal goes through components such as a filter, amplifier, and demodulator. Noise is therefore combined with the information-carrying signal.

Therefore, radio engineering and wireless link design should consider the transmission and propagation impacts through the link, on the one hand, and what is expected of the parameters involved, on the other, to meet the performance requirements.

The minimum acceptable received power is a design criterion, as is the received signal-to-noise power ratio. The signal-to-noise ratio may be expressed in different ways. A typical measure is the bit energy per noise-power-per-Hz.

4.4.2 Link Budget Analysis

Link budget analysis involves identification of the received power, given the transmitted power, the antenna gain, the different losses, and the required fade margin. It may be outlined in different though fundamentally similar ways. Furthermore, it typically goes beyond the signal power expression to identify the signal-to-noise ratio, and accordingly the design parameter requirements.

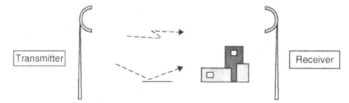

Figure 4-8: Simplified and Generic View of a Radio Link

For example, consider a downlink satellite link as in Figure 4-9.

$$P_r[\text{dB}] = P_t + G_t + G_r - Losses = EIRP_{satellite} + G_{r\,earth} - Losses \tag{27}$$

The noise at the receiver is taken to be

$$N = kT_e B \tag{28}$$

where k is Boltzmann's constant (1.38×10^{-23} Joules/Kelvin), B is the noise bandwidth in Hz, and T_e is the equivalent receiver noise temperature in °K. Alternatively, noise power per Hz is defined as

$$N_0 = kT_e \tag{29}$$

Now, the expression for the received signal power per noise power per Hz can be written as

$$\left(P_r / N_0\right)_{\text{dB}} = EIRP_{satellite} + \left(G_{r\,earth} / T_e\right)_{\text{dB}} - Losses + 10\log k \tag{30}$$

Figure 4-9: The Satellite Link Can Be Characterized by *EIRP* and *G,/T*

In this downlink signal-to-noise ratio expression for the satellite example, the ratio $G_{r\ earth}/T$ is a function of the receiver parameters and therefore a Figure of Merit.

4.4.3 RF Engineering – Modeling and Design of Radio Links

Radio communication may be used for a line-of-sight (LOS) application, such as point-to-point microwave or a satellite link with fixed earth receivers, or for a non-line-of-sight (NLOS) application, such as mobile communication. Both fixed wireless and mobile have distinct propagation and performance attributes, impacting coverage, capacity, and experience. Furthermore, the RF environment must be taken into consideration for planning and engineering, as conditions between outdoor and indoor, or between high-traffic-density areas vs. low ones are different.

The channel model for a wireless link should include all the propagation effects shown earlier. Simple models for the propagation mechanisms outlined there have lead to deterministic models for most of the propagation losses seen. However, realistic wireless channels require more complex models that account not only for all the effects, but also for propagation with multiple obstacles, non-line of sight, or that even take into account the movement of one or both sides of the communication link, such as mobile and satellite channels. The most common model for a wireless channel is a statistical model where the channel impulse response would be given by [Wer00]:

$$h(\tau,t) = \left[\frac{k}{d^n}g_{sh}(t)\right]^{\frac{1}{2}}c(\tau,t) \qquad (31)$$

The first term between brackets models the long-term or large-scale fading and the second term models the short-term or small-scale fading. The large-scale fading term accounts for the mean signal strength loss that occurs due to path loss and shadowing. The second term models the rapid fluctuations of the signal strength. The first term changes very slowly in time compared to the second term.

4.4.3.1 RF site surveys

In both the planning and operational phases of deploying an indoor or outdoor wireless network, RF site surveys must be performed to optimize network performance and to reduce implementation, operating, and maintenance costs. For an indoor wireless network, the basic steps are to obtain a facility diagram (blueprint, etc.), visually inspect the facility, identify user areas, determine preliminary access-point locations, verify final access-point locations, and document the findings [Gei02]. Often, an RF site survey involves walking around the facility and measuring the signal strength, either from wireless network access points or from potential sources of interference (e.g., microwave ovens), or from both. Approaches vary from the simplest case of walking around with just a signal strength meter or a laptop with a wireless network-interface card (NIC) and specialized RF signal-mapping software, all the way to a more sophisticated measurement setup with GPS location recording and a spectrum analyzer.

Network performance will be best if a walkaround survey is completed with a spectrum analyzer to identify possible interference sources at the start of the design phase. Software planning tools exist that allow the user to create or import building layouts and plan the wireless network using RF propagation simulators. These tools reduce testing costs by identifying initial locations for access points which can then be fine-tuned with fewer walkaround surveys. Several manufacturers offer special testing equipment such as spectrum analyzers, Wi-Fi analyzers, stimulus transmitters, power meters, and receivers.

Outdoor surveys for cellular networks are done in much the same way as indoor surveys, but they deal with terrain maps rather than building plans. Signal-strength measurements are done with RF drive surveys in a vehicle outfitted with antennas and measurement equipment. Software planning and analysis tools also exist for cellular networks to analyze multipath propagation and signal strengths for a particular coverage area. Much of the measurement equipment is similar to that for indoor networks, though more specialized. The basic measurement procedures are similar, too. They identify potential sources of interference, record signal strengths at various locations and then optimize the network coverage and performance. Site surveys must be done periodically as part of an ongoing network operations plan.

4.4.3.2 Diversity

In any wireless communication system, transmitted signals propagate over a variety of paths to the receiving antenna due to scattering and diffraction by man-made and natural objects such as the ground, buildings, vehicles, trees, hills, mountains, and other structures. This phenomenon is known as *multi-path propagation*. These propagation paths can add attenuation, distortion, delays, de-polarization, or phase changes to the signals. They can also be time-varying when communicating with a mobile phone in a vehicle. When these signals are coherently summed in the receiver, constructive or destructive interference occurs [Hou05]. With destructive interference, the power of the resulting signal can be significantly reduced. This phenomenon of destructive interference is called *fading*. *Deep fading* occurs when the combined signals are almost 180° out of phase. Fading can severely impact the performance of a wireless system [Vol07, Rap02].

To combat fading, two or more copies of the transmitted signal can be combined at the receiver to increase the signal power. The basic premise is that while some transmitted signals may experience fading, others may not. By using several copies of the signal and combining them at the receiver, the signal power can be increased and system performance enhanced. This is the basic concept of *diversity*. Diversity takes advantage of the statistical properties of the time-varying communication channel by assuming that the signals combined at the receiver experience fading independent of one another.

There are four primary types of diversity: frequency, time, space, and polarization. In frequency diversity, the information signal is modulated onto several different carrier frequencies, each separated by at least the coherence bandwidth of the fading channel, so that each signal will fade independently of the others. With time diversity, the desired signal is transmitted in several different time intervals separated by at least the coherence time of the fading channel. Again, this is done so that each signal will experience independent fading. Space diversity involves receiving the desired signal using several antennas, each separated in space at the receiver. The idea here is that signals reaching different antennas will fade independently, and so could be combined to maximize signal power. Spatial diversity can be effectively equivalent to a phased array, depending how the received signals are combined at the receiver [Vol07, Hou05, Rap02, Die00].

Polarization diversity involves transmitting the desired signal using two orthogonal polarizations because the polarizations have very a low correlation. Thus, it is unlikely that both communication channels will experience a deep fade simultaneously. Therefore, the received signals could be combined to maximize received signal power. Polarization diversity could effectively double the network capacity in a robust implementation [Vol07, Die00]. For optimal performance, there has to be sufficient electrical isolation between the two polarizations in both the transmit and receive antennas.

There are three basic techniques for diversity combining at the receiver: selection diversity, equal-gain combining, and maximum-ratio combining. In selection diversity, the signal having the highest received signal level is switched into the receiver. In equal-gain combining, all received signals are coherently summed with equal amplitude and phase. In maximum-ratio combining, a weighted summation of the signals is performed where the amplitudes are proportional to the signal-to-noise ratio (S/N) for each signal, and the phases are kept equal. When they are used with spatially separated antennas, equal-gain and maximum-ratio combining result in phased array antennas. The measure of diversity performance is the *diversity gain*, which is the improvement in signal level when diversity techniques are used versus the signal level when they are not [Die00].

There are three basic types of antenna diversity: space, pattern (or angle), and polarization. Space and polarization diversity were mentioned previously. Pattern (or angle) diversity involves discriminating among the propagation channels based on angle. For example, it could mean using antennas with different or orthogonal radiation patterns or using a multiple-beam antenna to receive the multipath signals that arrive from different angles [Die00].

4.4.3.3 System capacity and multiple access

The capacity of any wireless communication system is often most limited by the frequency spectrum. Capacity improvements generally focus on frequency *reuse* and more efficient use of the spectrum. Reuse refers to the ability to use some system resource to create multiple physical channels that occupy the same frequency spectrum. System capacity can be increased provided the phenomenon of *co-channel interference* (CCI) can be mitigated. CCI occurs when two transmitters transmit on the same frequency channel. If the two signal levels at a receiver are sufficiently high, the receiver cannot distinguish between the two signals, and CCI occurs. On the other hand, CCI will not occur if one signal level is much lower than the other.

The spectrum allocated for a wireless system is typically divided into many different frequency channels to allow multiple users to communicate. This concept is called *multiple access*. The diversity concept should not be confused with multiple access or reuse. Frequency reuse for cellular telephone systems is accomplished through spatial separation by using the same frequency in different geographical areas, or cells. Cells are separated by enough distance so that CCI is below a certain threshold. In spatial division multiple access (SDMA), frequencies can also be reused by using directional antennas. In this case, to avoid CCI the antennas should have low sidelobe levels and low front-to-back ratios [Die00].

System capacity can be increased through *cell splitting* and *sectoring*. Cell splitting subdivides a crowded cell into smaller cells where each new cell has its own base station with an antenna but with lower transmit power. Capacity is increased because the number of channels per unit area has increased. In sectoring, the base station uses several directional antennas that each transmit within one angular sector. Typically, a cell is partitioned into three 120° or six 60° sectors. Directional antennas help reduce CCI and hence increase the signal-to-interference ratio (SIR). The reduction in interference leads to an increase in system capacity. However, in a sectored system, the antennas must be designed with the correct beamwidth in the azimuth direction to make the implementation successful [Rap02].

The issue with antennas and multiple access is that the antenna must transmit or receive over the entire frequency band spanning all the frequency channels of the wireless system. Each frequency channel does not have a separate antenna. Since antennas are generally resonant structures with narrow bandwidths, they must be selected or designed with enough bandwidth to cover the relevant frequency spectrum. System capacity can be improved through multiple access by decreasing the bandwidth of each channel and/or increasing the number of users per channel.

4.4.3.4 Base station, indoor and mobile handset antennas

Cellular base stations primarily use two types of antennas: omnidirectional and, increasingly, directional (or sectored) antennas. Omnidirectional antennas are typically used in rural areas or at sites with lower capacity where sectoring is not required. Most cellular base stations use sectored antennas to meet large traffic and capacity demands, with the 120° sector generally the most popular. Base station antennas are often mounted on stand-alone towers, but they are sometimes mounted on buildings, water towers, power line towers, and other manmade structures. Occasionally, the base station tower is built such that it blends with the landscape.

Base station antennas are usually designed with moderate gains of anywhere from about 5 dBi to 24 dBi. Manufacturers sometimes specify the antenna gain in dBd, which is the gain referenced to a ½-wave dipole. Recall that the ½-wave dipole has a gain of ≈2.1 dBi; thus the gain in dBd will be 2.1 dB lower than the same specification in dBi. Base station antennas are typically designed for linear polarization (horizontal or vertical) and to radiate broadside, at a 90° angle to the antenna structure. When the antenna is mounted vertically on the tower, the main beam points in the horizontal direction. Some manufacturers offer dual polarized antennas (i.e., ±45° polarization) to take advantage of polarization-diversity combining [And07]. Parabolic reflector antennas are used, particularly in the 2.4-GHz band, for higher gain (≈ 20–24 dBi) outdoor wireless LAN applications. Flat panel antennas with ≈15–20 dBi of gain are also used for both outdoor WLAN applications and to solve some indoor WLAN issues. Sectorized cellular tower antennas are often tilted downward so that the main beam points below the horizontal.

Careful attention must be paid to the mechanical design of the antenna and its enclosure, which have to be protected from snow, rain, ice, and sun. Furthermore, the antenna and its mount must be rigid enough to withstand severe wind loads that might otherwise mechanically move the antenna and result in a network outage (loss of signal) or increased interference. Also, tower antennas must be protected from lightning strikes that could destroy the antenna and transmitter. See Chapter 5 for more on this topic.

Indoor antennas are omnidirectional or directional, and are used in such devices as access points and wireless routers for wireless local area networks (LANs). Omni antennas provide the widest coverage and allow the system to be designed with circular overlapping cells from multiple access points throughout the building. Indoor omni antennas generally have low gains of up to about 6 dBi, but using higher gain omnis can reduce the number of access points required, hence reducing system cost. Higher gain directional antennas (6 dBi to 20 dBi) can be used to cover large, narrow areas or in short links between buildings. Directional indoor antennas can also reduce the number of access points required in some applications. Manufacturers are always developing new antennas with diversity to help solve building coverage issues in WLAN applications.

Designing an integral handset antenna is complicated due to size constraints and overall system issues. For handsets, antennas are typically low-gain and omnidirectional due to the random orientation of the handset at any given time. Furthermore, the handset antenna must be small, and small antennas are notoriously inefficient radiators. A retractable monopole antenna is a good choice for a handset antenna, of which a portion may be wound into a coil for inductive loading. Other popular designs are based on a normal-mode helical antenna (NMHA), where a zigzag design is a popular option. This design can be easily printed on dielectric film and rolled into a round shape with a dielectric cover to form a short antenna stub for handsets.

The planar inverted F antenna (PIFA) is another good design to enclose in handsets; it is a combination of wire- and patch-antenna structures. It can also be designed for dual or tri-band use for global roaming handsets. Also in use are ceramic-chip antennas that more tightly confine the electromagnetic fields at the expense of much narrower bandwidth. Dielectric resonator antennas (DRAs) have also been designed. They work by exciting a mode inside a dielectric structure attached to the antenna. An example here is a dielectrically loaded, balanced, quadrifilar helix antenna. Finally, ferrites have been used in some applications to reduce the size of monopoles and other antennas [Vol07].

4.4.3.5 Antenna tilt

The mechanical tilt angle of a base-station antenna is a well-known additional degree of freedom used to optimize radio networks. Since most of these antennas are installed in a vertical orientation and since their main beam is in the broadside direction (i.e., horizontal), mechanically tilting the antenna moves its main beam below the horizontal plane. The advantage of tilting is that it reduces path loss, delay spread, transmitted power, and system interference. The tilt also strengthens signals from the home cell and reduces the CCI, thus increasing system capacity and improving network performance.

Antenna tilt is also an important specification in developing the network topology in the planning phase. Early network deployments used static tilt where the tilt angle was preset and optimized during initial network operation. Any changes in the tilt angle required a costly site visit by tower crews to re-position the antenna. Recent developments have focused on dynamic antenna tilt control [Pet04, Sio05, Cal06]. Base-station antenna manufacturers now offer remote electric tilt (RET) modules built into the mounting structure to provide remote tilt control from the operations center. Antenna tilt can also be controlled remotely through the Internet via secure channels [And07]. Furthermore, engineers can dynamically control antenna tilt to optimize network performance depending on traffic characteristics. For example, preset tilt angles could be used at different times of day to accommodate varying traffic loads (e.g., rush hour, overnight). The RET uses a stepper motor to mechanically tilt the antenna to the desired angle. A phased array base-station antenna could be used equally well, with the main beam tilted electronically by adjusting the phase of the individual array elements. Care must be taken to ensure that the number of bits used to quantize the phase shifters provides enough angular resolution for tilting the main beam.

4.4.3.6 MIMO

More recently *multiple input/multiple output* (or MIMO) techniques have been applied that place multiple antennas at the transmitter and receiver to improve wireless system capacity and performance. The basic idea behind MIMO is to exploit the space resource of the propagation channel and combine it with sophisticated signal processing to achieve significant gains in spectral efficiency. A MIMO demonstration by Lucent Technologies with 8 transmit and 12 receive antennas achieved a 1.2 Mb/s data rate in 30 kHz of bandwidth [Yli04]. A conventional radio link is single input/single output (SISO) and is subject to fading. A single input/multiple output (SIMO) channel offers diversity options at the receiver, such as space diversity, with multiple antennas. A multiple input/single output (MISO) channel offers diversity options at the transmitter, such as pattern or angle diversity. For example, a transmit phased array antenna can steer the main beam toward a particular receiver. Also, beamforming techniques can be employed at the transmitter to generate multiple beams.

Finally, for improved data rates a MIMO channel allows multiple data streams to be transmitted and recombined at multiple receive antennas. The system capacity improvement for a MIMO system is linearly proportional to the number of antennas. However, in a MIMO system, many handheld receivers do not have multiple antennas. In such cases, maximum gain is achieved through transmit diversity by transmitting the same signal on multiple antennas. The premise is that one of the propagation channels to the receiver will likely have less fading (or none at all) compared to the other channels. When the receiver does have multiple antennas (true MIMO), maximum capacity is achieved through parallel transmission of different data streams. A good analogy for true MIMO is parallel computing where different portions of a large numerical simulation are sent to different processors, achieving much higher compute performance than a single processor. A current area of research in wireless communications is multiple-user MIMO (MU-MIMO) where the basic MIMO concept is extended to multiple receivers, each with multiple antennas [Ges03].

Manuafacturers are also deploying aggressive beamforming technology, particularly for urban Wi-Fi networks to improve coverage and capacity over current configurations. The idea is to use multiple transmit antennas with digital beamforming technology to improve the wireless signal quality by maximizing antenna gain in the link. Essentially, this is an application of transmit diversity in a MISO link with a single Wi-Fi client. Through digital beamforming, interference is significantly reduced by focusing the antennas on a particular Wi-Fi client. SDMA can also improve wireless network capacity by leveraging the beamforming technology to enable simultaneous downlink channels with multiple users.

The standards (e.g., LTE and beyond – 3GPP) coupled with innovations define advanced dynamic antenna systems that can be used for performance (diversity), capacity (Multi-user MIMO), and/or coverage (beamforming).

4.4.3.7 Statistical fading models

Useful propagation models are mostly based on combining analytical and empirical methods. The empirical approach fits curves derived from measured data. In general, these curves are very useful because they take into account all known and unknown propagation factors through measured data. However, one has to be careful in applying these curves in other geographical areas or at other transmission frequencies. In most cases these curves are re-verified before being used to design the system. Over time, some classical propagation models have emerged that are now used to predict large-scale coverage for mobile communication systems designs. Two such models are discussed next.

4.4.3.7.1 Log-distance path-loss model

Models based on both analysis and experimental data reveal that the propagation loss increases as some power of distance. The exponent factor n varies from place to place and assumes different values in different environments. The range of n is from 2 for free-space propagation to about 6 in urban areas with tall buildings. Thus, choosing the right exponent, n, is a must for a realistic loss estimate. The average large-scale path loss for a Tx–Rx separation distance, d, with exponent n is given by

$$\overline{PL}(d) \propto \left(\frac{d}{d_0}\right)^n \tag{32}$$

or

$$\overline{PL}(d)[dB] = \overline{PL}(d_0) + 10n\log_{10}\frac{d}{d_0} \tag{33}$$

The bars in these equations denote the ensemble average of all possible path-loss values at distance d. The parameter d_0 is the reference distance at the far field of the antenna. The dB plot of this equation is a straight line on log-log paper where the slope is $10n$ or the loss increases by 10 dB with every decade increase in distance.

The recommended value of d_0 for a large-scale propagation model is 1 km for cellular and 100 m for microcellular systems [Lee85].

4.4.3.7.2 Log-normal shadowing

The above model can be modified to account for clutter around the area of interest. The value of n is chosen and the power variance factor is added to the equation to match the measured data. Measurements have shown that at any value of d, the path loss $PL(d)$ at a particular location is random and distributed log-normally (normal in dB) about the mean distance-dependent value [Cox84, Ber87]. The modified form of the equation is

$$PL(d)[dB] = \overline{PL}(d) + X_\sigma = \overline{PL}(d_0) + 10n\log_{10}\frac{d}{d_0} + X_\sigma \tag{34}$$

Here, X_σ is the zero mean Gaussian process having a standard deviation of σ, in dB. The equation is valid over a large number of measurement locations having a same Tx-Rx separation but with different clutter factors.

4.4.3.7.3 Outdoor propagation models

Radio transmission in the outdoors depends upon a precise estimate of the terrain undulation, building heights, and the surrounding greenery, such as plants, trees, etc. In urban areas, the average height of the buildings plays a critical role in path-loss determination, while in rural areas shadowing, scattering, and absorption by trees and other vegetation can play a vital role, especially at higher frequencies. A large number of models are available to predict path loss over irregular terrain. While all these models aim to predict signal strength at a particular receiving point or in a specific local area (called a sector), the methods vary widely in their approach, complexity, and accuracy. The prediction models differ in their applicability over different terrain and environmental conditions. Some are generally applicable while others are restricted to specific situations. However, no one model stands out as ideally suited to all environments. Thus the model must be chosen carefully. Most models aim to predict the median path loss, i.e., the loss not exceeded at half the locations and/or for half the time; knowledge of the signal statistics then allows the variability of the signal to be estimated. This makes it possible to determine the percent of the specified area where the received signal strength is adequate. From the large number of models we will review the models by Egli, Okumura, Hata, and COST-231.

The Egli Model

Egli's model for median path loss (path loss less than the predicted value at half the locations and half the time) is

$$L_{50} = G_t G_r \left[\frac{h_t h_r}{d^2} \right]^2 \beta \tag{35}$$

The factor β accounts for excess path loss and is given by $\beta = (40/f_{MHZ})^2$. Egli's interpretation of β at 40 MHz was to represent median path loss irrespective of the irregularity of the terrain. He further related the standard deviation of β to that of the terrain undulations by assuming the terrain height to be log-normally distributed about its median value [Reu74, Egl57]. Hence, he depicted these results in a set of curves that provide the value of β in dB referenced to the median value versus frequency in MHz.

The Okumura Model

One of the most widely used models for predicting path loss in urban areas is Okumura's. The model applies in the frequency range of 150 MHz to 1920 MHz (and extrapolates up to 3000 MHz). The transmitter-to-receiver separation can be from 1 to 100 km and the base antenna height can range from 30 to 1000m. The median attenuation equation is represented as

$$L_{50}(dB) = L_F + A_{mu}(f,d) - G(h_{te}) - G(h_{re}) - G_{AREA} \tag{36}$$

where L_{50} is the 50th percentile (i.e., median) value of propagation path loss, L_F is the free space loss, A_{mu} is the median attenuation relative to free space, $G(h_{te})$ and $G(h_{re})$ are the base-station and the mobile-antenna gain factors and G_{AREA} is the gain due to the environment. The gain factors of the base and mobile antennas are strictly due to their heights and exclude the effect of their antenna gain patterns. $A_{mu}(f,d)$ is represented by a set of parametric curves where the antenna heights, h_t and h_r have fixed values of 200 m and 3 m, respectively. The following equations provide corrections to the antenna gains if the base and mobile antenna heights are different than 200 m and 3 m.

$$G(h_{te}) = 20 \log_{10} \left(\frac{h_{te}}{200} \right) \quad 30\,m < h_{te} < 100\,m$$

$$G(h_{re}) = \begin{cases} 10 \log_{10} \left(\frac{h_{re}}{3} \right) & h_{re} < 3\,m \\ 20 \log_{10} \left(\frac{h_{re}}{3} \right) & 3\,m < h_{re} < 10\,m \end{cases} \tag{37}$$

G_{AREA} corrections are divided into three areas: open, quasi-open, and suburban. Correction factor values in dB are read as a function of the operating frequency. Other correction factors applicable to the Okumura model involve the terrain undulation height, average slope of the terrain, isolated ridge height, and mixed land-sea path parameter [Oku68].

The wholly empirical nature of the Okumura model means that the parameters used are limited to specific ranges determined by the measured data. If, in attempting to make a prediction, the user finds that one or more parameters fall outside the specified range then there is no alternative but to extrapolate the appropriate curve, which may give unrealistic results and care must be exercised.

The Hata Model

To make the Okumura method easy to apply, Hata established empirically-based mathematical relationships to describe the graphical information given by Okumura. Hata's equation applies directly to urban-area propagation as a standard formula and supplies corrections to other situations. The formula for urban surroundings is

$$L_{50,urban} = 69.55 + 26.16\log_{10}(f_c) - 13.82\log_{10}(f_c)$$
$$- 13.82\log_{10}(h_{te}) - a(h_{re}) + 44.9 - 6.55\log_{10}(h_{te}) + 10\log_{10}(d) \tag{38}$$

Here, the frequency range is from 150 to 1500 MHz, h_{te} is the effective transmitter (base station) antenna height (in meters) ranging from 30 to 200 m, h_{re} is the effective receiver (mobile) antenna height (in meters) ranging from 1 to 10 m. The factor d is the T-R separation distance (in km), and $a(h_{re})$ is the correction factor for the effective mobile antenna height, which is a function of the size of the coverage area. Hata provides mobile antenna correction factors in dB for small and medium cities as [Hat90]

$$a(h_{re}) = (1.11\log_{10} f_c - 0.7) - (1.56\log_{10} f_c - 0.8) \tag{39}$$

and for large cities as

$$a(h_{re}) = \begin{cases} 8.29(\log_{10}(1.54h_{re}))^2 - 1.1 & f_c \leq 300 \text{ MHz} \\ 3.2(\log_{10}(11.75h_{re}))^2 - 4.98 & f_c > 300 \text{ MHz} \end{cases} \tag{40}$$

Hata also modified this last equation for suburban and open rural areas. These equations are

$$L_{50}(dB) = \begin{cases} L_{50,urban} - 2[\log_{10}(f_c/28)]^2 - 5.4 & Surburban \\ L_{50,urban} - 4.78[\log_{10}(f_c)]^2 + 18.33\log_{10}(f_c) - 40.94 & Open \& Rural \end{cases} \tag{41}$$

Although Hata's model does not have any of the path-specific corrections of Okumura's model, these expressions have significant practical value. Results using the Hata model compare very closely with the original Okumura model, as long as d exceeds 1 km. Thus, the model provides good results for large cells but is not applicable to smaller cells of the PCS type.

The COST-231 Model

The high end of the frequency range suitable for the Hata model is 1500 MHz. This limits its application to the PCS range. Under the European COST-231 program, the Okumura results were analyzed at the upper end of its frequency range. Modifications were then made to the program so it can be applied between 1500 and 2000 MHz. The PL model is

$$L_{50}(\text{Urban}) = 46.3 + 33.9\log_{10}(f_c) - 13.82\log_{10}(h_{te}) - a(h_{re})$$
$$+ [44.9 - 6.55\log_{10}(h_{te})]\log_{10} d + C \tag{42}$$

Here, the values of $a(h_{re})$ are the same as in Hata model. The value of C is 0 dB for medium-size city and suburban areas and 3 dB for metropolitan areas. The model applies in the following limits of variables:

$$1{,}500 \text{ MHz} \le f_c \le 2{,}000 \text{ MHz}$$
$$30\,\text{m} \le h_{te} \le 20\,\text{m}$$
$$1\,\text{m} \le h_{re} \le 10\,\text{m}$$
$$1\,\text{km} \le d \le 20\,\text{km}$$

(43)

4.4.3.7.4 Outdoor terrain

Propagation curves must account for environmental clutter and terrain irregularities correctly so that the predicted field strength comes close to the measured value of the propagation loss. There are an infinite variety of terrain features, but for reference analysis "quasi-smooth" terrain is assumed. Whether in quasi-smooth or irregular terrain, buildings and trees near the vehicular antenna affect the received field strength differently. If we classify obstacles in areas according to building or trees, it will make the prediction of field strength in a given area very difficult. Thus, rather than classifying them into many minute groups, the clutter is generally classed according to the degree of congestion in a given area, i.e., open, suburban, and urban. An urban area has large buildings and is further subdivided into dense urban, moderate urban and geographically limited urban. The terrain classification is shown in Table 4-1.

Category	Characteristics
Open area	Very few structures, such as tall buildings and trees, are in the propagation path; the area is mostly farms, open fields, and rice fields
Suburban area	Residential, with small one- or two-story houses, with some obstacles near the mobile radio, but not very congested
Urban area	Heavily built-up areas with tall buildings and high-rise apartments

Table 4-1: Characteristics of Different Terrain

It must be noted however, these distinctions are approximate approaches to modeling propagation based on openness. Otherwise, the boundaries are not always clear and traffic density growth can be significant in any area.

4.4.3.7.5 Indoor structures
Walls

The modeling of radio-wave penetration into the outer walls of a building differs from the more familiar cellular case in many respects. The modeling here is based on transmitters mounted outside on a high-rise building while the mobile is located inside a different building and at varying heights. In a rural environment this may result in a LOS path to the upper floors of many buildings, whereas this is relatively a rare occurrence in buildings located in urban and suburban surroundings.

In typical cellular systems, base stations for macrocells are located on the roofs of tall buildings, and may be as much as 100 m or more above the local terrain; the mobile can be a few kilometers away. Since the cellular model setup is different than the model here (inside building), it is difficult to use the results from the cellular design directly.

Penetration loss experiments have shown that the signal in small areas within buildings is approximately Rayleigh distributed, with the scatter of the median having a log-normal distribution. In other words, the signal statistics within a building can be modeled as superimposed small-scale (Rayleigh) and large-scale (log-normal) processes.

Partition losses between floors

It has been reported that losses between floors are influenced by the construction materials in the outside walls, the number and size of windows, and the type of glass. The external surroundings also must be considered since there is evidence [Dev90] that energy can propagate outwards from a building, be reflected and scattered from adjacent buildings, and re-enter the building at a higher and/or lower level depending upon the location of the antenna. Experiments have also shown that attenuation between adjacent floors is greater than the incremental attenuation caused by each additional floor but that after five or six floors there is little further attenuation. The floor attenuation factor (FAF) and the standard deviation σ (dB) for three buildings have been tabulated by Seidel and Rappaport [Sei92].

Partition loss on same floor

Building partitions take a different form based on whether the building holds residences or offices. Residential buildings typically use wood-frame partitions with sheetrock and generally have wooden floors. On the other hand, office buildings either use moveable office partitions (soft) or permanent (hard) partitions as part of the building structure. The materials used vary so widely that a general model for partitions does not exist. However, values of the loss cause by different types of partition materials have been tabulated by various authors [Cox83, Vio88, Rap91].

4.4.3.7.6 Indoor and outdoor coverage calculations

A number of examples illustrating the computational aspect of path loss calculation in mobile surroundings are given below. Examples covered are for: free-space loss; loss due to reflection from the ground; propagation loss inside a building; application of the COST-231 model for path loss calculation; and diffraction loss due to a knife edge obstacle. These examples show the complexity which may be involved and the software tools developed accordingly.

Example 1: Compute the free-space loss with the following parameters: P_t = 20 Watt, $G_t = G_r = 0.0$ dB, f_c = 900 MHz. Find P_r (received power) at a distance of 1.5 km.

$$P_r(d) = \frac{P_t G_t G_r \lambda^2}{(4\pi)^2 d^2} = \frac{20 \times 1 \times 1 \times (1/3)^2}{(4\pi)^2 \times (1.5 \times 1000)^2} =$$

$$= \frac{2.22}{355.305 \times 10^6} = 0,006248 \times 10^3 \text{ mWatt} = -52.04 \text{ dBm}$$

Example 2: Two-path model: For the following conditions test whether the two-ray model can be applied or not: h_t = 40 m, h_r = 3 m, d = 300 m and d = 500 m. When $d > 10(h_t + h_r)$ we can say $d \gg h_t + h_r$. Applying the distance criteria, it fails for d = 300 m. For d = 500 m the criteria is satisfied and the two-ray model applies.

Example 3: Measurements inside a building follow a log-normal distribution about the mean. Assume the exponent to be 3.5. If a signal of 1 mW was received at d_0 = 3 m from the transmitter and at a distance of 30 m, and 20% of the measurements were stronger than 25 dBm, define the standard deviation σ, for the path loss model at d = 15 m.

$$\overline{P_r(d)} = -35\log10(d/d_0) + P_r(d_0) = -35\log_{10}(30/3) + 0 = -35 \text{ dBm}$$

$$\text{For } \upsilon = -35 \text{ dBm}, P_r[P_r(d) > \upsilon] = Q\left[\frac{\upsilon - \overline{P_r(d)}}{\sigma}\right] = Q\left[\frac{-25+35}{\sigma}\right]$$

$$= Q\left[\frac{10}{\sigma}\right] = 0.2; \text{ or } \frac{10}{\sigma} = 0.8416; \quad \sigma = 11.9 \text{ dB}$$

Example 4: Use the COST-231 model for a large city to find the received power at 2 km and 10 km when the received power at a distance of 1 km = 1μW. Assume f = 1800 MHz, h_t = 40 m, h_r = 3 m, $G_t = G_r =$ 0 dB.

$$L_{50}(\text{Urban at } d = 2 \text{ km}) = 46.3 + 33.9\log_{10}(1800) - 13.82\log_{10}(40)$$

$$- 3.2\left[\log_{10}(11.75 \times 3)\right]^2 + 4.98 + \left[44.9 - 6.55\log_{10}(40)\right]\log_{10}(2) + 3$$

$$= 145.19 \text{ dB}$$

Similarly, L_{50}(Urban at d = 1 km) = 134.83 dB. Thus, the power received at 2 km is

$$P_r(d = 2 \text{ km}) = P_r(d = 1 \text{ km}) + \left[L_{50}(d = 2) - L_{50}(d = 1)\right] = -30(\text{dBm}) - (145.19 - 134.83) = -40.36 \text{ dB}$$

Similar to the above calculation for a distance of 10 km, $L_{50}(d = 10 \text{ km})$ = 169.24 dB. Thus, the power received at 10 km is

$$P_r(d = 1 \text{ km}) + \left[L_{50}(d = 10) - L_{50}(d = 1)\right] = -30(169.24 - 134.83) = -64.41 \text{ dBm}$$

Example 5: Compute the diffraction loss due to a *knife-edge ridge* and the value of the power received for the configuration shown in Figure 4-10. Compute also the free-space loss as if the obstruction does not exist. The following parameters can be assumed: P_t = 10 watt, G_t = 10 dB, G_r = 0 dB and L = 1 dB at f = 900 MHz.

$$P_r(d) = \frac{P_t G_t G_r \lambda^2}{(4\pi)^2 d^2} = \frac{10 \times 1 \times 1 \times (1/3)^2}{(4\pi)^2 \times (6 \times 1000)^2} = \frac{1.11}{5684.9 \times 10^6} = 1.954 \times 10^{-10} \text{ Watt} = -67.1 \text{ dBm}$$

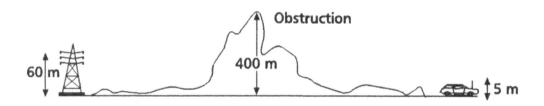

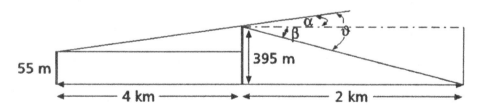

Figure 4-10: Diffraction Loss Computation Due to a 400m-High Ridge

$$\tan(\beta) = \frac{395}{2000}, \text{ thus, } \beta = 0.195 \text{ radians}; \ \tan(\alpha) = \frac{340}{4000}, \text{ thus, } \alpha = 0.085 \text{ radians}; \text{ or } \theta = \alpha + \beta = 0.275 \text{ radians}$$

$$\upsilon = \alpha \sqrt{\frac{2 \times d_1 \times d_2}{\lambda \times (d_1 + d_2)}} = 0.275 \sqrt{\frac{2 \times 2000 \times 4000}{0.333 \times (6000)}} = 0.275 \times 89.44 = 24.6$$

Using Lee's approximation for $\upsilon > 2.4$, *Diffraction loss* $= G_d = 20 \log \left(\frac{0.225}{24.6} \right) = -40.8$ dB

Received power $= -67.1 - 40.8 = -107.9$ dBm

Software Modeling Tools

Numerous simulation models are being developed for cellular and other mobile surroundings. These models are intended as design aids in different wireless environments such as indoor or urban outdoor.

4.5 Radio System Considerations

Radio systems, shown in a simplified block diagram in Figure 4-11, are used to transfer information from an information source to a remote destination. Typically, such a system consists of an information source, which can be audio, video, or computer data, or some combination thereof. A transmitter modulates this information onto a carrier frequency and the result is radiated into the channel.

The channel is, of course, not perfect. As the signal radiates from the transmitter, its energy density decreases. And the channel, which gains noise from natural and artificial sources, experiences "spreading loss" or a reduction in energy. What is more, other transmitters at various frequencies may also occupy the channel and add co-channel or adjacent-channel interference.

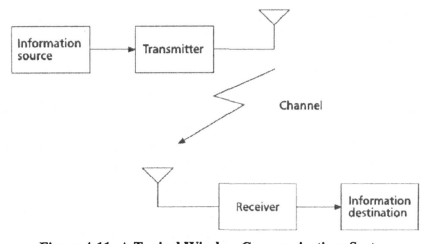

Figure 4-11: A Typical Wireless Communications System

The receiver recovers the information from the channel. While our perspective is of receivers for VHF/UHF wireless applications, the principles discussed here apply to any radio receiver. The ideal receiver has a number of characteristics.

- Sensitivity – Radio signals that reach the receiver are extremely weak, so the ideal receiver should be able to detect very small signals.

- Low noise – Noise and sensitivity are closely related because noise dictates the weakest detectable signal. At frequencies in the low-VHF range and below, receiver sensitivity tends to be limited by atmospheric noise sources. In the UHF range and above, sensitivity tends to be limited by internally generated amplifier noise.

- Selectivity – The ideal receiver should detect only the desired signal and reject all others.

- Stability – The receiver should remain tuned to the desired frequency, and its gain, noise, and demodulation characteristics should remain constant despite variations in supply voltage, temperature, vibration, and other environmental impairments.

- Low spurious responses – Certain receiver topologies result in undesired receiver responses at specific frequencies. It is possible to predict these frequencies and minimize these undesired responses.

- Low phase noise – Many receiver topologies employ one or more internally generated signals. Noise can affect the oscillators that generate these signals, resulting in phase noise. High phase noise can reduce the dynamic range of a receiver, as well as interfere with the detection process for phase-sensitive modulations such as quadrature amplitude modulation (QAM).

- Wide dynamic range – It is desirable that the receiver detect weak signals close in frequency to strong ones.

- Automatic gain control (AGC) – In some receiver systems, it is desirable that the output remain constant for varying inputs. The ideal AGC system will maintain the output constant over a wide range of input signal level and follow temporal input variations without overshoot or delay.

4.5.1 Receiver Topologies

The most widely used receiver topology is the superheterodyne, or superhet, whose block diagram is shown in Figure 4-12. The incoming signal from the desired channel passes through a pre-selector filter, RF amplifier, and image filter, and is applied to the mixer, where it is combined with the local oscillator (LO). The output of the mixer is nominally the sum and difference of the signal and LO frequencies. The mixer acts as a frequency converter. The intermediate-frequency (IF) filter selects either the sum or the difference frequency as the intermediate frequency. The IF amplifier provides additional gain at the IF, and its output is applied to the detector/demodulator, which extracts the original information.

This design has numerous advantages. Gain can be distributed between the RF or signal frequency and the IF to produce a very-high-gain yet stable receiver. And the receiver can be easily tuned to a new channel by changing the frequency of the LO. The IF filter generally has a fixed center frequency so it can be optimized for bandwidth and/or delay characteristics, and it does not have to be retuned when the receiver is tuned to a new channel.

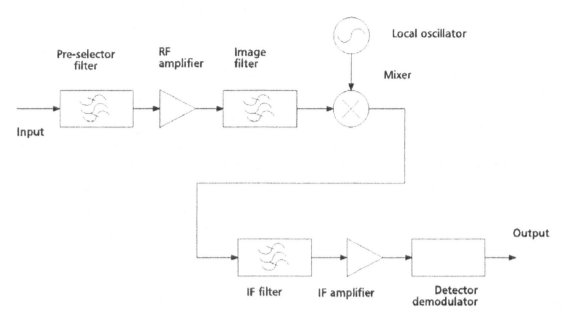

Figure 4-12: Block Diagram of a Superheterodyne Receiver

4.5.2 Receiver Sensitivity

At UHF frequencies, internally generated noise limits the weakest signal that can be detected. Noise generated by the RF amplifier is amplified by all the stages in the receiver so it determines the receiver sensitivity. Low-noise, high-gain amplifiers for the RF amplifier produce the highest sensitivity.

Receiver sensitivity is measured in a number of ways, but is typically expressed as noise figure (NF) and noise temperature (T). NF is defined as the ratio of the input signal-to-noise ratio (S/N) to the output signal-to-noise ratio. The output S/N is measured at the input of the demodulator. NF must be 1 or greater because the intervening stages can only degrade the S/N. NF is almost always defined in units of dB but calculations involving NF must be done as ratios.

Referenced to the antenna port, there is a signal input power, P_{Sin}, and a noise power, P_{Nin}. Any electronic device not at absolute zero creates noise over a wide bandwidth due to black-body radiation. The input noise power comes from losses in the antenna, atmospheric noise, and sky or ground noise. This noise is usually lumped into an equivalent antenna noise temperature, T_A and the input noise is given by $P_{Nin} = kT_AB$, where B is the system bandwidth and k is Boltzmann's constant.

Many active devices will add noise power much different from what their physical temperature might suggest. They are assigned an equivalent noise temperature, T_e. This temperature equates the device noise with a resistor having the equivalent physical temperature that produces the same amount of noise. The NF is calculated as

$$NF \equiv \frac{(S/N)_{in}}{(S/N)_{out}} = \frac{P_{Sin}/P_{Nin}}{GP_{Sin}/G(P_{Sin}+kT_eB)} = \frac{P_{Sin}/kT_AB}{GP_{Sin}/G(kT_AB+kT_eB)} = 1 + \frac{T_e}{T_A} \qquad (44)$$

where G is the gain of the receiver. Note that NF is not determined by the signal power, the gain, or the system bandwidth. For this reason, NF is usually calculated at the receiver input. The relation between NF and T_e is given by

$$T_e 0 = 290(NF - 1) \qquad\qquad\qquad (45)$$

Note that this assumes that T_A = 290K. For satellite systems, T_A may be 50K since the antenna will be pointed at "cold" sky while, for terrestrial systems, much of the antenna beam will intersect "hot" earth and T_A will be more like 290K. Historically, NF has been tied to terrestrial systems and hot-cold load measurements where one load is at room temperature.

While NF and noise-temperature analysis for cascaded stages is found in many textbooks, an excellent analysis that handles the NF of lossy devices and the correction necessary to handle additional the noise produced by spurious mixer products has been performed by Maas [Maa92]. The image filter and the pre-selector filter affect receiver noise performance differently. The image filter is needed to reduce the effect of RF amplifier noise at the mixer image response. The image response is explained next.

4.5.3 Mixing

The mixing or frequency conversion process introduces spurious responses. The main spurious response is called the image. Consider a simple example: a TV broadcast receiver is tuned to 627.25 MHz, which is the video carrier frequency for channel 40. The receiver has an IF of 45 MHz so the LO runs at 582.25 MHz. A difference frequency of 45 MHz is generated by mixing 627.25 MHz and 582.25 MHz. Notice that when 537.25 MHz, which is the video carrier for channel 25, is mixed with 582.25 MHz, the difference is also 45 MHz. The receiver responds to channel 25 just as it does to channel 40. Figure 4-13 shows the frequency relationships of the desired signal, the image, and the LO. The function of the pre-selector and image filters is to pass the desired signal but reject the image.

The image is only one of many possible spurious mixing responses. In practice, the mixer is a highly non-linear device that generates harmonics of the applied LO and signal frequencies. The second harmonic of the LO at 582.25 MHz is 1164.5 MHz. The second harmonic of 604.75 MHz is 1209.5 MHz. If a 604.75 MHz signal reaches the mixer, the sum and the difference of the harmonics are created. The difference is the IF of 45 MHz, and the receiver has a spurious response at 604.75 MHz. This response is sometimes called the "half IF spur" because it appears only 22.5 MHz, or one half the IF frequency, from the desired response. The image is twice the IF or, in this case, 90 MHz. It is much easier to design the image and pre-selector filters to remove the image than the half IF spur.

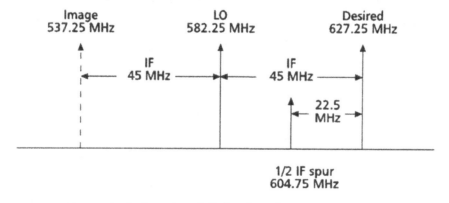

Figure 4-13: Spurious Mixing in a Superhet Receiver

Spurious mixing products are a serious problem in superhet receivers. Mixers generate not only the second harmonic but higher order harmonics as well. Balanced mixers that suppress even-order harmonics are widely used to reduce some of the spurious mixing products. It is also possible for superhet receivers to have more than one mixer, but predicting all the spurious mixing products can be difficult and tedious. Careful selection of the IFs, mixers, and filters reduces the impact of spurious mixing. Commercial software is available to aid in predicting spurious responses.

In addition, the LO may leak out through the image filter, the RF amplifier, and the pre-selector filter, and may radiate from the antenna input. This can create interference. An example of this is the LO in TV and FM broadcast receivers that tune through portions of the aircraft communications band. Thus, operation of these receivers is banned on commercial aircraft.

4.5.4 Dynamic Range

At UHF frequencies it is difficult to build a pre-selector filter narrow enough to pass only a single channel of information. In addition, it may be undesirable to build very narrowband filters if the receiver is to be tuned over a range of frequencies. If the range of desired signal frequencies cannot pass through the pre-selector/image filters, then these filters must be retuned when the LO is tuned to the new frequency. This is called tracking and it increases the receiver's complexity.

Since most pre-selector/image filters will pass more than the desired signal, many different signals will be amplified by the RF amplifier and applied to the mixer. All of these signals will produce LO sum and difference frequencies. The IF filter will pass only the desired intermediate frequency. However, real mixers will create harmonics of all the applied signals and produce sum and difference frequencies of these harmonics. The result may be a mixing product that falls in the IF filter's bandwidth.

Similarly, RF amplifiers are not perfectly linear and strong signals can cause them to saturate. This non-linearity generates harmonics that mix either in the amplifier itself or in the following mixer. While careful design can minimize internally generated mixing products, it is impossible to completely eliminate them with real mixers and amplifiers.

The linearity of amplifiers and mixers is specified in terms of third-order mixing products. Figure 4-14 shows a test setup for measuring these products. The output for an ideal amplifier would be only f_1 and f_2. But nonlinearities in actual amplifiers generate harmonics of f_1 and f_2 plus the mixing products of these harmonics. These mixing products are specified in terms of the harmonic number of the signals that generate them. For example, the third-order products are $2(f_1)-f_2$ and $2(f_2)-f_1$. The odd-order products are the most problematic since they fall close to the desired signal, making them difficult to filter out.

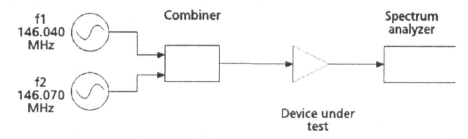

Figure 4-14: Measuring a Third-Order Intercept

Consider a channelized FM mobile-radio system; Figure 4-15 shows the output of a non-ideal amplifier. Channel 1 is 146.01 MHz, channel 2 is 146.04 MHz, channel 3 is 146.07 MHz, and channel 4 is 146.10

MHz. The second harmonic of channel 2 is 292.08 MHz. If this mixes with channel 3, the difference is 146.01 MHz, which is channel 1. Similarly, the second harmonic of channel 3 mixes with channel 2 to produce a signal at 146.04 MHz, which is channel 4. Figure 4-15 shows that third-order mixing of channels 2 and 3 create signals on channels 1 and 4. If the receiver is tuned to channel 1 or 4 and strong channel 2 and 3 signals are present, then the nonlinearity of the RF amplifier or mixer will create internally generated interference on channel 1 or 4. This will limit the receiver's ability to detect a weak signal on these channels.

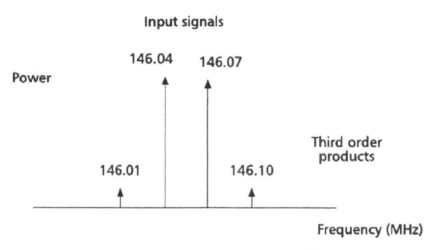

Figure 4-15: Amplifier Output in Figure 4-14

As the channel 2 and 3 signals grow stronger, more of the interfering channel 1 and 4 signals are generated. Figure 4-16 shows the relation between the power in the desired signal and the power in the third-order products. There is a linear relation between the output power and the input power of the desired signal until the device reaches saturation. At saturation the output power no longer increases for an increasing input. The apparent gain goes down. The maximum linear power output is defined as the output power where the gain is 1 dB less than for a small-signal. This is referred to as the 1-dB compression point.

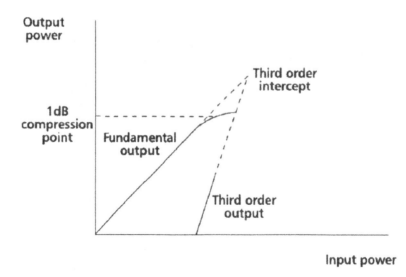

Figure 4-16: 1-dB Compression Point and Third-Order Intercept

Third order mixing products increase as the amplifier output increases. Due to the nature of the mixing, the power in the third-order products increases three times faster than the power of the desired signals. This can be seen in Figure 4-16. Third-order products would eventually overwhelm the desired signal if the amplifier did not go into saturation. The point where the power in the third-order products is equal to the power of the desired signal is called the third-order intercept. The intercept point is a figure of merit because the amplifier will go into compression long before the intercept power is reached.

The weakest signal the receiver can detect is limited by the internally generated noise. The equivalent power of this noise is the "noise floor." The strongest signals the receiver can tolerate are those that create a third-order mixing product equal to the noise floor. The ratio of the strongest signal to the noise floor is the spurious free dynamic range (SFDR).

$$SFDR = \frac{2}{3}(IIP3 - MDS) \tag{46}$$

where the noise floor is the minimum discernible signal, $MDS = kT_eB$ and $IIP3$ is the third-order intercept power referred to the receiver input. High-gain low-noise RF amplifiers maximize sensitivity by reducing MDS. However, maximizing SFDR requires a careful trade-off between the gain, noise performance, and third-order intercept of all the stages prior to the narrowest filter in the system. Building a sensitive receiver is relatively easy; building a sensitive receiver that can handle strong signals can be an engineering challenge. Building a sensitive receiver that can handle strong signals with a minimum of battery power is more challenging still.

4.5.5 Selectivity

Filters play a key role in rejecting adjacent-channel signals, reducing the effect of the receiver's spurious responses such as the image, and reducing the effect of non-linear mixing products. The ideal filter would have a lossless passband centered on the desired frequency that is only wide enough to pass the desired modulation, and have zero response for all other frequencies.

Real filters have a passband with some minimal loss and then a transition band where the response falls to a specified level. This is the stopband, as shown in Figure 4-17. It is impossible to go from zero attenuation to some large attenuation with zero frequency offset. More complex filters, i.e., those having more poles, have more attenuation in the stopband and can transition from passband to stopband faster. However, this fast transition comes at the expense of waveform distortion in the time domain [Bli76]. This is of particular importance in systems employing digital modulation. Filter design and specification is both an art and a science [Zve67].

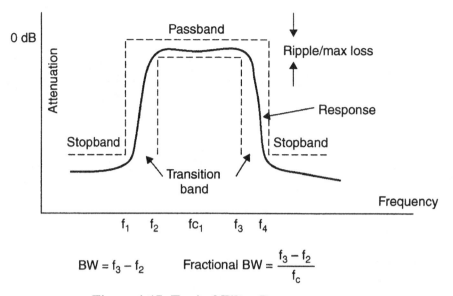

Figure 4-17: Typical Filter Response

An example of the bandwidth trade-off is shown in Figure 4-18. Using the fractional BW of Figure 4-17, the pre-selector filter is just over 1% wide. The pre-selector filter is chosen to do a credible job of rejecting spurious mixing products. A pre-selector filter much narrower will have too much loss. As bandwidth is reduced, filter loss tends to go up with an attendant increase in NF. However, it is necessary to select only one channel of the example given the previous section. The IF filter must be approximately 30 kHz wide to pass the desired FM signal. However, the pre-selector filter cannot protect the RF amplifier and mixer from signals that may limit dynamic range.

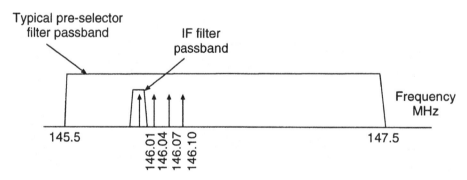

Figure 4-18: Relation between the Receiver Pre-Selector Filter and IF Filter

4.5.6 *Phase Noise*

All superhet receivers have one or more internally generated signals or local oscillators. These oscillators are not noise free. The amplifier in any oscillator generates noise, which modulates the resulting oscillator output. Noise in oscillators takes two forms: amplitude and phase noise. The effect of amplitude noise can be neglected in well-designed oscillators [Roh88], but noise effectively phase-modulates the oscillator frequency. Oscillator phase-noise performance is measured in terms of the power spectral density of the resulting modulation sidebands. The spectral density can be predicted from the Q of the oscillator resonator and from the noise power generated by the active device in the oscillator [Rap02]. Figure 4-19 shows how the output spectrum of a typical oscillator is affected by phase noise.

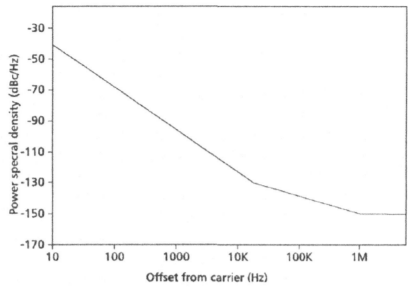

Figure 4-19: Oscillator Phase Noise

Phase noise degrades receiver dynamic range through what is called "reciprocal mixing." Figure 4-20 illustrates the effect of reciprocal mixing. The figure shows a noisy LO and a strong adjacent-channel signal outside the IF bandwidth. Without the LO noise, the IF filter would reject this signal.

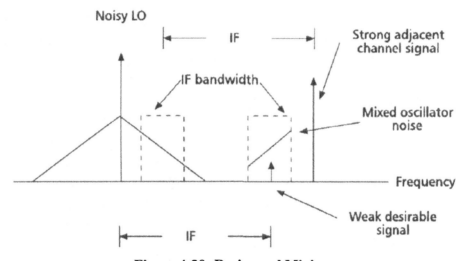

Figure 4-20: Reciprocal Mixing

The strong adjacent-channel signal acts as a local oscillator for the oscillator noise. This noise appears as a signal adjacent to the LO. Noise frequencies separated by the IF from the interfering signal will be mixed into the IF, as shown in Figure 4-20. The noise floor is increased by the oscillator phase noise, and this obscures the desired weak signal. In reality, all the signals in Figure 4-20 contain phase noise. Phase noise in the oscillator and in the strong adjacent channel will be mixed into the IF.

Phase noise produces additional receiver impairment in systems using phase-sensitive modulation, such as phase shift keying (PSK) or quadrature amplitude modulation (QAM). In digital PSK systems, it is necessary to regenerate the phase of the carrier signal as a demodulation reference. The phase noise of the receiver's LO is transferred to the incoming signal by mixing. This increases the phase uncertainty of the demodulation reference and leads to bit-decision errors. Phase noise has the effect of reducing the signal-to-noise ratio.

4.5.7 Direct Conversion

The superhet receiver converts the incoming signal to an intermediate frequency. It is possible to frequency-convert the incoming signal directly to its original baseband. The LO is tuned to the frequency of the incoming signal. The mixer output is the sum and the difference of the signal and the LO frequencies. Because the signal and LO frequencies are equal the difference is zero. The resulting output is the original information contained in modulation sidebands about the carrier frequency. This is called a direct-conversion (DC), homodyne, or zero-IF receiver. Figure 4-21 shows a block diagram of a direct-conversion receiver.

The DC receiver has the advantage of being simpler than the superhet. It is easier to build high-gain amplifiers at baseband than at the signal or IF frequency, so a DC receiver can be as sensitive as a superhet. This topology minimizes the number of components at RF frequencies. In a simple DC receiver, the signal can be applied directly to the mixer.

Since all the signal processing occurs at baseband, it is possible to integrate the entire receiver including the mixer. This is attractive for small low-power receivers in devices such as pagers or in receivers employing digital signal processing (DSP).

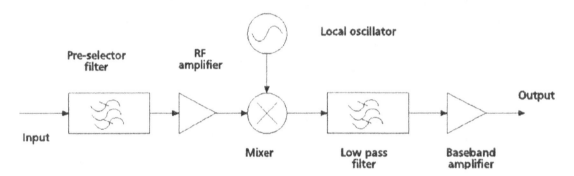

Figure 4-21: A Direct-Conversion or Homodyne Receiver

A DC receiver is "tuned" to itself so LO leakage into the input is a problem. The oscillator must be carefully shielded to prevent this radiation. Oscillator phase noise increases close to the LO frequency. The close-in oscillator phase noise is mixed directly into the baseband.

4.5.8 Detection

Modulation is the process of attaching information to a carrier wave that is suitable for transmission through the channel. The modulator can vary the carrier amplitude, frequency, or phase. Equation 47 shows the quantities that can be varied in the process of modulation.

$$s(t) = A(t)\cos\{2\pi f_c(t)t + \varphi(t)\} \tag{47}$$

Modulation can be analog or digital, and is widely covered in the literature [Cou93]. At the receiver, the process of demodulation, or detection, recovers the original information from the modulated carrier.

While demodulators vary, a commonly used demodulator is shown in Figure 4-22. This is a simple product detector that converts the signal down to baseband. The phase of the original signal must be recovered by additional circuitry. A more complex version of the detector is shown in Figure 4-23. An in-phase and quadrature baseband signal is generated and then digitized by the analog-to-digital converters (ADC). Digital signal processing is used to extract the modulating signal and the carrier phase. It is possible to reconstruct almost any signal with this demodulator. This type is widely used for complex modulations such as phase shift keying (PSK), QAM, and orthogonal frequency division multiplexing (OFDM).

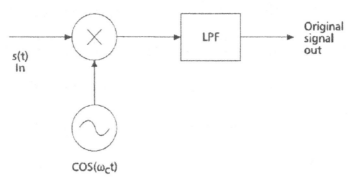

Figure 4-22: A Simple Product Detector

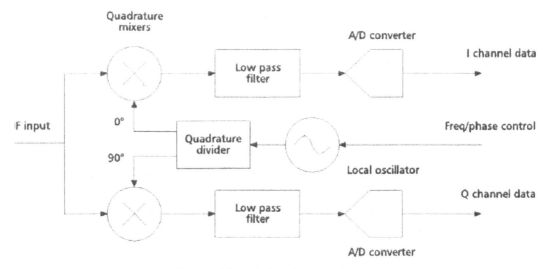

Figure 4-23: A Quadrature Detector

As noted earlier, noise figure and effective temperature are used to measure receiver performance. While these measures are independent of modulation, they do not give direct insight into the noise performance of various modulation types.

For linear analog modulation such as large carrier AM (still widely used in broadcasting) and single sideband (SSB), the S/N_{in} or carrier-to-noise (C/N) ratio at the antenna is the same as the post detection or output S/N (S/N_{out}). Digital modulation performance is often specified in terms of E_b/N_o (the energy per bit to noise density ratio) or E_s/N_o (energy per symbol to noise density ratio). The noise density is the noise power in 1 Hz of equivalent bandwidth. For biphase shift keying (BPSK) where the carrier wave is multiplied by 1 for a one bit and -1 for a zero bit, the E_b/N_o ratio is given by

$$\frac{E_b}{N_0} = \frac{CT_b}{kT_e} = \frac{C}{N} \tag{48}$$

where the bit period $T_b = 1/BW$. With E_b/N_o, it is possible to calculate the probability of a bit error or the bit-error rate (BER) for a particular S/N_{in} or (C/N) ratio. With higher order modulations where each phase or amplitude state represents more than one bit, E_b/N_o becomes E_s/N_o, or the energy per symbol to noise density ratio.

Consider quadrature phase shift keying (QPSK) where there are four phase states and each phase state represents two bits. QPSK and BPSK have the same BER when E_b/No is considered. However, QPSK has two bits per symbol so each bit receives half the power. QPSK requires 3 dB of additional carrier power to match the BER performance of BPSK. For BPSK, $E_b/N_0 = C/N$, and for QPSK, $E_b/N_0 = C/2N$.

QPSK is not a bad tradeoff because 3 dB in additional power results in 3 dB (×2) additional data that can be transmitted through a fixed bandwidth. Higher order modulations such as M-PSK, QAM, or OFDM do not provide this 1-for-1 increase in data throughput for a corresponding increase in power. They are not as power efficient as QPSK. These higher order modulations are used where bandwidth is at a premium so the power disadvantage can be tolerated.

4.6   Summary

Wireless systems are based on the transmission of radio signals through wireless links. These links connect to transmitters and receivers via antennas. The radio signal experiences losses due to basic propagation loss (free-space), which is a function of frequency and distance, and to the system, such as waveguides, multipath, and the environment. The environmental losses are typically due to the terrain, obstacles, and atmosphere, and exhibit their differences particularly as indoor vs. outdoor, urban vs. rural, LOS vs. NLOS, or fixed vs. mobile.

The operating frequency is fundamental in radio link design, particularly in terms of coverage and capacity. Higher frequencies have a more limited reach but typically offer more capacity. They also exhibit other properties, such as notable atmospheric losses at frequencies higher than around 10 GHz, or LOS requirements at frequencies higher than around 3 GHz. Networks have become more user-centric, potentially using multiple frequencies and multiple access mechanisms, increasingly in a multi-cell and heterogeneous way.

The goal is communication with the required performance. Link budget analysis is used to estimate the received power based on all parameters and the signal-to-noise ratio at the receiver. End-to-end radio engineering involves estimations using simplified models and software tools, as putting all the attributes together creates a complex system of contributing factors, some impractical, if not impossible, to determine. This is particularly true about fading channels, notably in mobile communications.

We started with individual elements and factors and moved to the characterization of links. Then we discussed radio systems, which include transmitters and receivers on the two ends of the links. These radio systems are used in the technologies and standards discussed in Chapter 1 on Wireless Access Technologies.

4.7 References

[Ber87] R.C. Bernhardt, *Macroscopic Diversity in Frequency Reuse Systems*, IEEE Journal on Selected Areas in Communications, vol. SAC 5, 1987.

[Bli76] H.J. Blinchikoff and A.I. Zverev, *Filtering in the Time and Frequency Domains*, Robert E. Krieger Publishing, 1976.

[Bul77] K. Bullington, *Radio Propagation for Vehicular Communications*, IEEE Transactions on Vehicular Technology, vol. VT 26, 1977.

[Cal06] G. Calcev and M. Dillon, *Antenna Tilt Control in CDMA Networks*, 2nd Annual International Wireless Internet Conference, 2006.

[Cou93] L.W. Couch, *Digital and Analog Communications Systems*, Macmillan, 1993.

[Cox83] D.C. Cox, R.R. Murray, and A.W. Norris, *Measurements of 800-MHz Radio Transmission into Buildings with Metallic Walls*, Bell System Technical Journal, vol. 62, no. 9, 1983.

[Cox84] D.C. Cox, R. Murray, and A. Norris, *800 MHz Attenuation Measured in and around Suburban Houses*, AT&T Bell Laboratories Technical Journal, vol. 673, no. 6, 1984.

[Dev90] D.M.J. Devasirvatham, C. Banerjee, M.J. Krain, and D.A. Rappaport, *Multi-Frequency Radiowave Propagation Measurements in the Portable Radio Environment*, IEEE International Conference on Communications, 1990.

[Die00] C.B. Dietrich, Jr., *Adaptive Arrays and Diversity Antenna Configurations for Handheld Wireless Communication Terminals*, PhD Thesis, Virginia Polytechnic Institute, 2000.

[Egl57] J.J. Egli, *Radio Propagation above 40 Mc over Irregular Terrain*, Proceedings of the IRE, vol. 45, no. 10, 1957.

[Gei02] J. Geier, *RF Site Survey Steps*, 2002, http://www.wi-fiplanet.com/tutorials/article.php/1116311.

[Ges03] D. Gesbert, *MIMO Space-time coded Wireless Systems*, University of Oslo, 2003, http://www.iet.ntnu.no/projects/beats/Documents/GesbertMIMOlecture.pdf.

[Han98] R.C. Hansen, *Phased Array Antennas*, John Wiley & Sons, 1998.

[Hat90] M. Hata, *Empirical Formula for Propagation Loss in Land Mobile Radio Services*, IEEE Transactions on Vehicular Technology, vol. VT 29, no. 3, 1990.

[Hay04] W. Hayward, *Introduction to Radio Frequency Design*, American Radio Relay League, 2004.

[Hou05] H. Hourani, *An Overview of Diversity Techniques in Wireless Communication Systems*, Helsinki University of Technology, S-72.333 Postgraduate Course in Radio Communications 2004-05.

[ITU08] International Telecommunications Union, *ITU Radio Regulations, Article 5*, Edition 2008.

[Kra02] J.D. Kraus, *Antennas, (3rd edition)*, McGraw Hill, 2002.

[Lee85] Lee, W.C.Y, *Mobile Communications Engineering*, McGraw Hill, 1985.

[Maa92] S. Maas, *Microwave Mixers, (2nd edition)*, Artech House, 1992.

[Meh94] A. Mehrotra, *Cellular Radio Performance Engineering*, Artech House, 1994.

[Oku68] T. Okumura, E. Ohmori, and K. Fukuda, *Field Strength and Its Variability in VHP and UHF Land Mobile Service*, Review of the Electrical Communication Laboratory, vol. 16, no. 9-10, 1968.

[Pet04] M. Pettersen, L. Braten, and A. Spilling, *Automatic Antenna Tilt Control for Capacity Enhancement in UMTS FDD*, Proceedings of the IEEE Vehicular Technology Conference, 2004.

[Rab08] J.M. Hernándo Rábanos, *Transmisión por Radio, (6th edition)*, Ed. Ramón Areces, 2008.

[Rap91] T.S. Rappaport, *The Wireless Revolution*, IEEE Communications Magazine, 1991.

[Rap02] T.S. Rappaport, *Wireless Communications: Principles and Practice, (2nd edition)*, Prentice Hall, 2002.

[Reu74] D.O. Reudink, *Properties of Mobile Radio Propagation above 400 MHz*, IEEE Transactions on Vehicular Technology, vol. 23, no. 2, 1974.

[Sei92] S.Y. Seidel and T.S. Rappaport, *914-MHz Path Loss Prediction Models for Indoor Wireless Communications in Multi-floored Buildings*, IEEE Transactions on Antennas and Propagation, vol. 40, no. 2, 1992.

[Sio05] I. Siomina, *P-CPICH Power and Antenna Tilt Optimization in UMTS Networks*, Proceedings of the Advanced Industrial Conference on Telecommunications, IEEE, 2005.

[Vio88] E.J. Violette, R.H. Espeland, and K.C. Alien, *Millimeter-Wave Propagation Characteristics and Channel Performance for Urban-Suburban Environments*, National Telecommunications and Information Administration. NTIA Report 88-239, 1988.

[Vol07] J.L. Volakis, *Antenna Engineering Handbook (4th edition)*, McGraw Hill, 2007.

[Wer00] D. Werner and F. Mittra, *Frontiers in Electromagnetics*, IEEE Press, 2000.

[Yli04] J. Ylitalo, *Tutorial #2: MIMO Communications with Applications to (B)3G and 4G Systems*, University of Oulu, http://www.cwc.oulu.fi/nrs04/slides/mimo_introduction.pdf, 2004.

[Zve67] A.I. Zverev, *Handbook of Filter Synthesis*, John Wiley & Sons, 1967.

4.8   Suggested Further Reading

C. Balanis, *Advanced Engineering Electromagnetics*, John Wiley & Sons, 1989.

C. Balanis, *Antenna Theory, Analysis and Design (3rd edition)*, John Wiley & Sons, 2005.

E. Biglieri, R. Calderbank, A. Constantinides, A. Goldsmith, A. Paulraj, and H. Poor, *MIMO Wireless Communications*, Cambridge University Press, 2007.

H. Bölcske, D. Gesbert, C.B. Papadias, and A.J. van der Veen, *Space-Time Wireless Systems: From Array Processing to MIMO Communications*, Cambridge University Press, 2006.

D. Dobkin, *RF Engineering for Wireless Networks: Hardware, Antennas, and Propagation*, Newnes Publishing, 2004.

R. Elliott, *Antenna Theory and Design*, IEEE Press, 2003.

F. Gross, *Smart Antennas for Wireless Communications*, McGraw Hill, 2005.

R. Harrington, *Time Harmonic Electromagnetic Fields*, McGraw Hill, 1961.

M.C. Jeruchim, P. Balaban, and K.S. Shanmugan, *Simulation of Communication Systems: Modeling, Methodology and Techniques (Information Technology: Transmission, Processing and Storage), 2nd edition*, Kluwer Academic Plenum Publishers, 2000.

R. Mailloux, *Phased Array Antenna Handbook (2nd edition)*, Artech House, 2005.

T. Milligan, *Modern Antenna Design*, John Wiley & Sons, 2005.

D. Miron, *Small Antenna Design*, Newnes Publishing, 2006.

C. Oestges and B. Clerckx, *MIMO Wireless Communications: From Real-World Propagation to Space-Time Code Design*, Academic Press, 2007.

A. Peterson, S. Ray, and R. Mittra, *Computational Methods for Electromagnetics*, IEEE Press, 1998.

S. Saunders and A. Aragon-Zavala, *Antennas and Propagation for Wireless Communication Systems (2nd edition)*, John Wiley & Sons, 2007.

C. Sletten, *Reflector and Lens Antennas*, Artech House, 1988.

J. Stratton, *Electromagnetic Theory*, McGraw Hill, 1941.

W.L. Stutzman and G. Thiele, *Antenna Theory and Design (2nd edition)*, John Wiley & Sons, 1998.

A. Taflove and S. Hagness, *Computational Electrodynamics: The Finite Difference Time Domain Method (3rd edition)*, Artech House, 2005.

R. Waterhouse, *Printed Antennas for Wireless Communications*, John Wiley & Sons, 2008.

K. Wong, *Planar Antennas for Wireless Communications*, John Wiley & Sons, 2003.

Chapter 5

Facilities Infrastructure

5.1 Introduction

This chapter outlines the information needed to specify, design, and implement wireless facilities and sites. The scope of this information is very broad, as can be seen from the list of topics below:

- AC and DC Power Systems
- Lightning Protection
- Electrical Protection Devices
- Heating, Ventilation, and Air Conditioning
- Equipment Racks, Rack Mounting Spaces, and Related Hardware
- Waveguides and Transmission Lines
- Tower Specifications and Standards
- Distributed Antenna Systems and Base Station Hotels
- Physical Security, Alarms, and Surveillance Systems
- Related Industry Standards Bodies

Therefore, this chapter can only touch on these topics at a high level. At the end of the chapter there are references to books, standards documents, and web sites where detailed information can be found. Because wireless site design and construction draw on many types of engineering backgrounds, information about possible sources of expertise in these areas is included throughout this chapter.

5.2 AC and DC Power Systems

Communications equipment needs electrical power to operate. It is important to ensure that this power is properly provisioned. In order to determine what power will be needed to operate a site, it is necessary to understand how much power is required at the site and how long the site will need to operate should the normal energy supply be interrupted.

5.2.1 Power System Specifics

A critical first step is to define the power/energy needs for the complete site. These include both the power consumption of the communications equipment and the power needed for support equipment supplying, for example, cooling or heating capabilities (see section 5.4). Both the maximum power consumption and the long term average power requirements must be calculated. The maximum power requirement must be known in order to size the site powering system, while the long-term average requirement must be known so that adequate long-term backup power or energy storage capacity is provided. Backup power needs to be evaluated and understood in terms how many hours of operation the backup system must provide. Unique customer needs may require that backup power be provided for an extended period of time. (See section 6.4.3, Example 3.)

Telecommunications systems are commonly powered from commercial utility AC power that has been converted on site to –48 VDC and/or ±24 VDC. In addition to the energy conversion equipment, energy storage must be provided on site to maintain the operation of the equipment in the event of a loss of utility power. The most common energy storage system is a bank of batteries; however, flywheels and other new energy storage technologies have been proposed for telecommunications applications. A second level of backup power may also be required and should be considered. For large installations, a diesel generator, turbine alternator, or other backup power source may be required. Any such source must include an automatic start capability in the event of a loss of commercial power, and, of course, a source or supply of fuel (gasoline, natural gas, etc.) must be provided at the site. Smaller sites may have provision for connection of a portable generator.

There are other options but the most common power plant for wireless base stations consists of centralized AC-to-DC rectifiers and battery storage, with DC distribution conductors carrying the power at the appropriate voltages to the communications equipment [Ree07]. Figure 5-1 shows a block diagram of a typical site powering system. Since most solid-state devices and circuits operate at voltages other than –48 VDC and ±24 VDC, internal power supplies within the communications equipment will typically include DC-to-DC converters which step these voltages down to the levels used by the electronic circuits. The power system for a site can obviously be simplified if it is possible to select communications equipment that is all designed to operate from the same DC distribution voltage. In some cases, bulk DC-to-DC converters may need to be provided as separate items of equipment to handle the cumulative low-voltage power needs of multiple pieces of communications equipment.

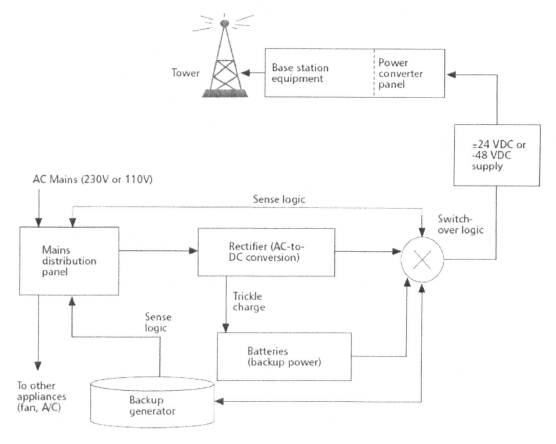

Figure 5-1: Typical Wireless Base Station Power System Layout

In addition to the DC power for the communications equipment, AC power must be distributed throughout the facility or site to power air conditioners, fans (although typically not the cooling fans that are incorporated into individual pieces of equipment), electrical heaters, and lighting, as well as to provide accessible outlets for powering test sets or other maintenance and repair tools. Whereas the DC voltages are not hazardous (but the current levels can be!), AC voltages can be hazardous. Therefore, industry standards, practices, and local codes must all be adhered to in the design and construction of these AC distribution networks. Proper shielding and grounding are also necessary to avoid introducing AC "hum" into any communications path.

In remote locations, alternative energy sources are being utilized more and more often. These include solar (photovoltaic) panels, wind turbines, fuel cells, and other new technologies. Because solar and wind energy are by their nature intermittent, the use of these energy sources requires a large battery bank or other energy storage facility on site to maintain the site operation in the event of long periods of cloudy weather or the absence of wind. Fuel cells, while not intermittent, require a supply of fuel, which must be replenished as it is consumed during the cell's operation. Whatever alternative energy source is used, the system must be sized, designed, and constructed to deliver both DC and AC power to meet the needs of both the communications equipment and its support equipment.

5.2.2 Additional Resources

A great deal of detail is involved in the design and implementation of AC and DC power systems. These systems usually connect to a commercial power company and involve potentially dangerous voltages and currents; therefore, they must be designed with care and they must conform to local codes. Because of these issues of safety and conformance, it is good practice to involve specialists with expertise in power system design and in locating the appropriate codes and standards that the power system must meet. Section 5.11 identifies some of the resources available, and suggests ways to locate such relevant expertise.

5.3   Electrical Protection

Engineered protection schemes are required at communications facilities to mitigate the effects of transient voltages [Car85]. The most significant and serious of these transients are the result of induced and conducted lightning currents; AC fault and induced currents, as well as power circuit switching operations also cause transient voltages. The latter include switching of power factor correction capacitors on the medium-voltage AC distribution system, along with any other make or break operation in a power system not occurring at a zero current point, such as motor contactors, fuse or circuit-breaker tripping, etc. Whenever an energized circuit is opened, the collapse of the magnetic field self-induces a transient voltage surge into the conductor.

Wireless communications facilities, with their towers and antennas, are frequently affected by direct lightning strikes due to their exposed locations and their shape. The frequency of lightning strikes depends on the geographical and topological location of these facilities as well as their elevation, height, and rooftop area. The correct lightning protection, grounding, and bonding systems are required at these facilities to mitigate the effects of direct lightning strikes.

Standards have been developed to enable such systems to be designed and installed effectively at communications buildings and towers.

5.3.1 Lightning Strikes as Sources of Transients

The largest protection threat to a wireless communications facility is a direct lightning strike to the radio tower. During a tower strike, all metallic entrances to the facility (AC power, telecommunications cables, water/gas lines, etc.) become potential lightning-current exit paths. This is due to the high ground potential rise (GPR), also known as earth potential rise (EPR), of the facility ground reference with respect to remote earth.

Figure 5-2 illustrates a typical site GPR/EPR during a lightning strike. It assumes a nominal 20 kA peak lightning current and a 5 ohm ground resistance. The figure of 20 kA is a median value for lightning currents; approximately 50% are lower and 50% are higher. Lightning originates from both a high source voltage and high source impedance. The facility ground resistance is negligible compared to the source impedance and has little effect on the delivered lightning current. Because of this, lightning is considered a constant current source. As shown, a nominal 20 kA strike through a nominal 5 ohm ground results in the facility ground plane rising 100 kV above remote earth

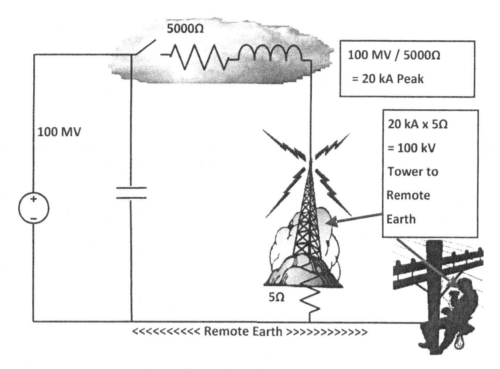

Figure 5-2: Ground/Earth Potential Rise Due to a 20 kA Lightning Strike

All metallic objects, including security fences and gates, should be solidly bonded to the facility ground reference. As a general rule, any two metallic objects within a person's arm-span should be bonded to the common facility ground reference for safety.

5.3.2 Lightning Strike Frequency

Lightning activity is unevenly distributed across the globe and its occurrence is usually measured by the annual thunder day (TD) occurrence. The TD is measured by the local weather station and indicates the presence of thunderstorms and lightning in the immediate area. The TD can vary from less than 10 in some parts of Western Europe to more than 200 in the equatorial parts of Africa, South America, and South East Asia.

Another measure of the lightning activity is the ground flash density (GFD) which can be derived from the TD. The GFD is also used to estimate the frequency of lightning strikes to a structure.

5.3.3 Lightning Strike Mechanism

Figure 5-3 shows the stages in the development of a lightning strike. The strike process usually starts in the thunder cloud several kilometers above the surface of the earth with the emergence of branching electrical discharges known as downward leaders. As the downward leaders descend to within a few hundred meters above the earth's surface, each downward leader will trigger a number of similar electrical discharges of opposite polarity, known as upward leaders, towards it from various nearby grounded objects such as buildings, towers, trees, and roof mounted equipment (e.g., antennas).

The lightning strike process is completed in about 20 ms when one of the downward leaders is connected to one of the upward leaders and forms the initial lightning return stroke. The return stroke is a high current discharge that starts from the grounded object and moves upwards into the cloud. All other unconnected downward and upward leaders will dissipate as soon as the return stroke occurs. The average lightning flash has several return strokes which gives it a flickering appearance.

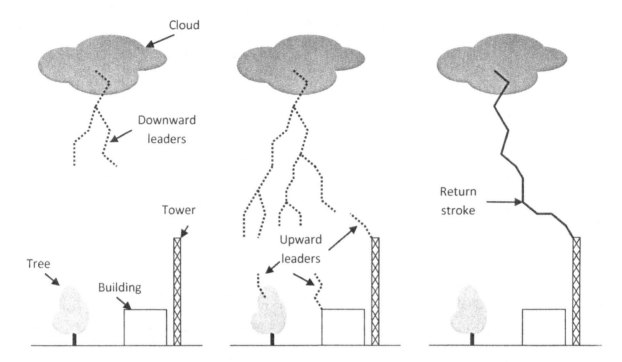

Figure 5-3: Example of a Lightning Strike Mechanism

The object whose upward leader made the connection with the downward leader becomes the primary target of the lightning strike and the degree of damage it suffers will largely depend on its material composition. Metallic objects usually suffer little or no damage, while the severity of damage to non-metallic objects will depend on the magnitude of the lightning stroke current which ranges from about 2 kA to more than 200 kA. As noted earlier, the median lightning stroke current is about 20 kA.

A very important consideration when dealing with lightning strikes is the location of the lightning stroke on the target structure. While it is generally true that taller structures are struck more frequently by lightning, it is not true that lightning always strikes the highest point of a structure. Studies have shown that more than 90% of lightning stroke attachments occur at high risk locations such as the corners and exposed points of a structure. Lightning stroke attachment points have also been observed to occur on the lower sides of very tall structures.

5.3.4 Surge Threat Evaluation

A tool commonly used in the U.S. to make lightning-risk site-selection decisions, or to determine if a given problem was lightning related, is the data from the U.S. National Lightning Detection Network (NLDN). The NLDN collects data on over 20 million lightning strikes to the continental U.S. each year, including time, longitude, latitude, and peak current. Data sets for a particular location and time are available online for a nominal charge at http://thunderstorm.vaisala.com/.

5.3.5 Lightning Protection System

The objective of installing a lightning protection system (LPS) is to prevent the lightning stroke from attaching itself to a vulnerable non-metallic part of a structure or to exposed equipment. A well designed, constructed, and maintained LPS will ensure that a lightning stroke will attach itself to the system and not to the protected structure or equipment, and hence will result in minimal or no damage to the protected structure or equipment. The LPS typically consists of the following sub-systems: an air termination system, a down conductor system, and a grounding system.

5.3.5.1 Air termination system

An air termination system consisting of air terminals (i.e., lightning conductors or Franklin rods) positioned at the high risk locations on a structure will prevent a lightning stroke from attaching itself to the structure. Air terminals consist of a combination of vertical and horizontal finials and conductors.

An all-metal structure does not require an air termination system since a lightning stroke will have a negligible impact at the point of attachment. However, non-metallic and sensitive devices and cables installed on the surface of the structure and exposed to direct lightning strikes will require protection in the form of an air termination system that is installed on the structure.

The high risk locations can be identified using the Rolling Sphere Method (RSM) that is described in the NFPA780 and IEC62305 standards. In this method, an imaginary sphere, whose radius is determined by the required risk level, is rolled around and over the exposed facility. Those parts of the facility that touch the surface of the imaginary sphere are determined to be the high risk locations. Recent studies suggest that over 90% of lightning strikes to structures occur at the exposed corners, edges and pointed features. In order to protect these high risk locations, air terminals should be erected on top of them where they can effectively collect the lightning strokes.

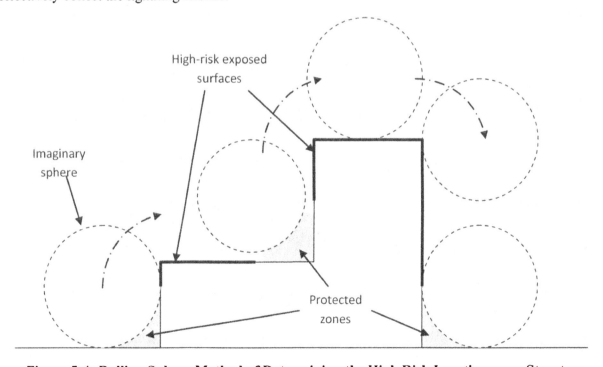

Figure 5-4: Rolling Sphere Method of Determining the High Risk Locations on a Structure

The RSM can also be applied in the protection of exposed communications facilities. The imaginary sphere is used to determine which equipment or parts of the facility are at risk of being struck by a lightning stroke (Figure 5-5). This risk is drastically reduced by the introduction of air terminals at strategic locations so that the imaginary sphere will only be able to touch the air terminals and not the sensitive parts of the communications system and facility (Figure 5-6). Sample air terminations are shown in Figure 5-7.

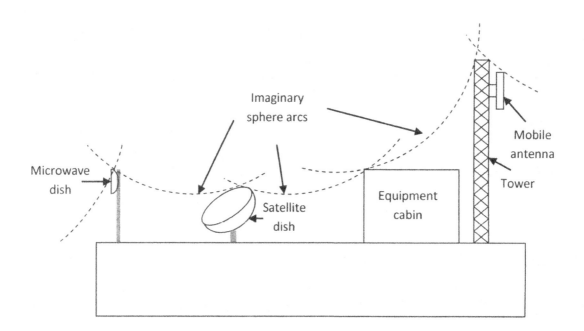

Figure 5-5: Identifying the High-Risk Locations on Roof-Mounted Equipment

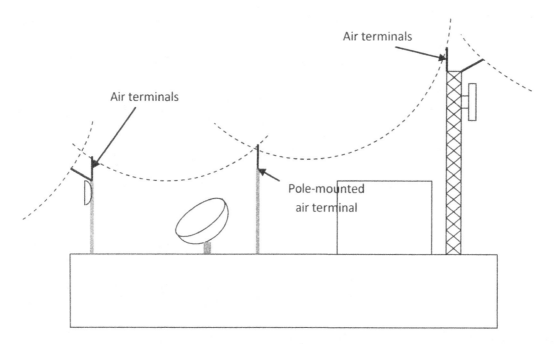

Figure 5-6: Protection of Roof Mounted Equipment Using an Air Termination System

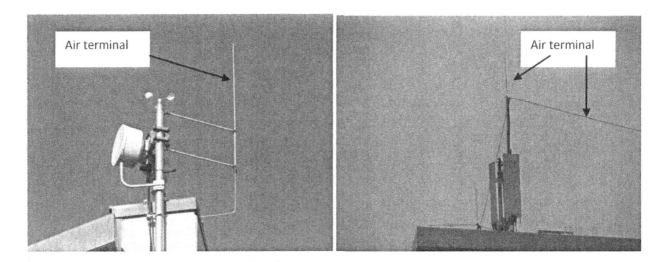

Figure 5-7: Examples of Air Termination Systems for Communications Antennas

Readers can obtain more information about air termination design methods from the current editions of the NFPA780 and IEC62305 lightning protection standards.

5.3.5.2 Down conductor system

A down conductor system consists of two or more conductors that link the air termination system to the grounding system. For a small structure, two down conductors routed down the opposite sides of the structure provide sufficient links. For larger or more complex structures, several down conductors should be added to link the air termination system to the grounding system. (See Figure 5-8 for examples.) The position and spacing between the down conductors should comply with the applicable standard(s) for the facility, location, and jurisdiction.

Modern buildings built with steel reinforced concrete can use the steel reinforcement bars as the down conductors. The air termination systems can be bonded to the steel bars at the corners of the building.

5.3.5.3 Grounding system

A grounding system consists of a number of conductor rods usually buried vertically in the soil and connected to each other by horizontal bare conductors. The metal reinforcement piles of modern buildings can be used as the grounding system. For older buildings and temporary structures built without metal reinforcement piles buried in the ground, the grounding system can be installed on the outer perimeter of the structure. Besides bonding to the down conductors, the grounding system should be bonded to all the communication and electrical grounding points in the facility.

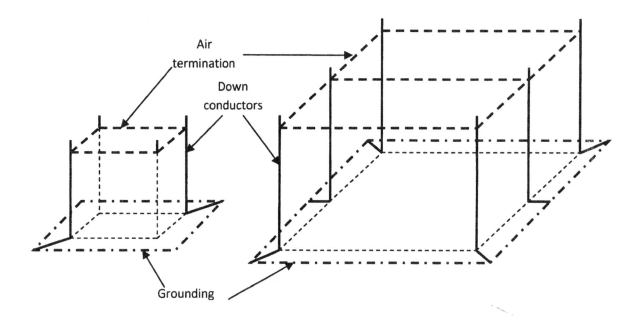

Figure 5-8: Examples of Down Conductor and Grounding Systems

5.3.6 Bonding

Proper bonding between various communication systems and to the electrical and lightning protection ground is essential to avoid lightning related damage during thunderstorms. A good way to ensure that all equipment is bonded together is to introduce a Master Ground Bar (MGB) in the building or facility. All communications, power, surge protection, electrical equipment, and frame ground terminals should be bonded to the MGB. The MGB should then be bonded to the building/lightning grounding system. If a tower is built nearby, the tower grounding system should also be bonded to the building/lightning grounding system.

For more details, readers should refer to the communication bonding and grounding standards published by the International Telecommunication Union (ITU) in their K-series documents.

5.3.7 Surge Protective Devices

Non-bonded metallic conductors, such as AC power and telecommunications cables, should be indirectly bonded to the facility ground through surge protective devices (SPDs). Since these conductors are referenced to remote earth, the high GPR/EPR will force their use as exit paths. The use of SPDs will prevent external flashover and will hold the incoming conductors close enough in potential to the facility ground reference to prevent equipment damage. AC SPDs (previously referred to as Transient Voltage Surge Suppressors [TVSSs]) are labeled with a maximum available fault current rating and what, if any, upstream over-current protection device is required. Care must be taken to ensure that the anticipated fault current at the point of SPD installation does not exceed the labeled rating on the SPD, and that any required upstream over-current protection is present. All SPDs used in communications facilities should be listed by a nationally recognized testing laboratory (NRTL) and used in accordance with the electrical code requirements. In the U.S., the National Electric Code (NFPA70, also known as NEC) article 285 covers SPDs for AC power circuits. A more detailed description of SPDs and their applications is contained in IEEE C62.43. (See also [Has00].)

European Council (EEC) Directives mandate that products offered for sale or placed in service in Europe must comply with the Member State's requirements as prescribed in all relevant CE marking directives. Related to specific topics, CE marking directives require EU Member States to appoint Notified (Type A Inspection Body) and Competent Bodies to evaluate and approve products against the requirements identified in each directive. Such bodies must be accredited against the essential requirements of standards EN 45004 and EN 45011, and the European Commission must be notified of their appointment. There must be demonstrable independence between a Notified Body and any activity associated with the design, manufacture, supply, installation, purchase, ownership, use, or maintenance of the items inspected. For example, in the UK, CE marking certification is performed by such private laboratories as Mira Ltd. The European Commission Low Voltage Directive, 2006/95/EC, essentially applies to consumer and capital electrical equipment (mains energized electrical appliances) designed for use within the voltage ranges from 50 VAC to 1,000 VAC and from 75 VDC to 1500 VDC. However, equipment subject to the Radio & Telecommunications Terminal Equipment (RTTE) Directive 1999/5/EC must meet the requirements of the Low Voltage Directive with no lower voltage limit.

Elsewhere, country-specific codes and requirements apply. For example, in Japan the applicable requirements are contained in the Electrical Appliance and Material Safety Law ("DENAN"; formerly the Electrical Appliance and Material Control Law "DENTORI"). In Malaysia, the national testing laboratory is SIRIM Berhad, and its electrical testing typically follows standards created and issued by the International Electrotechnical Commission (IEC).

All well-designed and constructed brands of AC SPDs utilize the same basic components and have approximately the same let-through voltage. The primary difference in AC SPD performance is not brand, but installation. AC SPDs are normally connected in parallel with the power circuit. The parallel or shunt lead length used to connect an AC SPD can significantly degrade performance. This is due to the inductive voltage drop along the leads at lightning frequencies. For example, a typical SPD for 120 VAC service has a let-through voltage of approximately 500 V on a standard 10 kA, 8/20 µs surge. Closely coupled connection leads, on the same surge, add an additional 600 V/m (200 V/foot). Widely separated leads have an even higher voltage drop. For acceptable installed performance, lead lengths should be kept as short and straight as practical, and as a rule of thumb should not exceed 1m (three feet). Outside North America, where 230 VAC is the common mains voltage, it is essential to follow practices based on applicable local standards. Several standards and work practices for safety in electrical testing are published by British standards bodies for work with low-voltage electrical systems of up to 1,000 VAC; similar publications are available in other countries and other regions and should be consulted when planning the AC protection system for any site.

SPDs for communications circuits are addressed in a variety of documents. In the U.S., the applicable requirements are included in NEC Article 800; in France, Association Française de Normalisation (AFNOR) French Standard UTE C15-443 covers such SPDs. There are multiple categories and listing requirements for communications SPDs. Different systems, installation topologies, and wiring methods have their own unique characteristics and require different protector performance to mitigate risks and operate without introducing hazards. Choosing the wrong protector can reduce (or even destroy) the effectiveness of the protection scheme and increase the risk of catastrophic protector failure, electric shock, fire, and injury.

For circuits directly connected to outside plant (OSP) conductors, the protector must protect against lighting and also against accidental contact of the OSP conductors with joint-use AC distribution conductors (referred to as a power-cross). In the U.S., the NEC requires the use of a listed primary telecommunications protector on exposed OSP circuits. These devices are NRTL-listed in the U.S. to Underwriters Laboratories UL497. Outside North America, where 230 VAC is the normal mains power, local jurisdictional requirements and codes should be followed to ensure proper protection in the event of a power-cross. The primary protector is normally supplied by the telecommunications service provider. Questions regarding the provided communications circuit classification should be addressed to the electrical protection engineer employed by this provider.

When a protector is directly connected to the metallic conductors between the listed primary protector and the first piece of electronics equipment, it is classified by the NEC as a secondary protector. Although the U.S. NEC does not require the use of secondary protection, it does require that if used, a secondary protector must be listed for that purpose. These devices are NRTL-listed in the U.S. to Underwriters Laboratories UL497A, which requires in-line fusing to prevent wiring fires during power-cross events. This fusing limits the surge capability of the secondary protector to less than that required for primary protection. Secondary protectors should only be used downstream of a listed primary protector. In other countries, local jurisdictional codes apply; see for example the *"Customer-Owned Outside Plant (CO-OSP) Design Manual, Third Edition,"* published by Building Industry Consulting Service International (BICSI).

Communications circuits that are not exposed to OSP surges or power-cross, are classified as isolated loops. Isolated loops include intra-site alarm and communications circuits, as well as many OSP-derived circuits. Often, the telecommunications service provider uses active listed electronics to convert from OSP circuits (such as DSL) to internal circuits (such as T1/E1 or Ethernet). In this case, the listed conversion electronics isolate the site's internal circuits from OSP surges and power-cross, and the internal circuits are classified as isolated loops.

Isolated loop protectors are NRTL listed in the U.S. to UL497B. Listing of an isolated loop protector does not require survival under the large GPR/EPR surges experienced during lightning strikes to the facility. Because of this, it is important to evaluate the surge capability of all isolated loop protectors used in communications facilities. Protectors at communications facilities rated for a multiple-pulse maximum-surge-current capability of 10 kA, 8/20 μs have proven robust enough to provide reliable service.

5.3.8 Additional Resources

For protection against direct lightning strikes, the lightning protection standard applicable in the USA is NFPA780 *"Standard for the Installation of Lightning Protection Systems"*. In IEC member countries, the applicable standard is IEC62305-3 *"Protection against Lightning – Part 3: Physical damage to structures and life hazards"*.

For protection against transients, the standard applicable in the USA is the IEEE Standard 1100 *"IEEE Recommended Practice for Powering and Grounding Electronic Equipment"*. In IEC member countries, the applicable standard is the IEC62305-4 *"Protection against Lightning – Part 4: Electrical and Electronic systems within structures"*. In addition, ITU member countries can also apply the relevant K-series documents for the protection of communication lines and equipment against transients.

A protection resource available in the U.S. is the Protection Engineers Group (PEG). PEG is an organization of protection engineers specializing in the electrical protection of communications facilities; it is sponsored by the Alliance for Telecommunications Industry Solutions (ATIS). For twenty years, PEG has held annual conferences where noted industry experts discuss communications facility protection issues. PEG has available for purchase CDs on topics from past conference presentations, as well as webinars in which communications facility protection topics are covered by noted industry experts. Information about PEG can be found at http://www.atis.org/peg/.

ATIS also sponsors a Network Electrical Protection (NEP) working group under its Sustainability in Telecom: Energy and Protection Committee. STEP-NEP has developed and maintains a number of standards and technical reports related to the electrical protection of telecommunications networks. Information about STEP can be found at http://www.atis.org/step/. In addition, standard BSR ATIS 0600334-200x (formerly ANSI T1.334-2002), prepared by the T1E1.7 working group, provides details on the electrical protection of communications towers and associated structures.

Similar technical committees and working groups exist in other countries, and their standards, technical reports, and best practices should be consulted when designing a facility in that country.

5.4   Heating, Ventilation, and Air Conditioning

Communications equipment typically requires a controlled environment to ensure reliable operation. For some equipment the temperature—and possibly the humidity, airflow, and/or air quality (e.g., levels of dust and potential pollutants)—must be tightly controlled. For other equipment, the environmental conditions can be allowed to vary over a wider range. It is important to understand the applicable requirements and to ensure that appropriate solutions are provided.

5.4.1   Heating, Ventilation, and Air Conditioning (HVAC) Requirements

Because communications equipment consumes power and dissipates heat, most such equipment requires some degree of environmental cooling to maintain its internal temperature within a range where it will operate reliably. High humidity, especially in conjunction with high temperatures, can be particularly damaging to sensitive electronic equipment; hence air conditioning (dehumidification), not simple ventilation, is typically required. In extreme environmental conditions, heating may be required to similarly maintain equipment at minimum operating temperatures. Whether heating or cooling (or both) is required, it is important to maintain a relatively stable temperature; wide swings from hot to cold and back can be particularly stressful on all types of communications equipment.

Batteries are commonly used as a backup power source and are typically quite sensitive to temperature, both as regards their energy storage capacity and their service life. Adequate backup power, including alternate power sources, may be required to maintain the controlled site environment if the potential site temperature can reach hot or cold extremes.

Storage batteries also require ventilation. Hydrogen and oxygen gases are generated as a function of overcharging. The amounts of both gases are directly proportional to the current that flows through the battery cells and the number of cells in series. There have been numerous events in which hydrogen gas explosions have damaged or even destroyed a communications facility. In most of the IEEE standards for stationary batteries, the maximum allowable gas concentration is 2% in order to provide a margin of safety against the lower flammability limit of hydrogen, which occurs at a concentration of 4% or above.

5.4.2 Additional Resources

When buildings are involved, it is often advisable to work with a mechanical engineer experienced in building systems to calculate the HVAC system requirements. An engineer with such experience can also provide information on the types and capacities of any cooling or heating equipment, based on the energy consumption and dissipation of the communications equipment that is to be located within the building and on the expected external environmental contributions to the required heating or cooling (e.g., heating from intense sunlight or cooling from strong icy winds).

5.5 Equipment Racks, Rack Mounting Spaces, and Related Hardware

Standards have been established to mount various types of equipment in common frameworks often referred to as equipment racks. Typically, these were originally developed many years ago by the telephone and broadcast industries and have subsequently migrated into all areas of electronic and industrial systems. A basic understanding of these standards, and where to find them, is essential to ensure that the site can accommodate the planned equipment, and that the equipment's needs for power, ventilation, technician access, etc., will be met. Complete site equipment requirements must be determined to ensure that there are no spacing issues. Both horizontal and vertical space requirements must be clearly defined at the outset. A helpful way of checking for spacing issues is to generate a complete set of blueprints for the proposed equipment layout.

5.5.1 Mounting Standards

The primary standard in the United States is included in ANSI/EIA 310-D and applies to equipment that mounts on a 48.3-cm (19-in.) wide rack. All equipment that mounts in such racks has a front panel 48.3 cm (19 in.) wide and is sized in increments of 4.4 cm (1-3/4 in.) high.

In telephone central offices and similar facilities, there is a comparable standard in which the front panels are 58.4 cm (23 in.) wide and sized in increments of 5.1 cm (2 in.) tall. This system also allows the equipment height to be sized in fractional units, i.e., in whole- and half-inch increments. Details on this standard rack format can be found in the catalogs of most providers of equipment racks for the telecommunications industry.

Both of these standards also include information on such things as the types of mounting hardware—for example, the diameter and thread pitch of the mounting screws for the equipment. There are also details about the installation of the racks themselves—for example, the size and spacing of the bolts by which the racks are anchored to the floor. Additional information applicable to a specific installation may include provisions for overhead bracing and for troughs or other guideways for containing and routing interconnect wiring and cable. In areas subject to significant earthquake risks, additional bracing, suspension from overhead beams, or other special design features may be required.

Because operating communications equipment generates heat and requires a flow of cooling air to maintain reliable operation, individual pieces of equipment may require space for heat baffles. Angled heat deflectors are often inserted between pieces of equipment. These typically direct the heat flow from the lower piece of equipment out toward the back of the rack, while simultaneously providing space for cooler air to flow in from the front of the rack and up into and through the higher piece of equipment. The space occupied by such heat deflectors must be planned for when calculating the amount of equipment to be installed in a rack.

While individual racks can be mounted abutting side by side in rows, adequate space must be allowed between rows. In the front of equipment racks, such space is obviously required to allow technicians to access the equipment for test, maintenance, repair, and replacement purposes. However, adequate space must also be allowed behind racks, since this is where most connections are made—to the power distribution bus, to the waveguide or cabling leading to the antennas, and to the cables or other communications connections that interconnect pieces of equipment, and that connect the site to the ground-based telecommunications network. The specifications for communications equipment often include information about the space that must be provided for such rear-panel connections.

5.5.2 *Additional Resources*

Unlike power distribution, electrical protection, or HVAC, there are few specialists in rack design and mounting. It is therefore important to understand the general nature of the equipment and its physical requirements for access and heat dissipation. Often there are also codes and regulations that define the access space that must be provided for personnel. All these factors must be considered when designing a facility, and it is advisable to review the equipment manufacturers' literature in some detail to understand how best to mount it and provide the necessary access. Indeed, the literature from rack and equipment manufacturers is often the best source for detailed information that will help ensure that the equipment can be mounted and accessed properly in the proposed racks and line-ups.

5.6   Waveguides and Transmission Lines

Waveguides and coaxial transmission lines are the most common devices for connecting transmitters and receivers to antennas. In this section we focus strictly on some of these devices' physical characteristics that need to be understood in planning the layout of a wireless facility.

5.6.1 *Physical Characteristics of Waveguides and Coaxial Cables*

Waveguide today is mostly elliptical and is handled much as a cable. The minimum bend radius is a very critical factor and is relatively large with respect to waveguide dimensions. Those dimensions and specific handling recommendations are available from the waveguide manufacturer. Strict adherence to these requirements is critical to the waveguide meeting its performance characteristics.

Coaxial cables are similar but not as critical as waveguide with regard to careful handling. However, strict adherence to the manufacturer's requirements is critical to meeting the specified performance characteristics. In general, coaxial cables have more transmission loss compared to a waveguide. As a rule of thumb, the more flexible the cable is, the greater the path loss will be. Conversely, cables with larger diameters have larger bending radii making it harder to turn them around corners.

Cables and waveguides must be supported at least every 1m (3 ft) on the horizontal and vertical when installed. Terminations may require short twistable-flexible waveguide sections at one or both ends to accommodate sharp angle bends that may be necessary for access to equipment or antennas. Large-diameter coaxial cables chosen to reduce transmission losses may require short sections of small-diameter highly flexible cables at the ends for the same reason.

Special provisions must be made when a waveguide or coaxial cable is to enter a building or equipment housing from outdoors. Such entry may require a bulkhead fitting or other feed-through device. The waveguides and coaxial cables must generally be bonded to ground prior to entering a building or equipment housing (see section 5.3). Special fittings are generally available from the manufacturer for this task. Coaxial cables are often fitted with surge protective devices at the bulkhead.

The maintenance of waveguides and coaxial cables is also critical to their ongoing performance as a link in a communications system. Dry air is often pumped into a waveguide to keep moisture away so as to reduce losses and resist deterioration of the waveguide's electrical performance. Pressurized waveguide needs to be checked periodically to determine if the system is holding pressure; once a year is the minimum suggested frequency. Overall system testing and inspection should also be done on an annual basis to identify and repair or, if possible, prevent, deterioration of the system.

More and more frequently, optical fiber cable is being used for backhaul from remote sites to switching offices and other centralized locations because of its high capacity. In such cases, additional equipment is required at the site to convert the optical signal to an electrical signal for wireless transmission, or to convert the electrical signal to an optical signal for transmission back to the central location. The power requirements (and heating and cooling needs) of such equipment must be factored into the calculations for the site. Optical cables may have a metallic strength member or may be of totally dielectric (non-electrically conducting) construction; it is important to identify which type(s) of cables are being used so that appropriate bonding and grounding measures can be taken. All-dielectric fiber optic cables are preferred as the transmission media serving tower locations due to their lack of a metallic path for conducting lightning surge currents.

5.6.2 *Additional Resources*

The most useful reference documents on this subject are usually the manufacturers' literature. They generally provide comprehensive guides to handling, terminating, mounting, and grounding the waveguide or coaxial cable, as well as the specific requirements for optical cable and optical/electrical signal conversion equipment. A number of the major manufacturers of coaxial cables and waveguides (including, for example, Andrew Corp Inc. [www.andrew.com] and RFS [www.rfsworld.com]) provide useful information at their websites.

5.7   Tower Specifications and Standards

Wireless communications towers are usually designed or selected by a professional structural engineer. However, the design depends on a complete understanding of the electrical and physical characteristics needed for the wireless system to perform properly. Standard designs from tower manufacturers may seem adequate and appropriate for a given site, but it is still wise to consult an experienced structural specialist to ensure that all of the tower requirements are met. Tower specifications are also required to guide the fabrication and installation of any tower that will support communications antennas.

5.7.1 Tower Requirements

There are two basic types of towers:

- Terrestrial communication towers
- Satellite communication towers

5.7.1.1 Tower requirements – terrestrial

Towers at wireless base stations not only support the antennas; they must support them in the proper positions and orientations to achieve the desired coverage, and they must do so under wide-ranging environmental conditions. By convention all terrestrial antennas can have a maximum upward tilt (elevation) of five degrees. This limits interference with satellite communications as well as incoming noise from atmospheric sources. Wind, rain, and snow and ice loads apply mechanical stresses to the tower. Dust, lightning, and salt fog (near the ocean) can erode, corrode, or damage the tower, resulting in a significant degradation of its strength. Uneven heating due to solar radiation can cause the tower to flex, altering the antenna orientation and adding stresses to the joints in the tower structure. Cyclic flexing due to daily heating and cooling cycles can be particularly fatiguing to metallic elements and joints. Antennas themselves are partly sheltered by radomes or similar protective structures and internal heating elements to counter some of the atmospheric effects. In addition, tower mounted amplifiers need ruggedized features as well. Towers must be robust enough to carry all imposed loads (antennas, equipment, and environmental) without significant distortion, and must do so for many years, or even decades, of service.

The tower design also must accommodate potential future loads. The full capacity tower loading must be analyzed to ensure that the tower is not overloaded beyond the physical weight it can withstand. This should include all future antennas that may be added, along with their associated cabling or waveguides and the potential ice, wind, or other loads that these antennas and cables/waveguides may impose. It is very common to reuse a good site, but often times the tower may already be overloaded. A new tower may be needed to replace the old one or may be erected nearby to supplement the older one. It is therefore good practice—and often cost-effective in the long run—to consider future needs and to over-engineer the tower when the site is initially constructed.

Two specific parameters in tower design address this problem: The EPA (Effective Plate Area) of the antennas and maximum swing at the top of the tower. EPA is the area offered by each antenna to wind loads that results in tower swing due to wind at a specified velocity. In existing towers, it is therefore important to verify the EPA of existing antennas on the tower in order to decide whether additional antennas can be installed. The swing (usually specified as two degree, four degree or six degree maximum swing at the top) is a parameter critical to microwave links as the alignment between a pair of antennas used for point-to-point links is affected by it. Tower manufacturers specify these numbers; in general the heavier the tower, the larger will be its EPA capacity and the higher its ability to tolerate swing. Typically an 18 m (60 ft) tall lightweight tower can tolerate a maximum sustained wind speed of 193 kph (100 mph) and a 3-sec wind gust of 225 kph (140 mph) while maintaining a maximum swing of 2 deg only if the EPA is no greater than $0.743m^2$ ($8 ft^2$). However, a heavier version of the same tower would support an

EPA of 3.25m^2 (35 ft^2) under the same wind load conditions. Therefore, towers located in windy climates must pay special attention to both EPA and swing under maximum wind load conditions. Tower vendors (including, for example, TESSCO, RFS, and others) typically provide data sheets and methods to calculate loads (see www.tessco.com or www.rfsworld.com).

Tower designs are categorized based upon the type of tower structure required for the specific project; for example, typical designs include monopole, three-legged, or four-legged lattice structures. Before a tower can be designed or selected, the total antenna load needs to be determined. The design of the structure depends upon the loading of both the antennas and transmission lines to be placed on the tower, as well as the wind design load. In addition, specifications for towers in areas where ice and snow occur need to be designed or selected to accommodate a defined ice/snow loading. Often it may be important to compare a monopole with guy wire support to a comparable three-legged or four-legged tower since they have different issues related to maintenance. Particularly in crowded areas, the use of a down-tilt mechanism for antennas is important and must be evaluated.

Proper tower design and installation also require a detailed geological study to determine the type of foundation necessary to support the structure. If the natural soil at the site is inadequate to support the specified fully loaded tower, materials may need to be brought to the site to strengthen the soil to the required standards. Alternatively, pilings or other supports may be required to distribute the tower load and transfer it to bedrock or other sufficiently strong underlayment. In addition, specific local requirements may apply, and these need to be determined, along with local codes and planning constraints.

5.7.1.2 Tower requirements – satellite

Satellite towers typically provide support to earth station antennas that are quite large in diameter. All satellite antennas are required to have an elevation angle above five degrees in order to avoid ground noise and interference from terrestrial communication systems. In addition to the antenna size, satellite towers often need to support tracking systems that move the antenna to follow the satellite location. In view of this, tracking antenna structures are inherently more complex since tracking accuracy is fundamental and structural stability is paramount. Many satellite towers support communications with geostationary satellites and only need an initial alignment. Alignment of antennas on such towers uses GPS and similar navigation equipment to locate the satellite and align the units. With the expansion of VSAT and broadcast satellite networks, a wide range of satellite dishes are used. INMARSAT terminals for example, provide GPRS feeds that can be used by handsets that can operate on both terrestrial and satellites links by the flip of a switch. However, setting up earth stations by first tracking the satellite and then aligning the antenna is more complex. A number of good references are available on this topic (for example, [Elb00]). Many satellite receiver antennas are located on roof tops and anchoring becomes important (concrete, cinder block, or similar materials are often used) in order to provide stability to satellite dishes. Often the low-noise front amplifier in the satellite receiver is mounted close to or on the same frame as the antenna.

5.7.2 Additional Resources

ANSI/TIA 222-G, *Structural Standards for Steel Antenna Towers and Antenna Supporting Structures*, is the document used in North America to specify towers. It is incorporated in most building codes by reference in the Structural Design section. This standard is continually being reviewed by experts in the design and erection of towers. The American Society of Civil Engineers (ASCE) is another good source for standards related to building and structure load requirements. Similar standards exist in other countries and are typically issued and maintained by appropriate civil and structural engineering organizations.

Because of the critical role that the tower plays in the performance of a wireless site, it is advisable to seek out the expertise of a structural engineer who will understand the behavior of a tower under the applied loads, as well as the applicability of various codes and standards in the selection of the tower design. Often the local power utility or similar organization is familiar with such codes, standards, and environmental concerns and may be able to provide such services.

5.8 Distributed Antenna Systems and Base Station Hotels

Due to the dense population in urban areas, it is often necessary for carriers and service providers to use multiple towers throughout a city. However, locating such towers in heavily built-up areas can be very difficult. And even with multiple towers, customers inside buildings often do not get strong, reliable signals. Therefore, an alternative approach to the traditional tower/antenna infrastructure has been developed in recent years in which many urban structures are served by a Distributed Antenna System (DAS) that is housed in a Base Station Hotel (BSH).

5.8.1 *In-Building Facilities Infrastructure*

The BSH is an indoor area, typically in the basement or near the roof of a building, where there is enough space for all the equipment needed at the site. The BSH serves as an alternative means to the conventional tower/hut infrastructure for providing indoor coverage. A centralized BSH uses DAS architecture for carrying signals to nearby buildings allowing for indoor distribution of the RF signals. Figure 5-9 shows the basic elements of a BSH/DAS system. Typically, a BSH consists of base station equipment from different service providers connected through a multi-channel amplifier. All incoming signals are combined onto a single carrier that is usually converted into an optical carrier signal transported over a single-mode fiber. The entire system at the BSH consists of multi-channel amplifiers, and the RF-to-optical converter is known as the Master node.

Fibers from the Master node are then distributed to different buildings in the area. Each building has receiving nodes known as the remote nodes that convert the signals back from optical to RF. The RF signal is then radiated within the building using multiple wall- and/or ceiling-mounted antennas that serve different rooms and areas. Typically, each Master node can support 24 to 32 remote nodes. The DAS system is quite common in many cities today and serves a variety of urban buildings ranging from airports and subway systems to office buildings and underground passageways that connect large malls. Unlike tower-based systems, DAS systems do not radiate signals over an open geographic area. Instead, they are designed to provide only local coverage within office buildings or restricted spaces where signals from towers clearly cannot reach or provide adequate signal strength. In terms of layout, the BSH is not very different from the base station layout described in Figure 5-1. The major difference is that the antenna tower does not exist, and, therefore, antenna diversity schemes cannot be used. Also, the output from the base station equipment is not carried as RF but is converted to an optical signal for easier transport and reduced transmission loss over significant distances in urban areas. Service providers typically connect their base stations located in a BSH to their network using traditional backhaul methods such as T1/E1 or Ethernet.

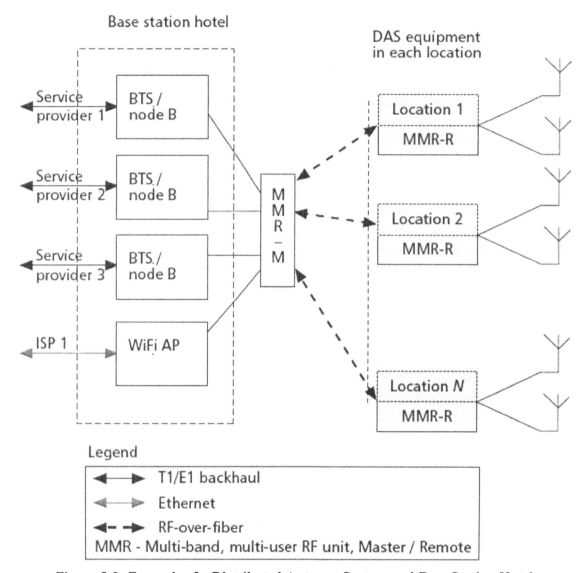

Figure 5-9: Example of a Distributed Antenna System and Base Station Hotel

5.8.1.1 Use of 60 GHz/80 GHz Gigabit Ethernet wireless for point-to-point links

With modern 3G/4G systems that support smart phones, cellular traffic has increased at an exponential rate. Therefore, all 4G networks will have at least a 100 Mbps Ethernet transport (compared to two T1/E1 lines from each base station used in earlier 2G/3G systems). Such a dramatic increase in the data rate offers new challenges to the design of both DASs as well as above-ground tower systems. In urban areas, providing high speed Ethernet backhaul to each site may become a challenge in itself, since laying fiber is both expensive and time consuming. There is a new technology of 60 GHz/80 GHz Gigabit Ethernet wireless links that has taken considerable market share from traditional fiber in recent times. Although these wireless links cover short distances (60 GHz links can go up to 1.5 km and 80 GHz can go up to 6 km), they provide flexibility and ease of installation. 80 GHz links are in licensed bands that may be used extensively by government agencies. Both systems provide cost-effective solutions not only to urban areas, but also across rivers and other types of difficult terrain where the laying of fiber becomes a major issue. Often the 80 GHz systems also provide legacy connections to TDM by providing SDH/SONET as well as Ethernet interfaces to existing networks. There are many vendors of such systems. Many manufacturers provide good reference publications on this technology (see for example BridgeWave Communications [www.bridgewave.com], Huber+Suhner [www.hubersuhner.com], and others).

With the increased use of handsets as the prime access device to the Internet, the use of DASs and BSHs to serve end users located in multi-story buildings and in large enclosed areas has become a priority. During working hours it is estimated that majority of business users (who use the majority of the bandwidth) are served by DAS networks. Therefore, the use of DAS networks is expected to continue to grow rapidly.

5.8.2 Additional Resources

Since both the BSH and the DAS network are typically located within buildings, the most useful resource is often the appropriate section of the local building code. Also, because it is often easier to obtain space in a new building or one undergoing extensive renovation, local architects and contractors can be helpful sources of information. This is particularly true when reliable wireless access is being promoted as a feature of the new or renovated building as a way to attract tenants. Calculations of vertical fiber run and horizontal RF cable run, where compromises are needed, and similar challenges are well discussed in text books that extensively cover this topic (see for example [Tol08]). Recently the concept of outdoor DAS systems has become popular among service providers. Essentially an outdoor DAS is used to fill gaps in service contours, particularly in areas of difficult terrain. Outdoor DAS equipment may be mounted on street lights or hung along utility poles to provide much-needed capacity fill in urban communities.

By providing signals from one or multiple service providers on a single antenna, a DAS can be perceived as a tower with multiple antennas. It spans multiple technologies, since the multi-channel amplifier banks used in a DAS can support GSM, UMTS, EVDO, WiFi or WiMAX channels. Currently, the major use of DASs is in airports, subway systems, and tunnels worldwide where there is no penetration of above-ground signals/coverage. DAS systems carry cellular and WiFi/WiMAX traffic, making them a neutral network to distribute different signals within a large building or facility. The growth area is outdoor DAS systems where a similar distribution is provided above ground, focusing on coverage gaps or specific communities where installation of a new base station does not provide the optimum solution.

Numerous vendors offer application notes on DAS and repeater systems, usually listed under the topics of coverage and capacity planning. Among the more easily accessible reference sources for information on DASs are CommScope/Andrew Solutions (www.andrew.com), Radio Frequency Systems (www.rfsworld.com), and Powerwave Technologies (www.powerwave.com).

5.9   Physical Security, Alarm and Surveillance Systems

Telecommunications facilities are often critical infrastructure and as such are potential targets for a variety of threats, natural or man-made. The degree of protection needed is a function of the criticality of the facility in question.

5.9.1   Importance of the Facility

A network plan generally will indicate the importance of a facility. A cell site that is one of many in an overlapping network is unlikely to be as critical as a single remote communications vault on a mountain top that covers 500 square miles or any site that includes public safety radio facilities. For sites that will carry high-priority traffic (e.g., security or public safety communications), complete site redundancy should be considered. Power backup and even equipment backup or redundancy will be useless if the entire physical site is lost or destroyed, whether due to a natural (e.g., a severe storm) or man-made (e.g., arson or other attack) catastrophe.

5.9.2   Team Effort

The security of a facility is normally a team effort in most organizations. The organization generally has a security group whose responsibility is the overall protection of all of the organization's assets. Although they may have unique and often site-specific requirements, the systems implemented to protect the communications facilities should interoperate with the general security plan of the organization.

5.9.3   Physical Security Considerations

Many issues must be considered when locating a facility. Flood, earthquake, fire hazard, and potential vandalism are just some of the factors that affect physical security. Locations exposed to extreme weather conditions such as hurricanes and tornadoes require additional protection. Remote sites in forests or wilderness areas should be protected from potential damage by local wildlife. In today's world, the possibility of terrorist attacks must be considered when locating and constructing critical communications facilities. Certain countries require that towers must not be located in regions that interfere with bird migration. In remote sites, it is common to see that birds nest on top of towers, particularly near antennas that have heating under radomes to prevent icing. Bird nests are sometimes known to cause network failures and their detection may require physical inspection, as alarm systems are not likely to detect the presence of birds.

5.9.4   Alarm Systems

Facilities should be provided with a remotely monitored alarm and surveillance system. Such a system should remotely report the status of all equipment and infrastructure at the facility. This should include unexpected entry to the facility, as well as temperature, humidity, and fire alarms. When sufficient bandwidth is available, video surveillance may be included.

5.9.5   Electronic Access System

Where electronic access systems are required, it is necessary to provide the bandwidth needed to implement site access authorization. The associated equipment should be powered via a battery backup system. This includes the actual door access devices as well as the access electronics. A secure backup plan should be provided to allow access in the event of the failure of the electronic access system.

5.9.6   Additional Resources

Site security and alarm systems are often overlooked in the process of planning a facility yet are important issues that must be addressed. A helpful resource may be the site specifications for other facilities maintained by the communications company, especially those in similar locations. Contact with the company's security organization is also recommended. Also of help is the literature from the manufacturers of various types of surveillance and monitoring systems whose capabilities are most applicable to the site.

5.10 National and International Standards and Specifications

Industry standard specifications and national and international standards make the job of site design and construction a bit easier. Many of the best practices make systems more reliable by drawing on the experience of others. In many cases the application of proven good practice helps speed approvals by local jurisdictions and may avoid unnecessary government regulation. Part of such good practice is to invite public participation through township planning committee meetings and presentations where the need for such a tower, the amount of power being radiated, and similar issues or concerns can be discussed in order to alleviate any anxiety in the minds of local residents. Holding such meetings during the early phase of planning often prevents frustration and delays.

5.10.1 Industry Standards Bodies Related to Facilities Infrastructure

Helpful insights can be gained by reviewing the relevant standards from the many organizations that develop and publish them. In addition, such organizations often publish handbooks or other documents that, although lacking the force of standards, provide useful background information and guidance on the rationale for and application of their standards. The following list includes the main organizations that have been involved in developing standards related to the facilities infrastructure.

ATIS – The Alliance for Telecommunications Industry Solutions (http://www.atis.org): The Protection Engineers Group (PEG) and the Sustainability in Telecom: Energy and Protection Committee (STEP) maintain a number of standards related to powering and protection of telecommunications facilities.

CDG – The CDMA Development Group (http://www.cdg.org) is a non-profit body that developed and has continued to evolve CDMA technology.

ETSI – The European Telecommunications Standards Institute (http://www.etsi.org) is an independent, not-for-profit standardization organization of telecommunications organizations in Europe, with worldwide projection.

GSMA – The GSM Association (http://www.gsmworld.com) is a nonprofit body that focuses on GSM-based technology.

IEC – The International Electrotechnical Commission (http://www.iec.ch) is a not-for-profit, non-governmental international standards development organization that prepares and publishes international standards covering all electrical, electronic, and related technologies.

Telcordia – In the United States, Telcordia (now part of Ericsson) publishes Generic Requirements documents that are used by many service providers as the basis for the procurement of reliable and standardized equipment. Among Telcordia's documents, the New Equipment and Building Standards (NEBS) have wide applicability to many of the topics related to the communications network infrastructure, including wireless site installations. NEBS addresses the physical aspects of the equipment installation, including test specifications for shock, vibration, temperature, etc.

3GPP – The European-based 3rd Generation Partnership Project (http://www.3gpp.org) is an international body that publishes wireless standards for W-CDMA as the successor to GSM. 3GPP is exclusive to the wireless industry and addresses all aspects of wireless infrastructure such as base stations, Mobile Switching Centers, base station controllers, and handsets.

In addition, the following organizations are among those that publish standards relevant to the construction of communications facilities, particularly within specific countries, regions, or jurisdictions.

Association Française de Normalisation (AFNOR), http://www.afnor.org/portail.asp?Lang=English

ATIS Protection Engineers Group, http://www.atis.org/peg/

ATIS Sustainability in Telecom: Energy and Protection Committee, http://www.atis.org/step

British Standards Institution, http://bsi-global.com

Building Industry Consulting Service International, http://www/bicsi.org

Bureau of Indian Standards, http://www.bis.org.in/

CDMA Development Group, http://www.cdg.org

Federal Agency on Technical Regulating and Metrology (Russia),
    http://www.gost.ru/wps/portal/pages.en.Main

German Institute for Standardization (DIN), http://www.din.de/cmd?level=tpl-
    bereich&cmsareaid=47565&languageid=en

Japanese Standards Association, http://www.jsa.or.jp/default_english.asp

National Lightning Detection Network, http://thunderstorm.vaisala.com/

NEMA Surge Protection Institute, http://www.nemasurge.com/

Standards Administration of China, http://www.standardsportal.org/splash/default.aspx

Standards Australia, http://www.standards.org.au/

5.11 Resources

The design of a wireless site requires the specification of numerous site components in addition to the communications equipment. These include the power system (primary and backup), equipment required to provide heating, cooling, and ventilation for the site, the tower, and in some cases, even the building or hut in which the communications equipment will be housed. The specification of such a broad range of components involves knowledge of many disciplines, applicable standards, and local codes. For this reason, it is advisable to seek expert assistance from others with experience in these areas.

5.11.1 Locating Professional Assistance

Many of the professionals who can assist in the design of the site infrastructure (power, HVAC, on-site fabrication, etc.) are connected to the building construction industry. Architects and engineers often are already known by the organizations involved. They are generally familiar with specialists in areas associated with communications facilities.

Local or national engineering associations are also good sources of contacts. Websites can help locate the local chapter or section of a professional association. The practical experience of workers in the applicable construction trades (for example, electricians) can be helpful, and they often may be familiar with the reputation and past performance of local architects and design engineers. Involving local experts can also be helpful in understanding any unique aspects of local codes, regulations, site inspections, and the permitting and approval process.

Additionally, the following websites may be helpful in locating and obtaining relevant standards and other resources related to system powering and applicable in their respective jurisdictions.

China

http://blogs.zdnet.com/BTL/?p=3913

http://nbr.org/publications/specialreport/pdf/SR10.pdf

http://www.fuzing.com/qrx/CCN/941/power-standards

Europe

http://ieeexplore.ieee.org/xpl/RecentCon.jsp?punumber=4375

http://www.bksv.com/Applications/SoundPower/SoundPowerStandards.aspx

http://www.powercords.co.uk/standard.htm

Japan

http://ieeexplore.ieee.org/Xplore/login.jsp?url=/iel1/39/10527/00491918.pdf?tp=&isnumber=&arnumber=491918

http://www.iop.org/EJ/abstract/0026-1394/17/1/006

Other Countries

http://sciencelinks.jp/j-east/article/199914/000019991499A0308609.php

http://www.equitech.com/support/worldpwr.html

http://www.kropla.com/electric2.htm

5.12 References

[Car85] E.P. Carter, *Telecommunication Electrical Protection*, AT&T Technologies, 1985.

[Elb00] B. Elbert, *The satellite communication ground segment and earth station handbook*, Artec House Space Technology and Applications Library, 2000.

[Has00] P. Hasse, *Overvoltage Protection of Low Voltage Systems, (2nd edition)*, IEE Power and Energy Series, 2000.

[Ree07] W. Reeve, *DC Power System Design for Telecommunications* Wiley-IEEE Press, 2007.

[Tol08] M. Tolstrup, *Indoor Radio Planning – A Practical Guide for GSM, DCS, UMTS and HSPA*, John Wiley & Sons, 2008.

Specific Applicable Standards

Application Guide for Electrical Relays for AC Systems, Bureau of Indian Standards IS 3842.

Cabinets, Racks, Panels, and Associated Equipment, ANSI/EIA 310-D.

Communications Circuits, NFPA70, 2008 NEC, Article 800.

Customer-Owned Outside Plant (CO-OSP) Design Manual (Third Edition), Building Industry Consulting Service International (BICSI).

Electrical Equipment Designed for Use within Certain Voltage Limits, (the "Low Voltage Directive"), European Commission Directive 2006/95/EC, August 2007.

Electrical Installations of Buildings, Section 442: Protection of Low Voltage Installations against Faults between High-Voltage Systems and Earth, German Institute for Standardization (DIN) VDE 0100-442:1997-11.

Electrical Installations of Buildings, Part 4-43: Protection for Safety–Protection against Overcurrent, JSA Standard JIS C 60364, 2006.

Electrical Installations of Buildings, Section 444: Protection against Electromagnetic Interferences (EMI) in Installations of Buildings, Federal Agency on Technical Regulating and Metrology (Russia) GOST R50571.25-2001.

Electrical Installations, Standards Australia AS/NZ 3000, 2007 (known as the "Australia/New Zealand Wiring Rules").

Electrical Protection of Communications Towers and Associated Structures, BSR ATIS 0600334-200x (formerly ANSI T1.334-2002).

Electromagnetic Compatibility and Electrical Safety – Generic Criteria for Network Telecommunications Equipment, Telcordia GR-1089-CORE, Issue 4, 2006.

General criteria for the operation of bodies performing inspection, EN 45004.

General criteria for certification bodies operating certification, EN 45011.

Generic Requirements for Surge Protective Devices (SPDs) on AC Power Circuits, Telcordia TR-NWT-0001011, Issue 1, 1992.

Guide for the Application of Surge Protectors Used in Low-Voltage (Equal to or Less Than 1000 V,rms or 1200 V,dc) Data, Communications, and Signaling Circuits, IEEE C62.43.

IEEE Recommended Practice for Powering and Grounding Electronic Equipment, IEEE Standard 1100.

Low Voltage Surge Protective Devices – Part 21: Surge Protective Devices Connected to Telecommunications and Signaling Networks – Performance Requirements and Testing Methods, AFNOR Groupe France Standard CEI 61643-21:2000.

Protection of Low-Voltage Electrical Installations against Overvoltages of Atmospheric Discharges and Switching – Selection and Erection of Surge Protective Devices, AFNOR Groupe France Standard UTE C15-443, August 2004.

Protection against Lightning – Part 3: Physical damage to structures and life hazards, IEC62305-3.

Protection against Lightning – Part 4: Electrical and Electronic systems within structures, IEC62305-4.

Protection against Interference – Bonding Configurations and Earthing inside a Telecommunication Building, ITU-T Recommendation K.27.

Protectors for Data Communications and Fire-Alarm Circuits, UL497B, 4th edition, 2004.

Protectors for Paired-Conductor Communications Circuits, UL497, 7th edition, 2001.

Radio & Telecommunications Terminal Equipment (RTTE) Directive, European Commission Directive 1999/5/EC.

Recommended Practice for Electric Power Distribution for Industrial Plant, IEEE Standard 141.

Recommended Practice for Grounding of Industrial and Commercial Power Systems, IEEE Standard 142.

Requirements for Electrical Installations – IEE Wiring Regulations, 17th edition, British Standards Institution BS7671:2008.

Secondary Protectors for Communications Circuits, UL497A, 3rd edition, 2001.

Standard for the Installation of Lightning Protection Systems, NFPA780.

Structural Standards for Steel Antenna Towers and Antenna Supporting Structures, ANSI/TIA-222.

Surge Protective Devices, UL1449, Issue 3, 2006.

Surge Protective Devices (SPDs), 1 kV or Less, NFPA70, 2008 NEC, Article 285.

Tower Design: Minimum Design Loads for Buildings and Other Structures, SEI/ASCE Standard No. 7-05, Structural Engineering Institute.

Chapter 6

Agreements, Standards, Policies, and Regulations

6.1 Introduction

Agreements, Standards, Policies and Regulations (ASPR) are the necessary mechanisms of the wireless industry used to anticipate, improve, and control the behaviors of entities that design, develop, implement, operate, and evolve wireless communications networks [ATI07]. [1] ASPR, sometimes abbreviated as "Policy", is one of the eight fundamental ingredients that make up the wireless communications infrastructure [Rau06, CQR01, NRI05a]. Similar to the critical role of hardware in providing the physical medium for electronic activity, and the critical role of software in providing the logic, automation, and control that shape communications features and services, ASPR plays a critical role in coordinating a wireless network's many diverse elements and stakeholders [NRS02, NRI03, NST06]. Without this coordination, equipment and devices from different suppliers would fail to interoperate; interfaces between network operators would be incompatible; the inconsistency of subscribers' experiences would be intolerable; and the result would be far less useful networks.

ASPR also helps ensure that wireless networks do not disturb their environments or other interests. In addition, they provide a framework where fairness can be managed among competitors. However, the primary reason for ASPR is to promote the reliability and quality of the wireless user's experience and the interoperability of networks. As will be seen, most implementations of ASPR do not require government involvement. Today's commercial legal frameworks are sufficient to facilitate most of the needed coordination. In this chapter, we are going to address these subjects:

- Agreements
- Standards
- Policies
- Regulations

---

[1] ASPR: "A term used to refer to the complete set of inter-entity arrangements that are necessary for communications services; these arrangements include national and international standards, federal, state, and local regulations or other legal arrangements, or any other agreement between entities – including industry cooperation and agreements and other interfaces between entities. ASPR is one of the elements of the Eight Ingredient Framework; in this Framework, it is sometimes abbreviated as 'Policy'." The term is used in many places in the Framework, as well as in the President's National Security Telecommunications Advisory Committee (NSTAC) Next Generation Task Force's *Final Report*, March 2006.

6.2 Agreements

6.2.1 Background

An agreement is a collection of mutually accepted terms and conditions that define the expectations among two or more parties in the wireless communications industry. Agreements may address issues such as: a network operator setting expectations for product quality with its equipment suppliers; legal arrangements between competitors to cooperatively share network resources during a time of crisis or regular operation; and service agreements with consumers or secondary operators regarding such things as network access coverage, pricing, and the types of services provided. It is not necessary for agreements to have government involvement, and commercial legal frameworks may suffice. Because the agreements are within the control of the parties involved, they can be implemented relatively quickly as a means of controlling the parties' behaviors.

6.2.2 Importance of Agreements

Due to the nature of communications, interoperability is a must, even in competitive environments. For example, cellular operators A and B are competing in the market, but users of operator A can execute phone calls with users of operator B. Generally speaking, agreements set a foundation for cooperative execution. Although stakeholders involved in wireless networks have different interests, agreements provide for tradeoffs so that each party's interests can be protected in an acceptable and even legal way.

Consumers are the most numerous and valuable stakeholders in the industry. Today's wireless communication industry supports billions of users globally. Users' interests vary around the common points of price, features, and quality. Service agreements are usually implemented in service contracts, which typically have options. Service providers are the stakeholders closest to the consumer in the supply chain. Some service providers do not own network equipment, relying instead on network operators through legal agreements to sell (or more precisely, resell) services that are run on networks operated by the latter. Other service providers are also network operators, which is the next role further back in the supply chain. Network operators, who typically obtain operating licenses from one or more government agencies and who follow regulations, design, build, operate, and maintain networks. Typically, they have agreements to provide roaming services to other operators' customers. In many cases, governments may force multiple network operators to share their co-located facilities in common physical sites, under regulations and mutual agreements, for the purpose of minimizing resources, promoting public safety, ensuring interoperability, etc.

As mentioned above, network operators also have agreements with their equipment suppliers. Agreements can cover the delivery schedule of new features and technical capabilities, the quality and reliability of products, the behavior of individuals who will be on site during equipment installation or upgrade, and penalties for failure in any of these areas.

6.2.3 Examples of Agreements

The following examples provide insight into the nature of agreements.

Example 1. Service provider-consumer agreements. One of the most common types of agreements is between a consumer and a wireless service provider. A simplified generic example of such an agreement is shown in Table 6-1, which shows a trio of service options with their corresponding differences in pricing.

Example 2. Mutual aid agreements. Competitors can reach an agreement well in advance of a crisis to offer equipment or other resources when human life is at stake. A wireless network operator can greatly increase its network resilience by setting up a legal framework in which an entity that may be a competitor on any other day would, during a crisis, make available its resources for a pre-arranged price [ARE07].

	Option A	Option B	Option C
Coverage	Regional	National	International
Services	Voice	Voice, text	Voice, text, Internet access
Minutes per month	100	1000	Unlimited
Price	25	100	250

Table 6-1: Sample Service Agreement Options

Example 3. Network operator-equipment supplier product-quality management agreements. To ensure that a product's quality and reliability are priorities of their suppliers, network operators may establish quality-management agreements with their suppliers. Such supplier agreements define a set of quality measures and acceptable product-performance levels. Reliability measurements may include the frequency of outages and system downtime averaged over a year. Quality measurements may cover modulation anomalies such as cross-modulation, interfering harmonics, and quantization impact. Equipment selection depends on the needs of the network, and the criteria included in an agreement must ensure that the necessary quality and other requirements will be met.

A network operator-equipment supplier agreement begins with the network operator specifying the performance requirements for systems and services in the initial purchasing agreements. The same criteria should also be used in evaluating the response of potential suppliers to the quality and reliability requirements. In addition, penalties may be introduced for suppliers whose equipment fails to perform within the agreed range. Methods to assess the quality of equipment and services can take advantage of industry standards as described below in section 6.3 [Tel05, TL11, NRS02].

6.2.4 Summary of Agreements

Agreements are mutually accepted terms that define the expectations of two or more parties. The establishment of an agreement is a relatively straightforward means of controlling the behaviors of the entities involved. In the wireless industry, agreements do not necessarily require government involvement. The quality of the wireless experience depends strongly on effective agreements being worked out between network operators and their equipment suppliers. Fortunately, the quality and reliability of network equipment and services can be managed by industry standards developed for that purpose.

6.3 Standards

6.3.1 Background

To allow interoperability among different equipment/device vendors who are likely competitors in the market, some commonly agreed technical specifications are required, along with mutually agreed technical conformance testing. Standards are therefore a widely accepted means of setting expectations for specific technical behaviors or specifications. Other ways of looking at standards in the communications industry include:

- an emphasis on quality: "a set of rules for ensuring quality" [ETS11];

- an emphasis on process: "a document established by consensus and approved by a recognized body that provides for common and repeated use, rules, guidelines, or characteristics for activities or their results, aimed at the achievement of the optimum degree of order in a given context" [ISO96], or "documents, established by consensus and approved by an accredited standards development organization that provides, for common and repeated use, rules, guidelines or characteristics for activities or their results, aimed at the achievement of the optimum degree of order and consistency in a given context" [IEE08]; and

- an emphasis on the voluntary nature regarding following standards: "technical specification approved by a recognized standardization body for repeated or continuous application, with which compliance is not compulsory ..." [ETS11].

Standards enable different networks to interoperate and users in different networks to communicate. They allow a large array of wireless handset manufacturers to produce different devices that operate on the same wireless network and offer the same or similar services. Standards enable different networks to interoperate with the network databases to confirm subscriber billing information. In addition, the logic of the standards development process can add value by helping identify the many possible circumstances that will have to be planned for and dealt with.

Standards are typically applied in a global, regional, or national way, depending on the agreement of authority for such standards. Example 1 in section 6.3.3 discusses a few illustrations. The International Telecommunication Union (ITU), as a part of the United Nations, considers global coordination. The International Standard Organization (ISO) deals with global standards. The IEEE as a professional society develops industrial common specifications (such as the family of documents developed under P802) by itself or under authorization of the ISO. The European Telecommunication Standards Institute (ETSI) is an example of a regional organization, which authorizes the development of standards by the 3rd Generation Partnership Project (3GPP) and other similar bodies. ANSI is usually considered to develop standards specifically in the United States.

6.3.2 Importance of Standards

Without the effective development and implementation of standards, the quality and reliability of services would not be as optimal as they are today. Network elements would fail to interoperate and, at best, user experiences would be inconsistent. Also, having multiple competitors to choose from would be less advantageous to the consumer because of inefficiencies in all of the unique interfaces that would have to be managed.

6.3.3 Examples

Additional insight regarding standards can be seen through examples.

Example 1. Wireless standards bodies. A number of standards development organizations play a critical role in writing regional and international standards. These include the following (with examples of the wireless-related areas they address, along with several key standards listed in Table 6-2):

- Alliance for Telecommunications Industry Solutions (ATIS): services and systems, network reliability, interconnection with emergency services;

- European Telecommunications Standards Institute (ETSI): electromagnetic compatibility, emergency communications, reconfigurable radio systems;

- International Telecommunication Union (ITU): radio regulations, network standards, radio standards, interference standards, electromagnetic compatibility, emergency communications, security, management of radio frequency and satellite orbits;

- International Standards Organization (ISO): compatibility with medical devices, cabling for wireless access points, radio frequency identification, device testing, performance testing methods;

- Institute of Electrical and Electronics Engineers (IEEE): wireless local area network (WLAN), quality and reliability, compatibility with hearing aids;

- Internet Engineering Task Force (IETF): mobility for IP, control and provisioning for wireless access points, mobile ad-hoc networks;

- Telecommunications Industry Association (TIA): terrestrial mobile multimedia multicast, steel antenna towers, vehicular telemetry; and

- The 3rd Generation Partnership Project (3GPP): a collaborative agreement among a number of communications standards bodies; its scope includes 3rd generation mobile systems, enhanced data rates for GSM, and GPRS.

ANSI C63.19-2006	*Methods of Measurement of Compatibility between Wireless Communications Devices and Hearing Aids* - addresses electromagnetic compatibility between ANSI C63.19:2006 mobile phones and hearing aids.
ETSI EN 200 220-1	Electromagnetic compatibility and Radio spectrum Matters (ERM); also covers short-range devices (SRD); radio equipment for the 25 MHz-to-1000 MHz frequency range ETSI EN 300 220-1 with power levels up to 500 mW; Pan I: Technical characteristics and test methods.
IEEE 802.11	*Wireless Local Area Networks* - standards for wireless LANs that cover IEEE 802.11 modulation techniques; sometimes referred to as Wi-Fi.
ITU-T X.805	*Security architecture for systems providing end-to-end communications* - provides ITU-T X.S05 a common framework for systematically addressing security issues.
TIA-222-G-1	*Structural Standards Abstract for Steel Antenna Towers and Antenna Supporting Structures* - outlines requirements for structural design and fabrication of structural antennas, support structures, mounts, components, guy assemblies, insulators and foundations.

Table 6-2: Examples of Standards Applicable to Wireless Networks and Services

Another important organization is the Association of Public-Safety Communications Officials – International (APCO), which supports the wireless communication services needs of police, fire, EMS, and other public safety agencies worldwide. In addition, APCO plays a critical role in the communications standards used by different public safety agencies throughout the world. For example, the TETRA (Terrestrial Trunked Radio) standard in Europe and the P25 (Project 25) standard in the U.S. are digital wireless communication standards developed by the organization [APC11].

Example 2. Describing quality measurements. When an equipment supplier and its many customers have a quality or reliability issue, different words can be used to describe the same situation, or the same words can describe different situations. By setting a common standard, the industry introduces efficiency and intelligence to classifying complex technical situations. In turn, this results in more efficient cost structures and, ultimately, a healthy competitive market and lower costs to consumers subscribing to wireless services.

Standards help manage the network operator-equipment supplier relationship. Typically, they include definitions and units of measurement. For example, the network reliability performance of a system operating in a live network environment may have reliability measurements expressed in terms of outage event frequency normalized by the number of systems deployed (events/system-year), and also as an expression of downtime (minutes/system-year). Other quality measurements include the number of software defects, the percentage of defective software patches, and the failure rate of hardware circuit packs.

Example 3. Priority communications. Traffic capacity is one of the fundamental limitations of wireless networks. Limits exist for both the network elements and transport facilities, and for the electromagnetic energy air interface. Because disaster situations are typically accompanied by a greater need for wireless communications, traffic congestion can significantly impair vital emergency communications. To address this, governments have encouraged the development of standards for priority communications. These standards have been developed and implemented voluntarily by the private sector, with government funding as an incentive.

Example 4. Radio interfaces. Figure 6-1 illustrates the process used for a decade in developing specifications for the radio interfaces of International Mobile Telecommunications-2000 (IMT-2000). The organizations at the left of the figure are all international standards development organizations (SDOs). The membership of each includes network operators, service providers, equipment suppliers, and other stakeholders in the global mobile wireless communications industry. These SDOs produce radio access network (RAN) standards that are continually being updated. A relatively new member of this group of standards is the mobile WiMAX standard that was developed by the IEEE 802.16 working group. Although not illustrated in Figure 6-1, there is a parallel standardization activity in ITU-T, which develops the network standards for each of the RAN technologies.

Example 5. Industrial Common Specifications. Another widely used approach in the wireless communications industry has been to develop a common industrial specification designed to be open for general applications, typically relieving patent constraints. Bluetooth has been developed by the Bluetooth SIG (Special Interest Group) which originally had nine promoters, led by Ericsson, working to deliver a common technology for wireless connections, e.g., the connection between an ear-phone and a handset.

The Bluetooth SIG develops technical specifications and conformance testing standards for common use within the industry.

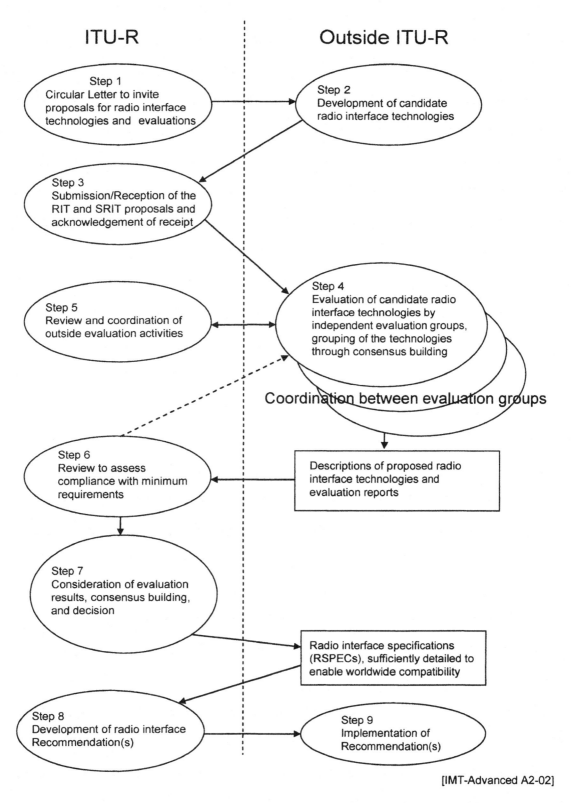

[IMT-Advanced A2-02]

Figure 6-1: Development of Mobile Communication Standards for IMT-2000 [ITU09]

To ensure international acceptance of these standards, they are brought into the ITU-R where they are incorporated into an ITU-R Recommendation. After the ITU-R verifies that the submitted standards meet requirements previously specified by the ITU-R, an ITU-R Recommendation is completed and submitted to the Radio Communication Assembly for final approval. The approved standards, which are often adopted by national standards and regulatory bodies, include the specific bands at which the mobile wireless communication devices can operate in their countries. This interaction between standards development and regulation will be discussed in more detail later. Clearly, spectrum must be made available for these systems. ITU-R working parties create study groups which produce recommendations, and the Radio Communication Assembly approves spectrum requirements. This is input into the regulatory process, also described later.

Last but not the least on the topic of standards, since it is hard to judge whether equipment or devices from different vendors following the same standard indeed operate as expected, conformance testing, usually accompanied by a certification process, has to be done by one or more reputable and independent organization(s). For example, IEEE 802.11 products have to obtain WiFi certification to ensure interoperability.

6.3.4 Summary

Standards are a widely accepted means of setting expectations for specific behaviors. Key aspects include their ability to promote quality, their development through consensus-based processes, and their voluntary implementation. Benefits of standards include increased efficiency, interoperability, and reliability, and consistency of the user experience. Compared to regulations, standards are quicker to develop and put into use. However, they take more time to develop than agreements and policies.

6.4 Policies

6.4.1 Background

Policies are the guiding principles or plans intended to influence decisions or actions. In the wireless industry, policies help stakeholders prepare for their own future decisions and actions. Policies address issues such as the expectations of the participants dealing with a multi-party outage; procedures for handling non-mandatory government requests for information; the use of environmentally friendly alternatives; corporate posture toward concerns regarding the long-term health effects of using personal wireless devices; and the emergency response to a catastrophe that has a widespread network impact. Policies can call for government involvement, but do not require it. This is because any entity can establish policies that other entities can expect will influence its behavior. Because the establishment of a policy is within the control of the entity that shapes it, policies are a very efficient means of providing guidance for how other entities should anticipate its behavior.

6.4.2 Importance and Interaction of Policies

Policies have the pivot role in the ASPR as Policy serves as an inclusive umbrella term that includes agreements, standards, and regulations. However, Policy has an important distinction: it also includes all the other mechanisms used to anticipate an entity's behavior that none of the other terms capture. For example, industry best practices are neither regulations nor standards, nor do they require an agreement among entities. Yet, industry-consensus best practices play a vital role in promoting the reliability, emergency preparedness, interoperability, and security of wireless networks. Another example is found in the areas of health, safety, and the environment. While some regions of the world have regulations to control behaviors in these areas, their approaches will fall short in some situations of being complete regarding what is appropriate. Many companies have policies that other entities, when interacting with them, can depend on, such as being able to expect that a company's facilities will be designed and operated with health and safety considerations in mind.

Getting into the new era of policies, at one level, Information and Communications Technology (ICT) seems best the province of engineers, geeks who program, techies who fiddle, and if they hit on a commercial success, more applause to them. They have made their mark and should reap the rewards. Often, at some point, however, governments intervene. There are many reasons why, but the most fundamental is that the distribution of information and ideas in a society is one of the foundations of political power. Once a techie phenomenon goes commercial and becomes popular, it has achieved two things – it has reached a lot of people and a lot of people love it. Communications services that have widespread reach and ideas that are popularly accepted attract the attention of politicians. In such circumstances, decisions made by businesses for largely business reasons often have major political effects and, therefore, it should not be surprising that there are significant political reactions [Wu08].

"Information is a kind of currency in politics; therefore, anything related to the production, distribution, and consumption of information is relevant to power politics. Looking back to the feudal history of Europe, Asia, and other regions, the source of wealth was land. On the control of land, power bases were built. With the coming of the industrial revolution, ownership of land was no longer the sole route to status, wealth, and power. Capital was another option. The essential change the industrial revolution wrought on the organization of society was to make important the distinction between those who own capital and those who do not. Those who own capital, profit. Those who do not must sell their labor. Today it is not just capital and land which are possible paths to wealth and power. Information is a third path, another basis of power that divides society into those who have it, control it, and understand it, and those who are at its mercy. Governments have long recognized that controlling information relates to controlling political power. Also, activists have long used technology to their advantage against governments." [Wu08]

There are many historical examples of how government has used communications and information technology for its own ends. Brazil deployed a military commission to build a telegraph network into the Amazon to assert its control over the new republic's far flung territory. Canada's telecommunications infrastructure came to symbolize its new nationhood, separate from the United Kingdom and distinct from the United States. Governments use television, especially public broadcasting, to create a sense of common national identity, such as Al-Jazeera for Qatar [Wu12].

As for those activists protesting government policy, again there is varied tradition. Elites in nineteenth century China used the telegraph to remonstrate against the Empress Dowager Cixi's treatment of reformers in the government. Texting on cell phones helped organize people power in the Philippines to overthrow President Estrada. Cable television in Taiwan carried opposition political news and information which contributed to the democratization of the government [Wu12].

Changes in communications technology and massive flows of new information result in significant political transformation if the people who use the technology re-imagine their own identities and if new institutions emerge in society to compete with the old ones. With access to new information and different views of the world, people can see themselves in altered light, with a new consciousness – sometimes cultural, sometimes political [Wu08].

Also, there is a connection between the commercialization of communications technology and the use of that technology as a political tool. Benedict Anderson in *Imagined Communities* identified the innovations in printing presses and the development of commercial newspapers as instrumental in constructing a national identity. Commercialization spreads technology very rapidly. In other words, every household may buy a television in order to watch soap operas, but once the television is an established service in everyday life, television programming can convey politically-relevant messages as well. This translates to the Internet today. Many people subscribe to stay in touch with their friends, to play video games, to catch up on the latest entertainment, etc. Once they are connected, that Internet service can be used for political purposes as well [Wu08].

Governments take into account all these fundamental issues when making policy decisions, and often goals are in tension with each other. Communications and information technology policy touch on objectives related to national security, economic regulation, international trade, cultural promotion, and/or ideological discourse [Wu08]. Each national and international government institution focuses on just some aspect of these objectives – for example, the World Trade Organization (WTO) focuses on trade, the International Telecommunications Union treats a variety of telecommunications issues, and the United Nations Economic, Scientific, and Cultural Organization (UNESCO) highlights the importance of culture. Understanding that information is a source of political power and that ideas are one of the strongest ties that bind communities and nations together explains many of the challenges and conflicts which arise in this arena. All new ideas for the benefit of human beings can be reflected in policies.

6.4.3 Examples
Additional insights into policies can be gained through examples.

Example 1. Industry-consensus best practices. The wireless industry has developed hundreds of consensus best practices. Companies that support such development are demonstrating a policy of expert information sharing with their industry peers on critical aspects of wireless networks in order to promote their reliability, emergency preparedness, and security. Because the implementation of a given set of best practices is not applicable in all situations, their implementation is voluntary. They are not applicable in all situations because of such things as differences in some emerging technologies, the use of an alternative or otherwise unique configuration, and the need for site-specific rules for special considerations. Industry-consensus wireless-network best practices can be accessed on the Internet at:

> http://www.bell-labs.com/USA/NRICbestpractices/   and
>
> http://www.bell-labs.com/EUROPE/bestpractices/.

Table 6-3 provides several examples of industry best practices.

Example 2. Crisis support. Many consumers buy a wireless device and service to enhance their personal security. Realizing how important mobile phones are, some wireless service providers have set policies that continue to allow people to depend on them during a crisis.

For example, service providers will set up phone banks for the victims of a catastrophe so they can make free phone calls. Alternatively, providers will allow subscribers an automatic two-to three-month delay in

7-7-0454	Network Operators and Service Providers should consider establishing technical and managerial escalation policies and procedures based on the service impact, restoration progress, and duration of the issue.
7-7-0495	Network Operators and Property Managers should consider pre-arranging contact information and access to restoral information with local power companies.
7-7-0578	Network Operators, Service Providers and Public Safety Agencies should actively engage in public education efforts aimed at informing the public of the capabilities and proper use of 911.
7-7-0472	Network Operators and Equipment Suppliers should consider connector choices and color coding to prevent inappropriate combinations of RF cables.

Table 6-3: Sample Best Practices for Wireless Networks

paying their bills due to the hardship. In still another example, some wireless service providers will provide public safety workers – and even insurance adjusters – with free mobile phones to support their critical work.

Perhaps one of the most impressive examples of cooperative response during a major catastrophe is when the industry comes together to support volunteers from among its ranks in the use of advanced wireless technology to enhance traditional search-and-rescue efforts. In such circumstances, a wireless device may be a victim's only hope for survival. The Wireless Emergency Response Team (WERT) headquartered in the Lehigh Valley, Pennsylvania, USA, is a non-profit organization that coordinates such volunteers.

WERT's vision is to "connect the best minds and resources of the wireless industry to the most vital needs of its subscribers in an emergency" [WER11]. With the combined resources of expert volunteers and portable network equipment, WERT has been able to simulate wireless networks that have been destroyed or greatly impaired. In addition, network operations experts have been able to analyze wireless emergency traffic during crises to identify emergency calls that may be receiving insufficient attention.

Example 3. Emergency back-up power. Communications are vital during a crisis. One of the most common crises encountered around the world is the loss of commercial electric power. Personal mobile devices depend on a small battery that can typically last up to several days or provide several hours of talk time. However, during a power outage, the mobile device can only make a call or send a text message if a nearby cell site and its backhaul network have emergency backup power. Recognizing the vital importance of communications during a crisis, most network operators typically deploy emergency backup power capabilities at cell sites. Typically, these capabilities consist of batteries or generators, or a combination of the two (see section 5.2). Some network operators also reach out to the local the commercial electricity provider to improve their coordination during the restoration activities after a disaster.

It has been noted that the minimum backup power appropriate for a given cell site depends on the circumstances of that particular site [CQR04].[2] Underscoring this, industry-consensus expert guidance

---

[2] 58% of Subject Matter Experts Agreed That the Minimum Number of Hours of Backup Power Depends on a Number of Factors (from Survey Questions, slide 5).

was provided to the U.S. Federal Communications Commission (FCC) and the private sector as part of an official U.S. Federal Advisory Committee Act (FACA), as follows:

> *Network operators should provide backup power (e.g., some combination of batteries, generator, and fuel cells) at cell sites and remote equipment locations, consistent with the site-specific constraints, criticality of the site, the expected load, and reliability of primary power [NRI05b].*

An advantage of voluntary policies is that they allow limited resources to be invested in technologies that can optimize the benefits provided to end users. For example, while a strategy to have a minimum number of x hours of backup battery power at all cell sites is simple to understand, it does not address needs beyond x hours. On the other hand, some wireless network operators have found that, at times, they can better meet the needs of their subscribers in a crisis by providing virtually unlimited generator-based power at a small number of interspersed sites. At these sites the reception gain and transmission power are temporarily increased to extend into adjacent geographic areas normally covered by other cell sites, which, because of the crisis, may have no power. This example highlights the value of policies whereby experts have developed solutions that optimize the benefits to consumers.

6.4.4 Summary

In the wireless industry, policies help stakeholders prepare for their own future decisions and actions. Policies do not require government involvement, as entities can themselves establish policies that other entities can expect will influence their behavior. However, there are government policies that impact wireless communications. Because establishing a policy is within an organization's own control, setting policies is an expedient way to provide guidance for how other entities should anticipate its behavior.

One of the most popular forms of a wireless industry policy is voluntary, industry-consensus best practices. Distinct from regulations and standards, best practices play a vital role in promoting the reliability, emergency preparedness, interoperability, and security of wireless networks. However, best practices are not applicable in all situations for several reasons, including differences in emerging technologies, the presence of alternative or otherwise unique configurations, and the need for site-specific rules for special considerations.

While some regions of the world have regulations governing behavior regarding health and safety, such approaches will fall short of what is appropriate in all situations. Accordingly, many companies have policies that other entities interacting with them can depend on, such as being able to expect that a company's facilities will be designed and operated with appropriate health and safety considerations. With wireless playing a key role in ICT and thus human life, the evolution of policies to meet human needs is critical for the industry and for human beings.

6.5   Regulations

6.5.1 Background

Regulations are rules established by government authorities; they are mandatory, although with some portion often optional. When a regulation is not obeyed, the violator can face negative consequences, such as having to submit to a formal hearing, paying a fine, or having its license suspended or revoked. Regulations are necessary to ensure coordination among entities in some circumstances.

Regulations are characteristically slow to develop relative to the pace of the wireless industry's growth and introduction of new technology. This is due in part to the inclusion of checks and balances appropriate to the development of public policy. Once established, regulations are also hard to change. Another characteristic is that regulatory policy struggles with areas of rapidly developing technology. This is due in part to the level of expertise and experience required to effectively deal with such areas; most regulators are primarily trained as lawyers, economists, or in other disciplines. Seldom are they engineers with technical expertise in the emerging technologies they are regulating. It is also difficult to develop regulations that can predict the impact of new technologies.

Finally, regulations use the force of the law to control behavior. This is constraining and undesirable from the perspective of wireless industry stakeholders, who for the most part are continually upgrading the capabilities they offer. Because of these inherent attributes of the regulatory process, regulations are often seen as the last resort for controlling the behavior of industry entities. Regulatory agencies sometimes assign unlicensed bands such as those in the 2.4 GHz as well as 5 GHz (WiMAX) bands. Due to basic guidelines but lack of regulation, these bands can become so polluted that in many metropolitan areas the noise floor increases sharply, making it difficult for many designers to use these bands effectively. [Big04, Boc07, Che08, Woo05]

The wireless industry typically must submit to several levels of government authority, such as federal and regional authorities and the European Union, its member states, and their local municipalities.

Because of the potential impact that regulations can have on their interests, private sector companies have organized themselves to work together to ensure their points of view are heard during the development of public policy. CTIA, the international association for the wireless telecommunications industry, is an example of such an organization [CTI08]. An example of an industry association that focuses on infrastructure (towers, cell sites, etc.) is PCIA, the wireless infrastructure association [PCI11].

6.5.2 Importance of Regulations

Regulations are needed when other means of coordinating behavior among entities are ineffective. Reasons can include gridlock caused by competitive forces, a lack of incentives to act, or the need for a framework that establishes a level competitive playing field. Another reason for relying on regulations is if the risks of unacceptable behavior can adversely affect public safety or other important social or nation-state security concerns.

Regulations are particularly critical in wireless communications and should be developed and enforced by an independent government regulator standing in a fair position to domestic/international industry, a nation's economic/service policies, and in accordance with the spirit of the WTO.

In the past two decades, based on the 1998 WTO Agreement on Basic Telecommunication Services, telecommunication deregulation has happened in many developed and developing economies, which introduced regulatory reform and changes in the telecommunication industry. The WTO Reference Paper [WTO96] addresses six areas of telecommunication deregulation:

- Competitive safeguards,
- Interconnection,
- Universal service,

- Public availability of license criteria,

- Independent regulators, and

- Allocation and use of scarce resources.

The WTO Agreement induced great impacts on telecommunication regulation and on the industry. For example, in Japan, telecommunication regulatory reform took place and the re-organization of NTT and thus the appearance of new cellular operators followed [Sud05]. In China, the Ministry of Industry and Information Technology (the primary telecommunication regulator), together with the Ministry of Finance and the National Development and Reform Commission, implemented the reform of major state-owned telecommunication operators, for the purpose of better competition and the advancement of technology [FuM10]. Telecommunication reform also offers a good opportunity to promote a new wireless communication technology known as TD-SCDMA (see section 1.2.4.4), one version of TDD technology in 3GPP. In India, the Telecom Regulatory Authority of India (TRAI) initiated a series of efforts to better use spectrum and to introduce new operators and thus modern cellular services into India, which has one of the fastest growing markets in the world [Pra09].

6.5.3 Examples
Additional insights regarding regulations can be seen through the following examples.

Example 1. Wireless frequency spectrum management. The wireless transmission portion of a call path relies on invisible electromagnetic energy within a set frequency range. Because the frequency spectrum is limited, managing its use is of great concern. Misuse can cause interference with other wireless carriers and electronics. Because of such unacceptable consequences, government authorities have stepped in and regulated aspects of the spectrum. For example they have:

- Issued licenses to users of specific frequency bands [FCC11a];

- Addressed cross-border situations, where the transmission and reception of the RF signal crosses into another country; and

- Prohibited frequency jammers.

Figure 6-2 shows the frequency allocations for the entire radio spectrum managed in the United States by the National Telecommunications and Information Administration (NTIA). The figure is admittedly very difficult to read, but that is precisely the reason for including it here: to indicate how complex the management of frequency spectrum is [FCC11b]. Other countries have similar databases and tools.

One familiar example of the negative impact of not managing the spectrum correctly is the interference heard on GSM telephones, speakers, car radios, home computer speakers, etc.

Figure 6-3 depicts an example of an international and national regulatory process. Specifically, it is for mobile wireless telecommunication services and is a continuation of Example 4, Radio Interfaces, described earlier in section 6.3.3. One should keep in mind that these processes are both iterative and interactive. Approved ITU-R Recommendations on spectrum requirements are part of the input to the World Radio Communication Conference (WRC), including the Conference Preparatory Meeting. A primary product from the WRC is the Radio Regulations document, which is a treaty-level document since the ITU is chartered by the United Nations.

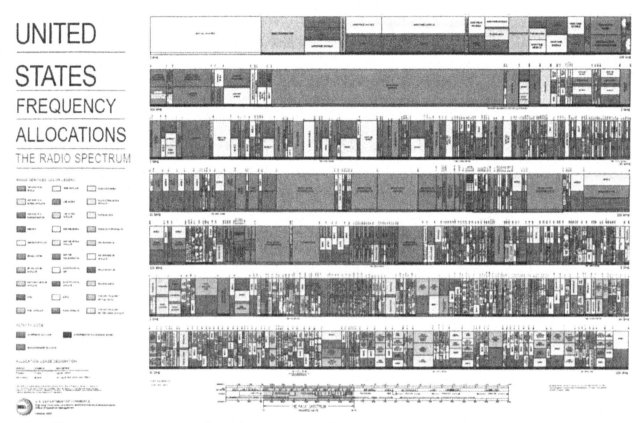

Figure 6-2: Frequency Allocations of the Radio Spectrum in the United States [NTI11]

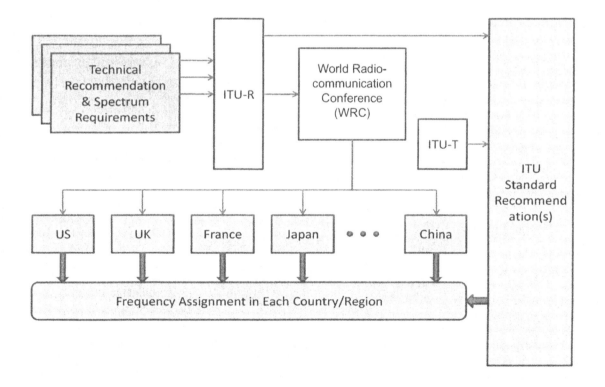

Figure 6-3: Block Diagram of the Process for Developing International and National Mobile Wireless Telecommunication Regulations

One of the most critical parts of the Radio Regulations is Article 5 of Volume I. This is the Table of Frequency Allocations, which allocates specific bands to specific services globally. New requirements for land mobile services, for example, are discussed during the WRC and changes are made to the Table of Frequency Allocations by international agreement. Because of the competing demands for spectrum, this is a very difficult process.

The ITU-R Radio Regulations provide an input to the regional and national regulatory bodies that make decisions on frequency assignments within the allocated bands. This process is indicated in Figure 6-3. Usually, the regulator in a nation determines the way(s) to release spectrum to wireless operators, typically by auction or "beauty contest", based on a wide range of considerations for the nation.

European Commission EC No 717/2007	Roaming on public mobile telephone networks: limiting roaming charges to consumers using mobile devices outside their home region
US FCC 47 CFR 4	Disruption to communications: requesting wireless service providers to report outages to FCC by meeting specific criteria
Denmark National IT and Telecom Agency Act no. 633	Joint utilization of masts for radio communication purposes: requiring owners of masts for radio communications to allow their parties to jointly use their masts
ITU-R Regulations	Frequency allocations: providing inputs to the regional and national regulatory bodies who make decisions on frequency assignments within the allocated bands

Table 6-4: Examples of Regulations Applicable to Wireless Networks and Services

Example 2. Promoting investment in technology. One of the benefits of regulating the spectrum appreciated by the commercial sector is the certainty it brings. When it is known that certain frequencies will be allocated to wireless services for a long time, companies can move forward with confidence that the enormous investment needed to develop equipment and build networks has a chance of yielding a financial return.

Similarly, in the U.S. the FCC has taken steps to "remove regulatory barriers and facilitate the development of secondary markets in spectrum usage rights among the Wireless Radio Services" [FCC03, Peh05]. The objective of these regulatory changes is to facilitate broader access to spectrum resources using spectrum-leasing arrangements.

6.5.4 Summary

Regulations are needed when the risk of an entity behaving unacceptably can lead to unacceptable consequences for public safety or other important social or nation-state security concerns that continually evolve. The wireless industry typically must submit to several levels of government authority.

Relative to the pace of the wireless industry's growth and introduction of new technology, regulations are characteristically slow to develop. Because of the inherent attributes of the regulatory process, regulations are often seen as the last resort for controlling the behavior of entities.

Violators of a regulation can face negative consequences. They could be required, for example, to submit to a formal hearing, pay a fine, or have their license suspended or revoked.

6.6 References

[APC11] Association of Public-Safety Communications Officials - International, http://www.apco911.org

[ARE07] *Availability and Robustness of Electronic Communications Infrastructures – The 'ARECI' Study – Final Report,* Formal Mutual Aid Agreements, Recommendation 2, 2007.

[ATI07] Alliance for Telecommunications Industry Solutions, *Telecom Glossary,* 2007.

[Big04] M. Biggs, A. Henley, T. Clarkson, *Occupancy Analysis of the 2.4 GHz ISM Band,* IEE Proceedings on Communications, vol. 151, no. 5, 2004.

[Boc07] E. Boch, *Wireless Metro WiMAX Backhaul: Licensed vs. Unlicensed,* Converge Network Digest, http://convergedigest.com/bp-bbw/bpl.asp?ID=391.

[Che08] K. C. Chen and J.R.B. deMarca, *Mobile WiMAX,* John Wiley & Sons, 2008.

[CQR04] IEEE Communications Society Technical Committee on Communications Quality & Reliability (CQR), *Proceedings of the Emergency Power Conference,* 2004.

[CQR01] IEEE Communications Society Technical Committee on Communications Quality & Reliability (CQR), *Proceedings of the 2001 CQR International Workshop,* 2001.

[CTI11] CTIA, The Wireless Association®, http://www.ctia.org.

[ETS11] European Telecommunications Standards Institute, http://www.etsi.org.

[FCC03] Federal Communications Commission (FCC), Report and Order and Further Notice of Proposed Rulemaking, *Promoting Efficient Use of Spectrum Through Elimination of Barriers to the Development of Secondary Markets,* 2003. http://wireless.fcc.gov/licensing/index.htm?job=secondary_markets.

[FCC11a] Federal Communications Commission (FCC) http://www.fcc.gov/spectrum.

[FCC11b] The FCC has tools for managing this information; see http://www.fcc.gov/oet/info/database.

[FuM10] H. Fu and Y. Mou, *An assessment of the 2008 telecommunication restructuring in China,* Telecommunications Policy, vol. 34, no. 10, 2010.

[IEE08] Institute of Electrical and Electronics Engineers (IEEE), *The Authoritative Dictionary of IEEE Standards Terms,* 2008.

[ISO96] ISO/IEC Guide 2:1996, definition 3.2.

[ITU09] International Telecommunications Union, http://itu.int/oth/R0A06000012/en.

[NRI03] FCC Network Reliability and Interoperability Council, NRIC VI, Focus Group 1, Subcommittee 1A, *Homeland Security - Physical Security - Final Report,* 2003.

[NRI05a] FCC Network Reliability and Interoperability Council NRIC VII, Focus Group 3B, Public Data Network Reliability, *Final Report,* 2005.

[NRI05b] FCC Network Reliability and Interoperability Council NRIC VII, Focus Group 3A, Wireless Network Reliability, *Final Report,* Best Practice 7-7-0492, 2005.

[NRS02] Alliance for Telecommunications Industry Solutions Network Reliability Steering Committee, *Annual Report*, 2002.

[NST06] President's National Security Telecommunications Advisory Committee, *Next Generation Networks Task Force Report*, 2006.

[NTI11] National Telecommunications and Information Administration, *United States Frequency Allocations – The Radio Spectrum*, http://www.ntia.doc.gov/files/ntia/publications/2003-allochrt.pdf.

[PCI11] Personal Communications Industry Association, http://www.pcia.com.

[Peh05] J.M. Peha, *Approaches to Spectrum Sharing*, IEEE Communications Magazine, vol. 43, no. 2, 2005.

[Pra09] R. Prasad and V. Sridhar, *Allocative efficiency of the mobile industry in India and its implications for spectrum policy*, Telecommunications Policy, vol. 33, no. 9, 2009.

[Rau06] K.F. Rauscher, R.E. Krock, & J.P. Runyon, *Eight Ingredients of Communications Infrastructure: A Systematic and Comprehensive Framework for Enhancing Network Reliability and Security*, Bell Labs Technical Journal, Vol. 11, No. 3, Wiley-InterScience, 2006.

[Sud05] Y. Suda, *Japan's Telecommunications Policy: Issues in Regulatory Reform for Interconnection*, Asian Survey, vol. 45, no. 2, 2005.

[Tel05] *GR 1929-CORE, Reliability and Quality Measurements for Telecommunications Systems (RQMS-Wireless)*, Telcordia Technologies, 2005.

[TL11] QuEST Forum, *TL 9000 Quality Management System, http://*www.tl9000.org.

[WER11] Wireless Emergency Response Team, http://www.wert-help.org.

[Woo05] R.W. Wooding and M. Gerrior, *Avoiding Interference in the 2.4GHz ISM Band*, Microwave Engineering Online, February 2005.

[WTO96] World Trade Organization Negotiating group on basic telecommunications, *Reference Paper*, http://www.wto.org/english/tratop_e/serv_e/telecom_e/tel23_e.htm.

[Wu08] I.S. Wu, *Information, Identity, and Institutions*, Georgetown University, Institute for the Study of Diplomacy, 2008, http://www12.georgetown.edu/sfs/isd/Wu_Information_Identity_and_Institutions.pdf.

[Wu12] I.S. Wu, *Network communities from the telegraph to the Internet: Using information as capital and ammunition*, book currently in preparation.

Chapter 7

Fundamental Knowledge

7.1 Introduction

Chapter 7 provides an overview of the fundamental knowledge deemed necessary for a practicing wireless communications engineer. The chapter content is intentionally made vast because wireless engineering is a demanding field of study which requires broad and interdisciplinary knowledge. This chapter is like a university curriculum, not a text book but instead a guide to study.

The topics covered in this chapter have different levels of importance. To help the reader find his/her way into this varying depth level, these are indicated with the keywords *low, medium,* and *high.*

A list of suggested reference books is included for in-depth study; the topic(s) for which each book is most relevant and useful are identified throughout the chapter. The reader can find similar information from many books available in the market. Therefore, this list is neither exhaustive nor unique.

Furthermore, supporting examples are included in this Chapter in this edition to help the reader understand the level of difficulty of the topic.

7.2 Electrical and RF Engineering

A wireless engineer should have excellent fundamental knowledge of electronic circuits and systems since electronics is the backbone of all wireless devices and systems. The size and cost of electronic devices keep decreasing while their performance keeps increasing.

Basically an engineer needs to understand the difference between low frequency, radio, and higher (microwave) frequency circuit operations as well as their interactions.

7.2.1 Electronics and Circuits

7.2.1.1 AC/DC circuit analysis

Analog and digital circuits at high or low frequencies form the backbone of the hardware used for wireless communications, either in the network or on the customer's premises. There are many basic topics to master.

Linear circuits: Voltage and current definitions and relationships, resistors (R), capacitors (C), inductors (L), voltage and current sources, and Ohms law. Specifically: analysis of RL, RC, and RLC circuits, natural and step response, series and parallel RLC circuits, resonant frequency and Q factor, higher order RLC circuits, the sinusoidal source, sinusoidal response, phasor analysis, frequency response, filters and frequency-selective circuits.

Kirchhoff's current and voltage laws: Node-voltage and mesh-current circuit analysis, Thevenin and Norton equivalent circuit representations, theory of superposition for linear circuit analysis, equivalence and reciprocity theorems.

Electric power and energy: Instantaneous, average, complex, real and reactive powers, root mean square (rms) value, power factor, maximum power transfer conditions.

Magnetically coupled circuits and transformers: Electric and magnetic fields, flux density, self and mutual inductance, current and voltage transformers.

Operational amplifiers (OPAMPs): Characteristics, open/closed loop gain, inverting/non-inverting amplifiers, OPAMPs as circuit-building blocks, unity-gain buffers, slew rate, and other specifications.

Boolean algebra: Fundamental theorems of Boolean algebra, Boolean functions (truth tables and canonical forms, representation forms, main properties).

Logic circuits: Discrete and integrated logic gates such as AND, OR, NAND, NOR and their interconnections, design of simple logic systems using transistors and gates.

In addition to the hardware fundamentals outlined above, most modern electronics depend on software. The hardware may include one or more microprocessors, field programmable gate arrays (FPGAs), analog or digital signal processing chips or application-specific integrated circuits (ASICs), peripherals, and input/output circuitry. Programmed with software, the microprocessors and FPGA chips are interfaced to other hardware. The wireless communication engineer may need to work with such electronic components, sometimes to develop and test new algorithms or simply to troubleshoot them.

The level of knowledge for this topic is <u>low</u>. Suggested references for detailed knowledge are [Lev79], [Mil87], and [Tho06].

7.2.1.2 Components of RF circuitry

The wireless engineer should have a basic knowledge of the components used in radio frequency circuits. Important RF circuit-design concepts include building blocks in RF systems, RF filter design, matching networks, active and passive device characteristics, and modeling.

Lumped elements: RLC circuit elements, design of lumped elements, lumped element modeling, fabrication, and applications.

Inductors: Basic definitions, inductor models, coupling between inductors, electrical representations, printed inductors on substrates, wire wound and bond wire inductors.

Capacitors: Capacitor parameters, chip capacitors, parallel-plate capacitors, voltage and current ratings, monolithic capacitors.

Resistors: Types of RF resistors, high-frequency resistor models.

RF transformers: Basic theory, wire-wrapped transformers, transmission-line transformers, ferrite transformers, parallel-conductor-winding transformers on silicon substrates, spiral transformers on GaAs (gallium arsenide) substrates.

RF building blocks: RF transceivers, impedance matching and impedance-matching circuits, low-noise amplifiers, mixers, filters, oscillators and power amplifiers, nonreciprocal components.

RF circuit design: Combining RF building blocks into functional communications circuits.

Parameters of multiport RF and microwave circuits: S-parameters, Z- and Y-parameters, ABCD parameters, conversion formulas.

Active RF components and modeling: RF diodes and transistors, modeling, measurement and characterization of active devices; the substrate effect, power loss in substrates, cutoff frequency (at which current gain is unity), $1/f$ noise and noise figure (NF), frequency synthesizers.

RF and microwave resonators: Lumped RLC, transmission-line and dielectric resonators, cavities, resonator coupling and excitation.

RF filters: Design of lumped-element filters (Butterworth, Chebyshev, elliptic, and Gaussian), filter transformations (low-pass, high-pass, band-pass, band-stop), implementation of RF and microwave filters, low-pass filters (stepped-impedance, stub), band-pass filters (coupled-line, coupled-resonator), resonator and filter configurations, LC resonators, special filters, filter implementation, coupled filters.

Oscillators: LC oscillator topologies, voltage-controlled oscillators, phase noise in oscillators, bipolar and CMOS LC oscillators.

Microwave circuit fabrication technologies: Microwave printed circuits, hybrid microwave integrated circuits, monolithic integrated circuits, device technologies, CMOS fabrication, micromachining.

Excitation and coupling in RF and microwave circuits: Aperture coupling, holes in waveguides, microstrip gaps and slots, ground-plane slots, proximity coupling, coupling via current loops, distributed coupling, hybrids, directional couplers, coupled-line filters.

The level of knowledge for this topic is <u>high</u>. A suggested reference for detailed knowledge is [Whi04].

7.2.1.3 Electronic design in practice

Basic printed circuit board design considerations: A printed circuit board, or PCB, mechanically supports and electrically connects electronic components using conductive pathways, or traces, etched from copper sheets laminated onto a nonconductive substrate. The PCB plays an important role in miniaturizing portable devices. The wireless engineer should know how a bare PCB is prepared, including its artwork (the printed circuit design), printed circuit assembly, the soldering process, and testing and quality control of the final product. Some of the key areas of PCBs are presented in Table 7-1.

PCB Types	Laminate Materials	Integrated Circuit (IC) Packaging Techniques
Single-sided/ double-sided Surface mount or through hole Single-layer or multi-layer Flexible or rigid	FR-4 (the most common PCB material), FR-2, polyimide, GETEK, BT-Epoxy, cyanate ester, Pyralux for flexible printed circuits, PTFE (polytetra-fluoroethylene), Rogers Bendflex, conductive ink	Dual in-line package (DIP), pin grid array (PGA), leadless chip carrier (LCC), surface mount with either gull-wing or j-leads, small-outline integrated circuit (SOIC), plastic leaded/leadless chip carrier (PLCC), plastic quad flat pack (PQFP), thin small-outline package (TSOP), land grid array, flip chip, ball grid array (BGA), multi-chip module, System in Package (SiP), hybrid integrated circuit (HIC), chip-on-board (COB)

Table 7-1: Key Topics in PCB Manufacture

Electromagnetic interference/electromagnetic compatibility: EMI and EMC issues are very important in preparing a PCB layout at microwave frequencies. Topics include the transmission line characteristics of traces, EMI characteristics of strip and micro-strip lines, RF loop current issues, EMI filtering using inductors and bypass capacitors, FCC Part 15 (covering intentional and unintentional EM transmitters and receivers), other regulations.

Grounding and electrostatics: Grounding and isolation of analog/digital and mixed grounds, providing ground planes, potentially harmful electrostatic discharge (ESD) issues, shielding against ESD and surges.

Standards bodies: Association of Connecting Electronics Industries, known as IPC; JEDEC Solid State Technology Association (formerly Joint Electron Device Engineering Council).

Specifications: Military Specification for Printed Wiring Board; Flexible or Rigid series (MIL-PRF-50884E, MIL-PRF-31032, MIL-PRF-55110) and Underwriters Laboratory Specifications for Printed Wiring Board (UL 796).

The level of knowledge for this topic is <u>low</u>. A suggested reference for detailed knowledge is [Kha05].

7.2.1.4 Power electronics

Amplifier circuits: The PN junction, BJT and FET, basic transistor circuits, biasing, current- and voltage-mode amplifiers, positive and negative feedback, gain-bandwidth product, input/output impedance, frequency response, 3-dB bandwidth, current and voltage gain, cascading amplifiers.

Power amplifiers: Characteristics of power amplifiers, power amplifier classes, high-efficiency power amplifiers, large-signal impedance matching, linearization techniques.

Power dividers and directional couplers: Properties of three- and four-port circuits, t-junction power and Wilkinson power dividers, waveguide directional couplers, quadrature hybrids, coupled-line directional couplers, Lange couplers.

Microwave amplifiers: Microwave transistors (bipolar & FET), gain and stability, noise linearity, input and output characteristics, noise floor, broadband amplifiers, multistage amplifiers, SFDR (spurious free dynamic range), BDR (blocking dynamic range), MDS (minimum detectable signal level), 1dB-compression point (input level at which the small-signal gain has dropped by 1 dB), IMD3 (third-order inter modulation), IIP3 (input-referred third-order intercept point), OIP3 (output-referred third-order intercept point).

The level of knowledge for this topic is <u>low</u>. Suggested references for detailed knowledge are [Mil87] and [Whi04].

7.2.1.5 Basic power supply design

A power supply refers to a system that supplies electrical energy to an output load or group of loads at certain specifications (output voltage, current, ripple, etc.). Mostly such supplies convert one form of electrical power to another form and voltage level. For electronic devices this typically involves converting 120- or 240-volt AC supplied by an electric utility company to a well-regulated lower-voltage DC. (See section 5.2 for a more detailed discussion.)

Topics to be aware of include power supply definitions, power adaptors, step-down transformers, rectifiers and inverters, linear regulators, switched-mode power supplies, voltage stability, ripple, surge protection, current/voltage regulation, continuous/pulse-mode supplies, uninterruptible power supplies (UPS), short-circuit protection, overpower (overload) protection, over-voltage protection.

The level of knowledge for this topic is <u>low</u>. Suggested references for detailed knowledge are [Mil87], [Ree07], and [Whi04].

7.2.2 Antennas and EM Wave Propagation

This is traditionally one of the main areas of knowledge required for a wireless engineer, which is why it is covered extensively in Chapter 4 of this book; the reader is advised to refer there for detailed information. We try here to provide a brief outline of the topic.

7.2.2.1 Transmission lines, antennas, electromagnetic waves, and applications

Electromagnetic waves: Dominant frequency bands, Maxwell's equations, electromagnetic wave equations, scalar and vector potentials, propagation properties, power density, polarization (horizontal and vertical).

Types of physical noise: Thermal noise (Johnson and Nyquist noise), shot noise, pink noise (flicker noise) or 1/f noise, Brownian noise, burst noise, phase noise, click noise, noise temperature, noise figure.

Radio wave propagation: Wave basics, the Friis formula, free space path loss, reflection, refraction, and diffraction; very-low-frequency (VLF), low-frequency (LF), medium-frequency (MF), and high frequency (HF) propagation; VHF and UHF propagation, microwave propagation; Snell's law, reflection and transmission coefficients, critical angle, Fresnel's equations, normal and vertical polarization, oblique incidence, Brewster's angle, reflection from and transmission through planar slabs.

Transmission lines: Transmission line components and equations.

Concepts and parameters of antennas: This topic is covered in Chapter 4.

Antenna theory: RF coupling, radiation, and related antenna theory concepts.

The level of knowledge for this topic is <u>high</u>. A suggested reference for detailed knowledge in wireless radio wave propagation is [Siz10] and for transmission lines is [Mag00].

7.2.2.2 EMI, EMC, and interference

Introduction to EMI and EMC: Sources of EMI, feedback mechanisms, protection schemes, techniques for EMI analysis and modeling, interconnections and wiring for EMI reduction, ground design, EMI filters, digital-circuit, common-mode and differential-mode emissions, common-mode filters.

Electromagnetic shielding: Basic theory, screening techniques, shielding materials and geometric structures, shielding for EMI protection, cable shielding and termination.

Intrinsic noise in circuit components: Noise generated by devices and active components, noise in digital circuits, noise suppression techniques and transient suppression.

Radiation in digital and microwave circuits: Electric and magnetic field coupling, transmission line effects and terminations, radiation by gaps, holes and loops in printed circuit boards, practical techniques to reduce coupling and radiation in electronic systems.

Grounding: EMC design of ground planes, ground grids, ground inductance, substrate coupling, closing current loops, AC power grounds and power-supply isolation, ground loops, equipment and enclosure grounding, mixed-signal printed circuit boards, return current paths, PCB partitioning and split ground planes, bridges.

Electrostatic discharge (ESD): Electrostatic fields, protection from ESD, metal enclosures, non-metallic enclosures, ESD immunity.

EMI/EMC measurements: Current and field probes, transmitted and radiated noise, electric field, magnetic field, power density, electromagnetic susceptibility, shielding effectiveness, absorption and reflection loss, open field and anechoic chamber tests.

EMC legislation and regulations: FCC Part 15 and CISPR (The International Special Committee on Radio Interference) regulations.

The level of knowledge for this topic is <u>high</u>. A suggested reference for detailed knowledge is [Pau06].

7.2.3 Measurements
7.2.3.1 Power calculations

Definition of electric power: The rate at which electric energy is dissipated by conversion into other forms of energy – electromagnetic, heat, sound and kinetic energy, etc. Power is measured in units of watts (W). One watt represents the electric power resulting from the dissipation of one joule (J) of electrical energy in one second.

Units of power: Watt (W), erg per second (erg/s), horsepower (hp), metric horsepower (PS), foot-pounds per minute (ft-lb/min), Btu per hour (Btu/h), conversion of units.

AC power: Instantaneous, average, peak and real power (in W), reactive power (VAR), complex power, apparent power (in VA), power factor.

Decibel measurements: Absolute (dBm, dBW, dBJ, dBf, dBk) or relative measurements (dBd, dBFS, dB-Hz, dBi, dBiC, dBO, dBrn, dBx, dBc and -dB, all values expressed relative to the carrier power); advantages of expressing the power level in dBm (or dBW).

The level of knowledge for this topic is <u>medium</u>. A suggested reference for detailed knowledge is [Bea00].

7.2.3.2 Measurements for RF circuits and sub systems

RF measurements require special training as at high frequencies most wires and PCB traces start behaving like antennas and capacitors.

Important measurements include: RF power, input and output impedance, scattering parameters (S-parameters) such as transmission and reflection coefficients, Smith chart characteristics, complex frequency response, gain bandwidth product, nonlinear characteristics such as 1 dB compression point, harmonic and inter modulation distortion (IMD), spurious free dynamic range, receiver sensitivity, phase noise and noise figure, measurement uncertainties and calibration.

The level of knowledge for this topic is medium. A suggested reference for detailed knowledge is [Sco08].

7.2.3.3 Operation of complex test instruments

Typical instruments include oscilloscopes, spectrum and network analyzers, TDRs, and signal generators.

Electrical measurement principles: Measurement errors and accuracy, error estimation, noise types and effects, instrument transformers and bridges.

Basic electrical instrumentation: Principles of operation, galvanometer, voltmeter, ammeter, wattmeter and power measurements, multimeters, series and parallel connections.

Computer-controlled measurements: The IEEE-488 interface and associated software.

Signal generators: Oscillators, RF and microwave sources, frequency, amplitude and power selection, internal/external AM/FM/PM/other advanced modulation, tracking generators.

Oscilloscopes: Principles of operation, internal/external triggering, markers, frequency and amplitude measurement, X-Y display, phase-angle measurements.

Time-domain interferometers: TDR principles of operation, cable testing, applications to nondestructive testing, fault testing of printed circuit boards with TDR.

Spectrum analysis: Basic functions (frequency and amplitude selection, bandwidth, span, markers, sweep and trace), measurement fundamentals (resolution and video bandwidth, sweep time, averaging), basic measurements (signal and channel power, occupied bandwidth, adjacent channel power, out-of-band spurious emissions, in-band/out-of-channel, carrier-to-interference ratio), interference measurements, measurement of specific signal types (AM, FM, SSB. GSM, CDMA, TDMA), AM/FM demodulation.

Network analysis: Basic functions (frequency and amplitude selection, bandwidth, span, markers, sweep and trace), calibration (TRL and full two-port calibrations, connector types, calibration kits), transmission measurements (insertion loss and gain, 3-dB bandwidth, pass-band flatness, out-of-band rejection, phase response, electrical length, phase distortion, group delay), reflection measurements (return loss, reflection coefficient, SWR, impedance, admittance, Smith chart displays), time domain measurements (time-domain reflection, gating time-domain response).

Protocol analyzers: Principle of operation and BER measurements.

The level of knowledge for this topic is <u>medium</u> for a general wireless engineer, although it may be necessary to develop an in-depth understanding during one's professional life. A suggested reference for detailed knowledge is [Wit02].

7.3 Communication Engineering

This section describes knowledge that is essential for the wireless engineer.

7.3.1 Signal Processing

7.3.1.1 Mathematics, including probability, statistics, and Boolean arithmetic

Multiple discrete random variables: Transforms, prediction, covariance, correlation.

Jointly distributed random variables: Definitions, calculations, independent RVs, sums.

Basic theorems: Weak law of large numbers, Central Limit theorem, strong law of large numbers.

Numerical statistics: Mean, median, standard deviation, variance, range and inter-quartile range.

Properties of signal and noise: Deterministic and random signals, continuous and discrete time signals, power and energy signals, periodic signals, operations like scaling and shifting, mean value and mean square value (power) of a signal, root-mean-square (RMS) value, the impulse (Delta) function and unit-step function, (non)stationary signals, Ergodicity, white noise, noise-equivalent bandwidth, noise probability density function, Gaussian noise.

Signal transmission with noise: Additive noise and signal-to-noise ratio (SNR) and SNR in various carrier-modulated and baseband systems.

Signal relationships: Auto-correlation, cross-correlation, orthogonal signals, convolution and de-convolution operations, convolution with the impulse function and the shifting property.

Sampling: Sampling frequency and theorem, Nyquist rate, quantization, digital to analog conversion (DAC), analog to digital conversion (ADC), aliasing and sample-rate conversion (decimation, up/down sampling interpolation).

The level of knowledge for this topic is <u>low</u>. A suggested reference for detailed knowledge is [Men11].

7.3.1.2 Fourier frequency spectrum and transforms

Fourier series: Trigonometric Fourier series and complex exponential Fourier series, properties of Fourier series, symmetric spectra for real signals, Parseval's relation.

Fourier transform: Time- and frequency-domain concepts, definition and properties of the Fourier transform; Parseval's theorem and energy spectral density, band-limited signal and noise, discrete and fast Fourier transform (DFT/FFT).

Other transforms and filters: Z-transform, digital filters (finite impulse response [FIR], infinite impulse response [IIR], lattice), high-pass filter (HPF), low-pass filter (LPF), band-pass filter (BPF), all-pass Filter (APF), pole-zero plots.

The level of knowledge for this topic is <u>high</u>. Suggested references for detailed knowledge are [Hay09] and [Pro07].

7.3.2   Information Theory

Today, information theory is becoming of greater and greater importance for the wireless engineer because of the need to compete with the enormous bandwidth of wired media. Thus the intelligent exploitation of the spectrum and successful techniques for ensuring the quality of information are more than necessary. This knowledge is largely outside the scope of this book; thus we indicate here the necessary background.

7.3.2.1 Communications and information theory (analog and digital)

Basics: Bit time, baud time, symbol time, inter-symbol interference (ISI), crest factor, peak-to-average power ratio (PAPR), complex envelope I & Q waveforms, Shannon theorem, coherent detection, non-coherent detection, geometrical representation, differential encoding and decoding, differential detection, constellation diagram, eye diagram, trellis diagram, signal-space representation, bit error rate (BER), symbol error rate (SER), Q function, antipodal signaling, constant envelope, probability upper bound, signal-to-noise ratio per bit (E_b/N_o), spectral analysis, occupied BW, effect of transmit non-linearity on modulation schemes and spectral re-growth.

Characterization of communication signals and systems: Signal space representations, spectral characteristics of signals, band-pass signals and systems, complex baseband representation.

The level of knowledge for this topic is <u>medium</u>. Suggested references for detailed knowledge are [Hay09] and [Pro07].

7.3.2.2 Modulation techniques for analog

Analog signal transmission and reception: Introduction to modulation, amplitude modulation (AM), angle modulation, effect of noise on analog communication systems (effect of noise on linear-modulation systems, carrier-phase estimation, effect of noise on angle modulation, effects of transmission losses and noise in analog communication systems).

Frequency and phase modulation: Relationship between FM and PM, instantaneous and peak frequency deviation, Carson's rule, narrowband and wideband FM, modulation index, Bessel functions, capture effect, phase nonlinearities, AM-to-PM conversion, pre-emphasis and de-emphasis.

The level of knowledge for this topic is <u>low</u>. A suggested reference for detailed knowledge is [Hay09].

7.3.2.3 Modulation techniques for digital

Digital transmission through additive white Gaussian noise (AWGN) channels: Geometric representation of signal waveforms, pulse-amplitude modulation, two-dimensional signal waveforms, multidimensional signal waveforms, optimum receiver for digitally modulated signals in AWGN, probability of error for signal detection in AWGN, performance analysis for wire-line and radio-communication channels, symbol synchronization.

Digital transmission through band-limited AWGN channels: Power spectrum of digitally modulated signals, signal design for band-limited channels, probability of error in detection of digital pulse amplitude modulation (PAM), digitally modulated signals with memory, system design in the presence of channel distortion, multicarrier modulation and orthogonal frequency division multiplexing (OFDM).

Digital amplitude modulation	On/off keying (OOK), amplitude shift keying (ASK), pulse amplitude modulation (PAM), quadrature amplitude shift keying (QASK), quadrature amplitude modulation (QAM), M-ary QAM
Digital frequency modulation	Frequency shift keying (FSK), binary frequency shift keying (BFSK), M-ary FSK, fast frequency shift keying (FFSK)
Digital phase modulation	Phase shift keying (PSK), binary phase shift keying (BPSK), differential phase shift keying (DPSK), quaternary phase shift keying (QPSK), offset QPSK (OQPSK), sinusoidal OQPSK, M-ary PSK
Continuous phase modulation	Minimum shift keying (MSK), Gaussian minimum shift keying (GMSK)
Pulse shaping	Raised cosine filtering, matched filter, correlation, union-bound approximations, pulse shaping, bandwidth efficiency (b/s/Hz)

Table 7-2: Schemes for Digital Modulation and Pulse Shaping

The level of knowledge for this topic is <u>high</u>. A suggested reference for detailed knowledge is [Pro07].

7.3.2.4 Coding techniques

Source coding compresses data to be transmitted, which increases the entropy in each symbol. Entropy is the actual information content in each symbol.

Topics: Mathematical models for information sources, measures of information, entropy and mutual information, asymptotic equipartition property, entropy rate, source-coding theorem and algorithms, coding for discrete and analog sources, optimum quantization, rate-distortion theory, waveform coding.

Channel coding reduces the channel capacity and the information rate through the channel but increases reliability. This is achieved by adding redundancy to the information symbol vector, resulting in a longer coded vector of symbols distinguishable at the output of the channel. Channel coding can be performed with block or convolutional codes.

Topics: channel models and channel capacity, bounds on communication.

Types of codes:

Linear block codes: The information sequence is divided into blocks of length k. Each block is mapped into channel inputs of length n $(n>k)$. The mapping is independent of previous blocks, that is, there is no memory from one block to another. Examples: Hamming code, BCH code, Reed-Solomon code, Reed-Muller code, Binary Golay code, low-density parity-check codes.

Convolutional codes: Each block of k bits is mapped into a block of n bits but these n bits are not only determined by the present k information bits but also by the previous information bits. This dependence can be captured by a finite state machine. Examples: Viterbi coding, punctured convolutional codes, trellis diagrams, turbo coding.

Error detection: Detecting the presence of error; parity check and cyclic redundancy check (CRC).

Error correction: Additional ability to reconstruct the original, error-free data.

Automatic repeat-request (ARQ): The receiver requests retransmission in case of error. Often, retransmission will begin if the transmitter does not receive an acknowledgement (ACK) of correctly received data in a reasonable time.

Forward error correction (FEC): The transmitter encodes data with an error-correcting code and sends the coded message. The receiver decodes what it receives into the "most likely" data. (FEC can be combined with ARQ.)

The level of knowledge for this topic is <u>high</u>. A suggested reference for detailed knowledge is [Pro07].

7.3.3   Communication Systems and Networks

A wireless system is always an integral part of a broader (even global) communications infrastructure. And actually it is a rather new extension of the previously existing wired network. Thus understanding the essential rules of the operation of the global infrastructure is quite necessary for the wireless engineer.

7.3.3.1 Wired networks, including signalling, switching, and transmission

Plain old telephone service (POTS): Local loop, pulse-code modulation, quantization and quantization noise, public-switched telephone network (PSTN), signaling systems (SS7), T and E carrier systems (the backbone of telephony networks, see Table 7-3 below), circuit switching.

T-carrier and E-carrier systems	North America	Europe (CEPT)[1]
Level zero (channel data rate)	64 kb/s (DS0)	64 kb/s
First level	1.544 Mb/s (DS1) (24 user channels) (T1)	2.048 Mb/s (32 user channels) (E1)
Intermediate level (U.S. hierarchy only)	3.152 Mb/s (DS1C) (48 Ch.)	
Second level	6.312 Mb/s (DS2) (96 Ch.)	8.448 Mb/s (128 Ch.) (E2)
Third level	44.736 Mb/s (DS3) (672 Ch.) (T3)	34.368 Mb/s (512 Ch.) (E3)
Fourth level	274.176 Mb/s (DS4) (4032 Ch.)	139.264 Mb/s (2048 Ch.) (E4)
Fifth level	400.352 Mb/s (DS5) (5760 Ch.)	565.148 Mb/s (8192 Ch.) (E5)

Table 7-3: Specifications of T and E Carrier Systems

Data communications over PSTN: X.25 and HDLC, voice-band modems, DSL (digital subscriber line) and its derivatives (ADSL, DSL++, and VDSL, etc.), cable modems, DOCSIS (Data Over Cable Service Interface Specifications),

Local area networks evolution: ALOHA, token ring, Ethernet, CSMA/CD, CSMA/CA, E1/T1 multiplexing, OC-n/SDH, ATM, FDDI, frame relay.

Internet: Encapsulation, fragmentation and reassembly, connection control, ordered delivery, flow control and error control, IPV4 and IPV6, voice over IP, IPTV.

Optical networks: SONET (SDH) and various PON standards, optical Ethernet.

7.3.3.2 Quality of Service (QoS) and DiffServ

The QoS model described here covers the features of DiffServ but not those of integrated services (IntServ). DiffServ is a connectionless service that provides for service differentiation for aggregates of flows specified by the DiffServ code point (DSCP) field of the IP header [Bla98].

DiffServ implementations: Behavior aggregate (simple classification using only the DiffServ code point); multi-field (more complex classification based on IP and layer-4 header fields); components of DiffServ classification functions:

Policy and policy scope: Provide a way to classify packets according to the operator's policies.

Hierarchical policy configuration specification: Allows policies to be constructed of one or more classifications and actions; classifications may be used by more than one policy.

Policy actions: Determine what is done once a packet is classified (drop, remark, police, etc.).

Filtering: Determines the inspection depth of the packet, that is, how far beyond the beginning of the packet it is inspected to determine what to do with it.

Mapping tables: Allow the operator to map between packet designations, such as DSCP, precedence and CoS bits, to allow for remarking and administrative domain changes.

---

[1] CEPT: European Conference of Postal and Telecommunications Administrations

Per-hop behavior (PHB): Network nodes that implement differentiated service enhancements to IP use a code point in the IP header to select a per-hop behavior (PHB). The Internet Engineering Task Force (IETF) has proposed two PHB groups, expedited forwarding and assured forwarding. The third PHB group (best effort) has no service differentiation.

> Expedited forwarding (EF): Used to build a low-loss, low-latency, low-jitter, assured bandwidth, end-to-end service through DS domains.

> Assured forwarding (AF): The AF PHB group delivers IP packets in four independently forwarded AF classes (AF1, AF2, AF3, AF4). Within each class, an IP packet can be assigned one of three different levels of drop precedence.

> Best effort (BE): This PHB group acts like the normal IP with no consideration of the IP TOS byte.

Scheduling mechanisms: Strict priority (SP) and weighted fair queuing (WFQ) are used to schedule IP packets from traffic class queues. The default router configuration does not assign any forwarded traffic to the network control class. If required, configuration of DiffServ policy or modification of the DSCP trust and the DSCP-to-class map table can be used to give forwarded network control traffic the same priority as the generated protocol traffic.

Active queue management (AQM): Different AQM techniques monitor traffic load in an effort to anticipate and mitigate congestion at network bottlenecks. It can be enabled for AF and BE traffic classes and is achieved through packet dropping. It will also be used to avoid global synchronization.

The level of knowledge for this topic is low. A suggested reference for detailed knowledge is [Sta02].

7.3.3.3 Concepts of queuing theory and traffic analysis

Traffic analysis constitutes a useful background for the wireless engineer to study the performance of any wireless system and provide the appropriate quality of service. Knowledge of the following topics is required.

Combinatorial analysis: Counting principles, permutations, combinations, sampling schemes, binomial and multinomial coefficients.

Probability models and axioms: Sample spaces, events, probability axioms.

Conditioning and independence: Conditional probabilities, Bayes' Rule, independent events.

Random variables: Discrete and continuous random variables, probability mass functions, expectations, conditional expectation.

Probability theory: Random variables, transformations, use of histograms, Markov chains and queuing theory; definition of traffic intensity and its measure, definition of queues, the Markovian case and the solution at regime, main queue analysis parameters, Little's theorem, M/M/s and similar queue analysis, the Erlang-B formula, distribution of the queuing delay.

M/G/1 queuing theory and applications: Solution of the state probability distribution for M/G/1 queues, different possibilities of embedding instants, differentiated service times, queuing theory applied to local area networks with analysis of token and polling schemes, analysis of different types of CSMA, and analysis of PRMA.

Networks of queues and closed networks: Traffic rate equations, the Burke theorem and Jackson theorem with examples, Markovian models used for traffic generation: on/off sources, fluid flow traffic models, traffic loss analysis, multi-rate traffic analysis, traffic load measurement, sampling methods, traffic model selection, traffic types and grades of service.

The level of knowledge for this topic is low. A suggested reference for detailed knowledge is [Hay04].

7.3.3.4 Frequency allocations and reuse
Much of this required knowledge is discussed in Chapter 1.

Cellular concepts and frequency reuse: Cellular systems, frequency reuse factor and cell clusters, co-channel interference, adjacent channel interference, channel reuse ratio, power control, cell splitting, sectoring, cell coverage area, outage probability, cellular base station, hexagonal geometry, center-excited cells, edge-excited cells, frequency reuse factor in FDMA, CDMA and OFDMA networks, the 60-degree coordinate system, soft/hard handoff, handoff prioritization, dwelling time.

Regulatory bodies: Working groups of the International Telecommunication Union (ITU) and Federal Communications Commission (FCC), 3rd Generation Partnership Project (3GPP), European Telecommunications Standards Institute (ETSI), and others.

The level of knowledge for this topic is high. A suggested reference for detailed knowledge is [Rap02].

7.3.3.5 Wireless multiple-access schemes
This topic is considered essential and is covered in detail in Chapter 1, to which the reader is referred for in-depth information and suggested references.

7.3.3.6 Concepts in wireless optical communications (infrared, Fi-Wi)
Many communications experts believe today that the future will require the coexistence and interoperability of wireless (for mobility) and optical (for bandwidth) systems. Thus a wireless engineer needs a background on the interfacing of these systems. Essential topics in this area include:

Basics of optical fibers: Optical wave propagation, waveguide operation, single/multi-mode optical fiber, attenuation and dispersion in fiber, optical sources (lasers and LEDs), detectors (PIN and APD), WDM concepts, subcarrier multiplexing and analog transmission, optical networks.

Optical wireless: Free space optics and indoor wireless optic systems, infrared sources and detectors, attenuation, sensitivity to environmental conditions, multiplexing, modulation schemes.

Fiber-Wireless (Fi-Wi): Radio over Fiber (RoF) concept and various Fi-Wi architectures.

The level of knowledge for this topic is low. A suggested reference for detailed knowledge is [Ini08].

7.3.3.7 Satellite communications
Satellite communications constitute a very special part of the wireless engineering curriculum. It is clearly understood that we are moving toward an integrated infrastructure solution (terrestrial and space); thus a general knowledge of satellite communications is essential for the wireless communications engineer.

Most multiple access techniques used in cellular radio systems are also applicable to satellite systems, as for example FDMA, TDMA, Aloha, etc. Further, the security solutions of Chapter 2 can be extended to cover the space infrastructure, while the channel propagation modeling and the antennas used constitute very specific categories of general mobile communications.

This topic covers the basics of:

Satellite architectures, that is *orbits*: Geosynchronous (GEO), medium-Earth orbit (MEO), low-Earth orbit (LEO), space and earth segment, types of systems (global, regional, national).

Frequency bands:

Band	Frequency	Service
L	1–2 GHz	Mobile
S	2.5–4 GHz	Mobile
C	3.7–8 GHz	Fixed
X	7.25–12 GHz	Military
Ku	12–18 GHz	Fixed
Ka	18–30.4 GHz	Fixed
V	37.5–50.2 GHz	Fixed

Satellite access: Access and contention schemes (FDMA, TDMA, CDMA, DAMA).

Payload architectures: Transparent bent pipe and onboard processing.

Performance measures: Link availability, bit error rate, throughput, delay, grade of service.

Rain fade mitigation techniques: UPC, ALC, ACM, site diversity.

Satellite protocols: ATM, IP, ATM over IP.

User-terminal features: Antennas, LNAs, PAs, cost, performance.

Interference calculations: Co-channel interference from adjacent satellites, crosstalk interference, cross polarization interference, cross modulation products.

Transmitter power: Types (SSPA and TWTA), performance (linearization and back-off).

Adaptive coding modulation techniques: Channel coding, advanced modulation.

Losses: Free space, pointing, absorption, rain fade, depolarization.

Onboard transponders and power amplifiers: Size (bandwidth and power), performance.

Modem performance: Speed, error control, modulation efficiency, performance threshold.

Earth coverage: GEO and non-GEO.

Frequency reuse: Spatial, polarization, both.

Satellite services: Telephony, video, data, ISDN, emergency, other (e.g., GPS), services for satellite communications, limitations, advantages and disadvantages.

Link budget calculations: Bandwidth, power utilization.

Inter-satellite links.

LEO Handover: Handover issues for LEO constellations, handover algorithms.

Mobility: Mobility models.

Channel modeling: Channel models for satellite communications (log-normal, Nakagami, Aluini).

Antennas: Types, gain, side-lobe performance, noise temperature, efficiency, tracking, installation.

Standards: UMTS-S, DVB-S, DVB-S2.

The level of knowledge for this topic is <u>medium</u>. A suggested reference for detailed knowledge is [ITU02].

7.4 Engineering Management

Engineers in professional life will address issues far beyond the fundamental knowledge received during their academic studies. This is especially true when advancing upwards through the levels of management and leadership in the work place. This section includes some necessary background for an efficient and ethical professional life.

The reader will realize that the knowledge in this field covers only the very basic concepts. Engineers who devote their professional lives to managerial rather than technical positions will need to broaden their qualifications by further studies (e.g., an MBA, MSc in Economics, etc.).

7.4.1 Project Management Methods and Processes

Engineering design processes: Prescriptive and descriptive processes, elements of the design process.

Requirements specification: Concept generation and evaluation, properties of engineering requirements, techniques for identifying requirements, properties of requirements specifications, requirements validation, the engineering-marketing tradeoff matrix.

Functional decomposition and design: Bottom-up and top-down design approaches, design architecture, functional specifications/requirements, state diagrams/state machines, flowcharts, data flow diagrams, entity relationship diagrams, the Unified Modeling Language (UML).

Project management: Work breakdown structures, network diagrams, Gantt charts, cost estimation, elements of effective presentation, reliability prediction.

Integration and verification: Properties of test cases and units, integration and acceptance tests.

Standards: Identifying and locating industry technical standards, codes and other applicable requirements.

Design and configuration for ease of maintenance.

Documentation and configuration control schemes.

The level of knowledge for this topic is <u>medium.</u> A suggested reference for detailed knowledge is [Gra02].

7.4.2 Fundamental Engineering Economics

Statistics theory: Jointly distributed random variables, weak law of large numbers, Central Limit theorem, strong law of large numbers, mean, median, standard deviation, variance, range, inter-quartile range, confidence intervals, point estimation, test procedures, errors, and large sample test for population mean and population proportion, scatter plots, Pearson correlation coefficient, simple linear regression.

Engineering costs and cost estimation: Types of costs and cost estimates, models for cost estimation, cash flow diagrams.

Interest and equivalence: Types of interest, computing cash flows, equivalence calculations, interest formulas.

Analysis techniques: Many different types abound, including present worth, annual cash flow, rate of return, future worth, benefit-cost ratio, sensitivity and break-even and replacement analyses.

Depreciation: Basic aspects of depreciation, depreciation methods, modified accelerated cost recovery system, asset disposal, depletion.

Other essential topics include taxation and after-tax economic analysis, inflation effects, capital budgeting, and accounting principles.

The level of knowledge for this topic is <u>low</u>. A suggested reference for detailed knowledge is [New04].

7.4.3 IEEE Code of Ethics

As they go about their work, engineers should all be aware of the IEEE's Code of Ethics. It should guide them professionally in all they do. Membership in the IEEE in any grade carries the obligation to abide by the code. Following is the code in its entirety. It can also be found on the IEEE website [IEE11].

We, the members of the IEEE, in recognition of the importance of our technologies in affecting the quality of life throughout the world and in accepting a personal obligation to our profession, its members and the communities we serve, do hereby commit ourselves to the highest ethical and professional conduct and agree:

1. To accept responsibility in making decisions consistent with the safety, health and welfare of the public, and to disclose promptly factors that might endanger the public or the environment;

2. To avoid real or perceived conflicts of interest whenever possible, and to disclose them to affected parties when they do exist;

3. To be honest and realistic in stating claims or estimates based on available data;

4. To reject bribery in all its forms;

5. To improve the understanding of technology, its appropriate application, and potential consequences;

6. To maintain and improve our technical competence and to undertake technological tasks for others only if qualified by training or experience, or after full disclosure of pertinent limitations;

7. To seek, accept, and offer honest criticism of technical work, to acknowledge and correct errors, and to credit properly the contributions of others;

8. To treat fairly all persons regardless of such factors as race, religion, gender, disability, age, or national origin;

9. To avoid injuring others, their property, reputation, or employment by false or malicious action;

10. To assist colleagues and co-workers in their professional development and to support them in following this code of ethics.

7.5 References

[Bea00] H.W. Beaty, *Handbook of Electric Power Calculations*, McGraw-Hill, 2000.

[Bla98] S. Blake, D. Black, M. Carlson, E. Davies, Z. Wang, and W. Weiss, *An Architecture for Differentiated Services*, Internet Engineering Task Force RFC 2475, December 1998.

[Gra02] C.F. Gray and E.W. Larson, *Project Management*, McGraw-Hill, 2002.

[Hay04] J. Hayes and T.V.J. Ganesh Babu, *Modeling and Analysis of Telecommunications Networks*, Wiley Interscience, 2004.

[Hay09] S. Haykin and M. Moher, *Communication Systems, 5th Edition*, John Wiley & Sons, 2009.

[IEE11] The Institute of Electrical and Electronics Engineers (IEEE), *The IEEE Code of Ethics*, http://www.ieee.org/portal/pages/portals/aboutus/ethics/code.html.

[Ini08] R.R. Inifuez, S.M. Idrus, and Z. Sun, *Optical Wireless Communications*, CRC Press, 2008.

[ITU02] International Telecommunications Union (ITU), *Handbook of Satellite Communications*, Wiley Interscience, 2002.

[Kha05] R.S. Khandpur, *Printed Circuit Boards*, McGraw-Hill, 2005.

[Lev79] K. Levitz and H. Levitz, *Logic and Boolean Algebra*, Barrons Educational Series, Inc., 1979.

[Mag00] P.C. Magnusson, G.C. Alexander, V.K. Tripathi, and A. Weisshaar, *Transmission Lines and Wave Propagation, 4th Edition*, CRC Press, 2000.

[Men11] W. Mendenhall, R.J. Beaver, B.M. Beaver, and S.E. Ahmed, *Introduction to Probability and Statistics*, Nelson Education, 2011.

[Mil87] J, Millman and A, Grabel, *Microelectronics*, McGraw-Hill, 1987.

[New04] D.G. Newnan, T.G. Eschenbach, and J.P. Lavelle, *Engineering Economic Analysis*, Oxford University Press, 2004

[Pau06] C.R. Paul, *Introduction to Electromagnetic Compatibility*, John Wiley & Sons, 2006.

[Pro07] J.G. Proakis and M. Salehi, *Digital Communications*, McGraw-Hill Higher Education, 2007.

[Rap02] T. Rappaport, *Wireless Communications, Principles and Practice*, Prentice Hall, 2002.

[Ree07] W. Reeve, *DC Power System Design for Telecommunications*, Wiley-IEEE Press, 2007.

[Sco08] A.W. Scott and R. Frobenius, *RF Measurements for Cellular Phones and Wireless Data Systems*, John Wiley and Sons Inc., 2008.

[Siz10] H. Sizun, *Radio Wave Propagation for Telecommunication Applications*, Springer-Verlag 2010.

[Sta02] W. Stallings, *High Speed Networks and Internets*, Pearson, 2002.

[Tho06] R.E. Thomas, *The Analysis and Design of Linear Circuits*, John Wiley & Sons, 2006.

[Whi04] J.F. White, *High Frequency Techniques – An Introduction to RF and Microwave Engineering* John Wiley & Sons, 2004.

[Wit02] R.A. Witte, *Electronic Test Instruments: Analog and Digital Measurements*, Prentice Hall, 2002.

Appendix A

Glossary of Acronyms

3DES	Triple DES Encryption		**ATPC**	Automatic Transmit Power Control
3G	3rd Generation Mobile Telecommunications		**AuC**	Authentication Center
			AUT	Antenna Under Test
3GPP	3rd Generation Partnership Project		**AUTN**	Network Authentication Token
3GPP2	3G Partnership Project 2		**AUTS**	Token used in resynchronization
			AWGN	Additive White Gaussian Noise
A5	Encryption algorithm in GSM		**AWS**	Advanced Wireless Services
AAA	Authentication, Authorization, and Accounting			
			BCCH	Broadcast Control Channel
AAD	Additional Authentication Data		**BCMCS**	Broadcast and Multicast Services
ACK	Acknowledge		**BE**	Best Effort
ACM	Address Complete Message		**BER**	Bit Error Rate
ACM	Adaptive Code Modulation		**BGP**	Border Gateway Protocol
ADC	Analog to Digital Converter		**BSC**	Base Station Controller
AES	Advanced Encryption Standard		**BSS**	Basic Service Set
AF	DiffServ Assured Forwarding		**BTS**	Base Transceiver Station
AFD	Average Fade Distortion			
AGC	Automatic Gain Control		**CB**	Certification Body
ALC	Automatic Level Control		**CBC**	Cipher Block Chaining
AM	Amplitude Modulation		**CBC-MAC**	Cipher Block Chaining Message Authentication Code
AMC	Adaptive Modulation and Coding			
AMPS	Advanced Mobile Phone System		**CC**	Call Control
ANM	Answer Message		**CCCH**	Common Control Channel
ANSI	American National Standards Institute		**CCI**	Co-Channel Interference
			CCM	Counter with CBC-MAC
AODR	Ad hoc On Demand Routing		**CCMP**	Counter with CBC-MAC Protocol
AP	Access Point		**CCSA**	China Communications Standards Association
APD	Avalanche Photo Diode			
AR	Axial Ratio in Elliptical Polarization		**CDMA**	Code Division Multiple Access
			CGM	Conjugate Gradient Method
ARIB	Association of Radio Industries and Businesses		**CID**	Connection ID
			CIR	Carrier to Interference Ratio
ARQ	Automatic Repeat-Request		**CM**	Connection Management
AS	Application Server		**CMOS**	Complementary Metal Oxide Semiconductor
ASCII	American Standard Code for Information Interchange			
			COFDM	Coded Orthogonal Frequency Division Multiplexing
ASK	Amplitude Shift Keying			
ASN	Access Service Network		**COMP128**	Authentication Algorithm in GSM
ASN.1	Abstract Syntax Notation One		**CP**	Circular Polarization
ASP	Application Service Provider		**CP**	Cyclic prefix
ATIS	Alliance for Telecommunications Industry Solutions (US)		**CPC**	Cyclic Prefix Code
			CQI	Channel Quality Indicator
ATM	Asynchronous Transfer Mode		**CRC**	Cyclic Redundancy Check

CRC-32	Cyclic Redundancy Check, 32 bits
CS	Coding Scheme
CSA	Canadian Standards Association
CSCF	Call Session Control Function
CSMA/CA	Carrier Sense Multiple Access with Collision Avoidance
CSMA/CD	Carrier Sense Multiple Access with Collision Detection
CSN	Connectivity Service Network
CST	Computer Simulation Technology
CTIA	International Association for the Wireless Telecommunications Industry
CTS	Clear to Send
DARPA	Defense Advanced Research Projects Agency
dBi	Decibel Isotropic
dBm	Decibel Milliwatts
dBr	Decibel Relative
DCF	Distributed Coordination Function
DCH	Dedicated Channel
DDoS	Distributed Denial of Service
DECT	Digital Enhanced Cordless Telephony
DES	Data Encryption Standard
DiffServ	Differentiated Services
DIFS	Distributed Inter-frame Space
DL	Downlink
DMB	Digital Multimedia Broadcasting
DNS	Domain Name System
DoS	Denial of Service
DPCCH	Dedicated Physical Control Channel
DPSK	Differential Phase Shift Keying
DQPSK	Differential Quadrature (or Quarternary) Phase Shift Keying
DRA	Direct Resonator Antenna
DRC	Data Rate Control
DS-CDMA	Direct Source Code Division Multiple Access
DSL	Digital Subscriber Line
DSR	Dynamic Source Routing
DSS1	Digital Subscriber Signaling System No. 1
DSSS	Direct Sequence Spread Spectrum
DVB-H	Digital Video Broadcast – Handheld
DWDM	Dense Wavelength Division Multiplexing
EAP	Extensible Authentication Protocol
EAP-FAST	EAP Flexible Authentication via Secure Tunneling

EAPoL	EAP over LAN
EAP-TLS	EAP Transport Layer Security
EAP-TTLS	EAP Tunneled TLS
E-DCH	Enhanced Dedicated Channel
EDGE	Enhanced Data Rates for GSM Evolution
EF	DiffServ Expedited Forwarding
EGC	Equal Gain Combining
EGPRS	Enhanced GPRS
EIA	Electronic Industries Alliance
EIR	Equipment Identity Register
EM	Electromagnetic
EMC	Electromagnetic Compatibility
EP	Elliptical Polarization
ERP	Effective Radiated Power
ESD	Electrostatic Discharge
ESS	Extended Service Set
ET	Error Tracking
eTOM	Enhanced Telecom Operations Map
ETSI	European Telecommunications Standards Institute
FA	Foreign Agent
FACA	Federal Advisory Committee Act (US)
FBSS	Fast Base Station Switching
FCAPS	Fault Configuration Accounting Performance and Security
FCC	Federal Communications Commission (US)
FDD	Frequency Division Duplex
FDDI	Fiber Distributed Data Interface
FDMA	Frequency Division Multiple Access
FDTD	Finite Difference Time Domain
FEM	Finite Element Method
FFT	Fast Fourier Transform
FHSS	Frequency Hop Spread Spectrum
Fi-Wi	Fiber-Wireless
FR 2, 4	Flame Resistant, ANSI Class 2 or 4
FSK	Frequency Shift Keying
FSO	Free Space Optics
FSS	Frequency Selective Surfaces
G.711	ITU-T Standard for Audio Pulse Code Modulation
GEO	Geostationary Earth Orbit
GGSN	Gateway GPS Support Node
GKH	Group Key Hierarchy
GMSC	Gateway Mobile Switching Center
GMSK	Gaussian Minimum Shift Keying
GPRS	General Packet Radio Service
GPS	Global Positioning System
GSM	Global System for Mobile Communications

GTC	Generic Token Card
H.263	Low Bit-Rate Video Compression Standard
H.264	Next-Generation Video Compression Format (aka MPEG-4 AVC)
HA	Home Agent
HARQ	Hybrid Automatic Repeat Request
HDLC	High-level Data Link Control
HE	Home Environment
HFSS	High Frequency Structure Simulator
HHO	Hard Handoff
Hi-Cap	High Capacity
HLR	Home Location Register
HLR/AuC	Home Location Register/ Authentication Center
HN	Home Network
HO	Handoff
HSDPA	High Speed Downlink Packet Access
HS-DSCH	High Speed Downlink Shared Channel
HSPA	High Speed Packet Access
HSS	Home Subscriber Server
HSUPA	High Speed Uplink Packet Access
HTTP	Hypertext Transfer Protocol
IBSS	Independent Basic Service Set
ICMP	Internet Control Message Protocol
I-CSCF	Interrogating CSCF
ICV	Integrity Check Value
IDEN	Integrated Digital Enhanced Network
IDU	Indoor Unit
IEC	International Electrotechnical Commission
IECEE	IEC System for Conformity Testing and Certification of Electrotechnical Equipment and Components
IETF	Internet Engineering Task Force
IF	Intermediate Frequency
IFFT	Inverse Fast Fourier Transform
IK	Integrity Key
IKE	Internet Key Exchange
IMS	IP Multimedia System
IMSI	International Mobile Subscriber Identity
IMT-2000	ITU Standard: International Mobile Telecommunications for 2000
IP	Internet Protocol
IP v4	Internet Protocol Version 4

IP v6	Internet Protocol Version 6
IP-CAN	IP Connectivity Access Network
IPSec	Internet Protocol Security
IS-95	Interim Standard 95 for CDMA
IS-136	Interim Standard 136 for TDMA
ISAKMP	Internet Security Association and Key Management Protocol
ISDN	Integrated Services Digital Network
ISI	Inter-Symbol Interference
ISIS	Intermediate System to Intermediate System
ISM	Industrial, Scientific and Medical Radio Frequency Band
ISO	International Organization for Standardization
ISUP	ISDN User Part
ISUP IAM	ISUP Initial Address Message
I-TCP	Indirect TCP
ITIL	Information Technology Infrastructure Library
ITU	International Telecommunication Union
ITU-R	ITU Radiocommunication Sector
ITU-T	ITU Telecommunication Standardization Sector
KA	Knowledge Area
KC	Ciphering Key
KCK	EAPoL Key Communication Key
KEK	EAPoL Key Encryption Key
LAN	Local Area Network
LDPC	Low-Density Parity Check
LEO	Low Earth Orbit
LHCP	Left Hand Circular Polarization
LMS	Least Mean Square
LO	Local Oscillator
Lo-Cap	Low Capacity
LOS	Line Of Sight
LP	Linear Polarization
LR-WPAN	Low Rate Wireless Personal Area Network
LS-CMA	Least Squares Constant Modulus Algorithm
LTE	Long Term Evolution
MAC	Media Access Control
MAC	Message Authentication Code
MAC-S	Authentication Token Used in Resynchronization
MAN	Metropolitan Area Network
MAP	Mobile Application Part
MBMS	Multimedia Broadcast/Multicast Service

MCW	Multi Codeword
MD5	Message Digest 5
MDHO	Macro Diversity Handover
MDS	Minimum Discernible Signal
Media-FLO	Forward Link Only Data Transmission
MGCF	Media Gateway Control Function
MGW	Media Gateway
MIB	Management Information Base
MIC	Message Integrity Code
MIMO	Multiple Input Multiple Output
MIP	Mobile IP
MISO	Multiple Input Single Output
MM	Mobility Management
MMUSC	Multiparty Multimedia User Session Control
MoM	Method of Moments
MOS	Mean Opinion Score
MPDU	MAC Protocol Data Unit
MPEG	Moving Picture Expert Group
MPLS	Multiprotocol Label Switching
MR	Mesh Router
MRC	Maximum Ratio Combining
MRF	Media Resource Function
MS	Mobile Station
MSC	Mobile Switching Center
MSC/VLR	MSC Visitor Location Register
MSK	Minimum Shift Keying
MSS	Maximum Segment Size
MTBF	Mean Time Between Failures
MTTR	Mean Time To Repair
MU-MIMO	Multiple User MIMO
NACK	Negative Acknowledge
NAS	Network Access Server
NAV	Network Allocation Vector
NCRP	National Council on Radiation Protection & Measurements
NEBS	Network Equipment-Building System
NEC	National Electrical Code (NFPA 70)
NF	Noise Figure
NFC	Near Field Communication
NGMC	Next Generation Mobile Committee
NGMN	Next Generation Mobile Network
NGN	Next Generation Network
NIC	Network Interface Card
NIST	National Institute of Standards and Technology (US)
NLOS	Non Line Of Sight
NMHA	Normal Mode Helical Antenna
NRSC	National Radio Systems Committee

NRSC	Network Reliability Steering Committee (ATIS – US)
NRZ	Non-Return to Zero
NSP	Network Service Provider
NSS	Network Subsystem
NSTAC	National Security Telecommunications Advisory Committee (US)
OATS	Open Area Test Site
ODU	Outdoor Unit
OFDM	Orthogonal Frequency-Division Multiplexing
OFDMA	Orthogonal Frequency Division Multiple Access
OGC	Office of Government Commerce (UK)
OLSR	Optimized Link State Routing
OSA	Opportunistic Spectrum Address
OSI	Open Systems Interconnection
OSPF	Open Shortest Path First
OSS/BSS	Operational and Business Support Systems
OTA	Over The Air
OTP	One Time Password
PA	Power Amplifier
PAN	Personal Area Network
PAPR	Peak to Average Power Ratio
PBCCH	Packet Broadcast Control Channel
PCM	Pulse Code Modulation
P-CSCF	Proxy CSCF
PDC	Personal Digital Cellular
PDSN	Packet Data Serving Node
PDU	Protocol Data Unit
PEAP	Protected EAP
PHY	Physical (layer)
PIFA	Planar Inverted F Antenna
PIN	Personal Identification Number
PIN	Positive Intrinsic Negative (photodiode)
PKH	Pairwise Key Hierarchy
PL	Path Loss
PLMN	Public Land Mobile Network
PN	Pseudo Noise
PO	Physical Optics
PON	Passive Optical Network
PPP	Point to Point Protocol
PRMA	Packet Reservation Media Access
PSK	Phase Shift Keying
PSTN	Public Switched Telephone Network
QAM	Quadrature Amplitude Modulation
QoS	Quality of Service

QPSK	Quadrature Phase Shift Keying
RAB	Radio Access Bearer
RACH	Random Access Channel
RADIUS	Remote Access Dial In User Server
RAN	Radio Access Network
RAND	Random
RC4	RC4 Cipher Algorithm
RET	Remote Electrical Tilt
RF	Radio Frequency
RFC	Request for Change
RFC	Request for Comments
RFID	Radio Frequency Identification
RHCP	Right Hand Circular Polarization
RIP	Routing Information Protocol
RLC	Radio Link Control
RLS	Recursive Least Squares
RMON	Remote Network MONitoring
RNC	Radio Network Controller
ROAMOPS	IETF Roaming Operations
ROF	Radio Over Fiber
ROHC	RObust Header Compression
RR	Radio Resource
RRC	Radio Resource Control
RSA	Rivest, Shamir, and Adelman algorithm for cryptography
RSN	Robust Security Network
RSNA	Robust Security Network Association
RTP	Real Time Protocol
RTS	Request To Send
RTT	Round Trip Time
S/N	Signal to Noise Ratio
SA	Security Association
SAR	Specific Absorption Rate
SCCP	Signaling Connection Control Protocol
SCP	(ETSI) Smart Card Platform
S-CSCF	Serving CSCF
SCTP	Stream Control Transmission Protocol
SCW	Single Codeword
SDCCH	Stand-alone Dedicated Control Channel
SDH	Synchronous Digital Hierarchy
SDMA	Space Division Multiple Access
SDR	Software Defined Radio
SEGF	Security Gateway Function
SET	Secure Electronic Transaction
SF	Spreading Factor
SFDR	Spurious Free Dynamic Range
SFID	Service Flow ID
SGSN	Serving GPRS Support Node
SGW	Signaling Gateway

SHA	Secure Hash Algorithm
SID	Secure Identification Number
SIFS	Short Inter-Frame Space
SIG	Special Interest Group (of WWRF)
SigTran	Signal Transport
SIM	Subscriber Identity Module
SIMO	Single Input Multiple Output
SIP	Session Initiation Protocol
SIR	Signal to Interference Ratio
SISO	Single Input Single Output
SLF	Subscriber Location Function
SMI	Structure of Management Information
SMS	Short Message Service
SM-SC	Short Message Service Center
SMTP	Simple Message Transfer Protocol
SNMP	Simple Network Management Protocol
SNR	Signal to Noise Ratio
SPC	Single Parity Check
SQN	Sequence Number
SRES	Signed Response
SRTP	Secure RTP
SS7	Signaling System No. 7
SSB	Single Sideband
SSID	Service Set Identifier
SSPA	Solid State Power Amplifier
STA	Station
STM	Synchronous Transfer Mode
SYN	Synchronization
T2P	Traffic to Pilot
TCAP	Transaction Capabilities Application Part
TCH	Traffic Channel
TCH/FS	Traffic Channel Full Rate Speech
TCH/HS	Traffic Channel Half Rate Speech
TCP	Transmission Control Protocol
TCP/IP	TCP/Internet Protocol
TD-CDMA	Time Division CDMA
TDD	Time Division Duplex
TDD-HCR	TDD High Chip Rate
TDD-LCR	TDD Low Chip Rate
TDMA	Time Division Multiple Access
TDOA	Time Difference Of Arrival
TD-SCDMA	Time Division Synchronous CDMA
TIA	Telecommunications Industry Association
TK	Temporal Key
TKIP	Temporal Key Integrity Protocol
TMF	TeleManagement Forum

TRAP	TDMA-based Randomly Accessed Polling
Triple DES	Triple DES Encryption (3DES)
TS	Time Slot
TSC	TKIP Sequence Counter
TSG	3 GPP Technical Specification Group
TSG CT	Core Network and Terminals TSG
TSG GERAN	GSM EDGE Radio Access Network TSG
TSG RAN	Radio Access Network TSG
TSG SA	Service & Systems Aspects TSG
TTA	Telecommunications Technology Association of Korea
TTC	Telecommunication Technology Committee (Japan)
TWTA	Traveling Wave Tube Amplifier
UDP	User Datagram Protocol
UE	User Equipment
UL	Underwriters Laboratories (US)
UMB	Ultra Mobile Broadband
UMTS	Universal Mobile Telecommunications System
UMTS AKA	UMTS Authentication and Key Agreement
UPC	Uplink Power Control
UPS	Uninterruptible Power Supply
USGS	United States Geological Survey
USIM	UMTS SIM
UTRA	UMTS Terrestrial Radio Access
UTRAN	UMTS Terrestrial Radio Access Network
UWB	Ultra Wideband
VLR	Visitor Location Register
VN	Visited Network
VoIP	Voice over Internet Protocol
VSAT	Very Small Aperture Terminal
VSWR	Voltage Standing Wave Ratio
WAN	Wide Area Network
W-CDMA	Wideband CDMA
WCET	Wireless Communication Engineering Technologies
WCP	Wireless Communications Professional
WEP	Wireless Encryption Protocol
WERT	Wireless Emergency Response Team
WG	Working Group (e.g., of WWRF)
WiFi	Wireless Fidelity
WiMAX	Worldwide Interoperability for Microwave Access

WINNER	Wireless World Initiative New Radio
WLAN	Wireless Local Area Network
WMAN	Wireless Metropolitan Area Network
WMN	Wireless Mesh Network
WPA	WiFi Protected Access
WPAN	Wireless Personal AreaNetwork
WRC	World Radiocommunication Conference
WWRF	Wireless World Research Forum
XG	Next Generation
XKMS	XML Key Management Services
XMAC	Cryptographic primitive in the 3GSM Key Generation Process
XOR	Exclusive Or
ZRP	Zone Routing Protocol

Appendix B

Summary of Knowledge Areas

B.1 Wireless Access Technologies

Tasks and knowledge related to wireless access networks, especially the physical, MAC, and link layers. Analyze building blocks, multiple access, mobility management, and spectrum implications in wireless access system design; analyze design considerations to optimize capacity/coverage; design and analyze a wireless access system; analyze the required bandwidth for a wireless system and tradeoffs; analyze wireless access technology standards, their features, and evolution.

Tasks:

1. Analyze multiple access schemes for various technologies.

2. Analyze spectrum implications in wireless access system design (examples might include applications, TDD/FDD, inter-modulation, LOS/NLOS, coverage/capacity).

3. Analyze design considerations and perform system design to eliminate coverage holes and to optimize capacity/coverage in urban/indoor areas.

4. Design and analyze a wireless access system (examples might include AP placement and channel selection) according to given bandwidth requirements, coverage, and other considerations.

5. Test devices with respect to interference issues in various operating environments (examples might include TDMA, CDMA, WCDMA, WLAN, 802.15).

6. Perform interference analysis (examples: co-site interference in TDMA, CDMA, WCDMA, WLAN, 802.15, and GSM; effect of interference on capacity in cellular, WLAN, WAN, *ad hoc* and sensor networks).

7. Compute the required bandwidth for a wireless system given certain network conditions (examples might include BER, flow count, and protocols in use).

8. Analyze the tradeoffs (examples might include bandwidth versus BER) of various error detection and correction techniques.

9. Analyze the tradeoffs and capacity implications of mitigation techniques for time-varying channels, including channel estimation; time- and frequency-recovery and tracking; modulation/demodulation; pre-coding; and power control schemes (examples: scheduling algorithms, bandwidth versus power efficiency analysis).

10. Calculate frequency re-use factor.

11. Design fundamental elements/attributes of wireless network systems (examples might include cellular, 802.16, WLAN, and satellite).

12. Analyze the steps involved in the process of handover/handoff for various wireless systems (examples might include UMTS, CDMA2000, 802.16, and WLAN).

13. Analyze the tradeoff between the size of a paging area and the location update frequency.

Knowledge of:

1. multiple access and multiplexing schemes (examples might include TDMA, CDMA, OFDMA, FDMA, and SDMA)

2. technology standards and their evolution (examples might include WCDMA, CDMA2000, LTE, 802.11, 802.15, and 802.16)

3. error detection and correction, ARQ, HARQ, Turbo Coding, link-adaptation, modulation/demodulation, and pre-coding techniques

4. objectives of channel-estimation and power-control schemes and their operation

5. handover/handoff/mobility management, including inter-technology handover/handoff

6. paging functions

7. the major components of a wireless network topology

8. LEOS, MEOS and geostationary satellites, their bands, and their usage for broadcasting

B.2   Network and Service Architecture

Tasks and knowledge related to network infrastructure, including core networks; service frameworks such as IMS; and application architectures such as voice, video streaming, and messaging. All-IP services architecture as in 3GPP Rel 6 and beyond, including Enhanced Packet Services (EPS) as in 3GPP Rel 8 LTE (Long Term Evolution) and EPC (Enhanced Packet Core). Analyze service platforms, IP addressing schemes for various technologies; design and test quality of service (QoS); select and test a load-balancing scheme; analyze IP routing and ad hoc routing and mesh protocols; perform capacity planning, error tracking, and trace analysis; analyze the evolution of mobile networks to enable IP multimedia.

Tasks:

1. Analyze service platforms including service enablers (examples might include messaging, positioning, and location), service creation/delivery (examples might include Open Service Access and Parlay), and service-oriented architecture (SOA). Design and engineer various VAS (CRBT, SMS, VMS, Alerts, etc.) services on wireless network CORE. Design optimum network services for data traffic.

2. Analyze IP addressing schemes for various technologies (examples might include Mobile IP, RObust Header Compression [ROHC] as in VoIP over HSPA or LTE, IPv4, and IPv6).

3. Design and test quality of service (QoS) (examples might include design and plan for adequate resources, selecting priority schemes, prioritization of differentiated services, queuing strategies, mapping of QoS classes between network and transport layers and call admission control) for VoIP and IMS-based services. Calculate Capacity and Grade of Service (GOS) for a cellular network e.g., GSM/WCDMA/LTE networks. Provision QoS for different applications per 3GPP standards, e.g. through QCI, ARP, etc. for LTE/EPC networks.

4. Select and test a load-balancing scheme.

5. Analyze IP routing (examples might include interpreting an IP routing table).

6. Analyze *ad hoc* routing and mesh protocols, and suitability for various deployment scenarios.

7. Perform capacity planning using traffic engineering principles.

8. Perform error tracking and trace analysis on protocol control messages for specific systems.

9. Analyze the evolution of mobile networks to enable IP multimedia services (including circuit-switched to packet-switched network evolution).

10. Analyze intra- and inter-domain roaming (examples might include roaming within a country or in different countries in 3GPP networks). Analyze service continuity across domains (e.g., VoIP in LTE and circuit-switched voice in GSM/W-CDMA networks).

11. Analyze the functioning of TCP/IP major transport protocols (examples might include TCP, UDP, and RTP) in the context of wireless communications and limitations of PING/Ack.

12. Develop a simple block diagram-level design for a network operations center (examples might include digital cellular, web-based mobile content, multimedia broadcast, and SMS).

Knowledge of:

1. IMS (IP multimedia subsystems) and its architecture, including session control and switching plane; knowledge of different VAS in wireless domain

2. VoIP/IP-multimedia protocols

3. wireless service enablers evolution, including call processing architecture/framework, feature development/enhancement, as well as applications such as presence, location, etc. policy rules, decisions, charging and enforcement

4. location and positioning techniques

5. load balancing principles in the context of wireless communications, and methods to avoid single point of failure through active/active or active standby, and concept of self organizing networks (SON)

6. IP routing and mobile IP networking and addressing schemes including WLAN systems. IP evolution in wireless access – backhaul and packet core connectivity

7. error tracking and trace analysis techniques for dropped cells, access failures and other network related problem reports

8. circuit switched and packet switched data and packet cellular networks; the differences between them; knowledge of various data capable technologies – 1xRTT, EVDO, GPRS/EDGE, LTE

9. roaming and roaming controls

10. TCP/IP including transport protocols including WLAN systems

11. Access Point Name and its functionality

12. heterogeneous architecture for single-hop and multi-hop wireless networks

B.3 Network Management and Security

Tasks and knowledge related to fault, configuration, account, performance, maintenance, security management, management availability, and operation support systems (examples include network service assurance and provisioning). Design a fault monitoring system and a performance monitoring system; develop/specify types and methods of alarm reporting; compute availability and reliability metrics; assess the potential impacts of known security attacks; plan corresponding solutions to known security attacks.

Tasks:

1. Design a fault monitoring system (examples might include using SNMP TRAP/NOTIFICATION, and using 2G OAM&P standards at Network Element Layer [NEL], Equipment Management Layer [EML], and Network Management Layer [NML]).

2. Design a performance monitoring system (examples might include using SNMP GET/SET and performance measurement on radio layer, BTS and RNC, usage and traffic analysis and accounting, monitoring SAACH frame error rate in 3GPP networks). •

3. Develop/specify types and methods of alarm reporting for an installation, and other OAM&P.

4. Compute availability and reliability metrics from both the "network performance" and "system designer" perspectives (related to equipment failure).

5. Assess the potential impacts of known security attacks on wireless systems (examples might include virus, worm, DoS, network sniffing, flooding and impersonation; additional examples might include SIM/USIM card cloning, attempting bank transaction using prepaid cellular handsets, integrity of SMS, multi subscription of USIM card etc).

6. Plan corresponding solutions to known security attacks (examples might include stolen SIM card, stolen PIN, use of different handsets using the same SIM card etc).

7. Monitor, log, and audit security-related data (including tasks such as streaming system logs to third party box for analysis and reporting).

8. Analyze security vulnerabilities and prepare/recommend corrective actions; develop comprehensive test plan for network security testing.

9. Design and plan a migration to a new network management scheme (including impacts on OSS, BSS, and billing); design proper access levels (user management) and its implementation.

10. Analyze wireless accounting and billing schemes including inter-operator accounting.

11. Design and establish VPN communications from client to host.

Knowledge of:

1. quality of service (QoS) monitoring and control

2. fault management

3. configuration management including licensing mechanisms, feature addition/integration, system initialization and installation, policy-based management, role-base access control, level of security offered OTA by standard cellular and wireless systems, and architectures for service management

4. authentication, authorization, and accounting (AAA) principles and mechanisms and APN security; CAVE, A3/A8 and other authentication algorithms – separating mobile from subscription data; cellular authentication schemes based on HLR, VLR, SIM card

5. types of security attacks on wireless networks (examples might include use of stolen SIM card, fraudulent techniques to use handsets in non-designated areas)

6. protocols to secure wireless networks (examples might include Application Security, Web security and Secure Socket Layer, VPN, RADIUS, DIAMETER, HLR/VLR and encryption methods based on cellular algorithms), and Self Organizing/Optimizing networks for next generation networks

7. security-violation events logging and monitoring, attempts towards billing fraud, SIM card manipulation and detection, etc. and different security testing tools

8. security issue management and resolution (examples might include management of A-key, OTAP and HLR/VLR updates, monitoring handoff and reauthentication during call)

9. network management protocols (examples might include simple network management protocol [SNMP], network scanning for BTS identification, interface measurements, data quality measurements, video quality measurements, verification of test mobile phones, acquisition of calibration data for planning tools)

10. performance metrics pertinent to various access networks (examples might include Carrier to Interface (C/I) matrix; recommended changes to neighbor list to ensure appropriate cell handovers and others)

11. IP security, Encapsulation Security Payload (ESP), Internet Key Exchange, and digital signature; root authentication keys in removable UIM, Data Subscriber Authentication – DSA over the air interface

12. MIB, RMON, and Internet Control Messaging Protocol (ICMP)

13. intrusion detection systems, DDoS attacks, and traceback techniques; GSM security IMSI/TMSI, RAND, SRES-HLR and AuC checking methods, network controlled policies, on-line and off-line charging for pre-paid subscribers

14. operational process models (examples might include ITIL and eTOM); writing A-key into mobile – manual and over-the-air procedures

15. hot billing during call, hot billing after call, and similar cases

16. OTAP (Over the air provisioning methods), USIM (Universal SIM) card architecture, Kasumi security algorithm

17. mobile money transaction methods, near-field communications and security

B.4   Radio Engineering and Antennas

Tasks and knowledge related to: antennas, RF engineering, transmission, reception, propagation, channel modeling, and signal processing. Evaluate system performance and reliability; calculate path loss; evaluate the effects of different fading and empirical path loss models; calculate and evaluate the effects on the received signal of path-related impairments; determine parameters related to antennas or antenna arrays; generate and evaluate coverage and interference prediction maps; develop and analyze procedure to optimize the coverage of a radio; make RF system measurements.

Tasks:

1. Calculate link budgets to evaluate system performance and reliability based on received signal level and fade margin (examples might include satellite, microwave link, base station to mobile station, wireless LAN and PAN); calculate path loss for various RF transmission systems (examples might include between isotropic or dipole reference antennas, base station to mobile station, base station to repeater, earth station to satellite, LOS/NLOS paths, and clutter losses).

2. Calculate the capacity of various multiple-antenna schemes, and analyze the tradeoffs involved in selecting from among alternative schemes (calculations might include analysis of pre-coding techniques).

3. Evaluate the effects of different fading models (examples might include Rayleigh and lognormal) and empirical path loss models on the received signal strength in various signal propagation environments (examples might include flat terrain, rolling hills, urbanized areas, and indoor environments [such as buildings or tunnels] with losses caused by walls, ceilings, and other obstructions).

4. Calculate and evaluate the effects on the received signal of path-related impairments, such as Fresnel Zone blockage, delay spread, and Doppler shift of a signal received by a moving receiver.

5. Calculate the polarization mismatch loss for various antenna systems (examples might include fixed microwave systems, cellular and mobile radio systems, and satellite systems).

6. Evaluate receive diversity gain for selection, equal gain, and maximal ratio diversity system configurations.

7. Determine parameters related to antennas or antenna arrays (examples might include pattern, beamwidth, gain, SAR-reduction features, distance from an antenna or array at which far field conditions apply, spacing, beam forming, tilt, and sectorization) and analyze the effects of these parameters on coverage.

8. Determine appropriate antenna location at base station sites to prevent inter- and intra-system interference effects, taking into account required radiation patterns and mutual coupling effects.

9. Generate and evaluate coverage and interference prediction maps (examples might include maps for cellular, mobile radio, and WLAN systems).

10. Develop and analyze a procedure to optimize the coverage of a radio system using propagation modeling and "drive test" measurements.

11. Develop a block diagram of an RF system (examples might include cellular, land mobile, and WLAN) employing standard modules (examples might include filters, couplers, circulators, and mixers) and/or using lumped or distributed matching networks, microstrips, and stripline.

12. Make and analyze RF system measurements (examples might include swept return loss to determine antenna system performance, transmitter output power [peak or average, as appropriate], signal-to-noise ratio at a receiver front end, and co-channel and adjacent-channel interference for specific types of signal spectra).

Knowledge of:

1. different types of losses (examples might include transmission line loss, antenna gain, connector losses, and path loss)

2. procedures to calculate antenna gain and free space path loss

3. statistical fading models and distance-power (path loss) relationships in different propagation environments

4. the effects of outdoor terrain and indoor structures such as walls, floors, and ceilings on signal propagation

5. common deterministic, statistical, and empirical propagation models (examples might include free space, Okumura, Longley-Rice, and ray-tracing) and software modeling tools (examples might include EDX Signal, ATDI, PathLoss, and similar radio network planning tools) used to implement them

6. topographical maps and digital terrain databases

7. indoor and outdoor coverage calculation and verification techniques

8. Es/N0, Eb/N0, RSSI, NF, and other system parameters

9. the relationship between receiver noise figure, noise temperature, and receiver sensitivity and the relationship between sensitivity under static conditions and the degradation of effective receiver sensitivity caused by signal fading in different propagation conditions

10. external noise sources and their impact on the S/N ratios of received signals, and techniques for measuring the impact of external noise

11. basic antenna system design and use including antenna types (examples might include omnidirectional, panel, parabolic, dipole array, indoor antennas), antenna patterns, gain and EIRP, EIS, ERP, TIS, TRP, antenna size, antenna polarization, receive and transmit diversity, antenna correlation coefficients (examples might include MIMO antenna systems), and proper antenna installation to provide for coverage, interference mitigation, and frequency reuse

12. adaptive antenna methods and techniques, including null-steering, selection diversity, optimal-ratio combining, adaptive antennas, spatial multiplexing, space-time coding, and MIMO techniques

13. subscriber unit, mobile, and device antennas and their performance characteristics, including SAR-reduction characteristics

14. use of test equipment such as network analyzers, spectrum analyzers, and TDRs

15. co-channel and adjacent channel interference analysis and measurement methods and techniques; multi-user detection and interference-cancellation schemes and their limitations

16. filters, power dividers, combiners, and directional couplers

17. signal processing techniques, including matched filtering, adaptive filtering, adaptive equalization, and Rake processing

B.5 Facilities Infrastructure

Tasks and knowledge related to the specification, design, implementation, and operation of facilities and sites. Determine power consumption; analyze electrical protection requirements and design the electrical protection layout for a wireless telecommunications facility; determine the required antennas for the facility and their positions; develop a specification for the required structure for a wireless base station facility; determine the required cable, antennas, and materials to implement an in-building wireless network; evaluate equipment compliance with industry standards, codes, and site requirements.

Tasks:

1. Determine the power consumption of a unit of communications equipment (examples might include tower amplifier modules, pressurization systems for waveguides).

2. Determine the power consumption for a facility containing communications equipment (examples might include base station amplifier racks, microwave system rack etc.).

3. Design a DC power plant to support the facility for a given required reserve time.

4. Analyze the electrical protection requirements (includes grounding/earthing, bonding, shielding, and lightning protection) and design the electrical protection layout for a wireless telecommunications facility.

5. Design a wireless communication facility layout plan with considerations for heating, air conditioning, ventilation, and structural issues.

6. Determine the required antennas for the facility, including specification of the antenna system from RAN to Antenna. Identify and size common types of antenna, amplifiers, and cable for a given scenario.

7. Determine the required antenna positions on a structure (examples might include towers located in remote/extreme conditions such as mountain tops, arctic areas, etc.).

8. Design the waveguide/transmission line layout between the communications electronics and the antenna(s).

9. Coordinate with other users when implementing a communications system in a shared location.

10. Develop a specification for the required structure for a wireless base station facility based on the required antenna sizes and elevations above ground.

11. Determine the required cable, antennas, distributed antenna systems, and materials to implement an in-building wireless network.

12. Determine the required number of racks on which to mount the equipment and the rack layout and placement, taking into account the maintainability of the equipment.

13. Evaluate equipment compliance with industry standards, codes, and site requirements such as NEBS/ETSI specifications as well as ANSI, IEC, local/city regulations, right of way, and other applicable standards.

14. Design a site-specific alarm and surveillance system.

Knowledge of:

1. procedures to determine the power consumption of wireless communications equipment (examples might include satellite earth station facility, ship/small island based facility, etc.)

2. how to determine the power required to support a site (examples might include solar and wind based support for tower sites, considerations of bird nests, heated radome, etc.)

3. the application of AC and DC power systems (examples might include urban towers based on roof tops, tunnels and bridges)

4. the application of alternative energy sources to wireless communications facilities (examples might include use of solar, wind power, or bio-mass in rural areas)

5. heating, ventilation, and air conditioning (HVAC) requirements

6. equipment racks, rack mounting spaces, and related hardware

7. electrical protection (including grounding/earthing, bonding, shielding, and lightning protection)

8. basic waveguides and transmission lines (examples might include elliptical waveguides, multiple cables runs, pressurization and sealing of connectors, etc.)

9. tower specifications and standards (examples might include wind load calculation based on Effective Plate Area, alignment kits, and elevation angle measurements)

10. physical security requirements

11. alarm and surveillance systems

12. effects of environmental exposure (examples might include corrosion, temperature, and UV susceptibility)

13. NEBS/ETSI specifications as well as ANSI, IEC, and other applicable standards, codes, and other relevant site-specific requirements

14. where to find expertise in structural engineering, fire suppression, and other building systems

B.6   Agreements, Standards, Policies, and Regulations

Tasks and knowledge related to externally imposed compliance requirements and conformance testing, including interoperability. Assess service and equipment quality; prepare specifications for purchasing services and equipment and evaluate the responses; verify compliance with regulatory requirements; select and analyze frequency assignments; perform standardized homologation tests as required by regulatory or standardization bodies; evaluate compliance with health, safety, and environmental requirements; perform conformance/interoperability analyses of systems and components; analyze the use of licensed vs. unlicensed spectrum; obtain licenses and permits.

Tasks:

1. Assess service and equipment quality and recommendations to standardization bodies for new requirements/features.

2. Prepare specifications for purchasing services and equipment, and evaluate the responses, including relevant country-specific standards (examples might include preparing request for

proposals for introducing new services/licenses and evaluating submitted proposals for implementation of universal services projects).

3. Verify compliance with regulatory requirements (examples might include licensing, standards, rules, and regulations).

4. Select and analyze frequency assignments.

5. Perform standardized homologation tests as required by regulatory or standardization bodies.

6. Evaluate compliance with health, safety, and environmental requirements.

7. Perform conformance/interoperability analyses of systems and components, including self organizing and self optimizing networks for NGN.

8. Analyze the use of licensed vs. unlicensed spectrum.

9. Obtain and draft licenses and permits where required, including software, hardware, product licenses (open source, GNU, IP, patent laws), as well as dispute settlement.

10. Perform market analysis, study of market indicators, and pricing of telecom services.

Knowledge of:

1. regulatory requirements and telecom laws (examples might include international, national, and local); emerging standards and network evolution (examples might include convergence of networks, IMT-advanced); regulatory pillars (examples might include transparency, free competition) and mandates (examples might include consumer protection, universal service); international organizations and corresponding structure and functions (examples might include the role of ITU and its subdivisions)

2. spectrum licensing (examples might include leasing options, primary and secondary assignments in license)

3. spectrum characteristics, availability, and management including formal methods of measurements to report non-compliance to regulatory bodies

4. local and site-specific rules/codes (examples might include the National Electric Code in the US and analogous codes in other countries) and engineering regulations (examples might include when engineering work needs to be sealed by a Professional Engineer)

5. electrical and RF safety (examples might include UL, EC, CSA, and IEEE C.95)

6. frequency assignment databases and online tools (examples might include verification of registered users in the area, experimental bands and their usages)

7. modulation anomalies (examples might include cross modulation, modulation products, harmonics, and quantization impact)

8. health, safety, and environmental issues (examples for RF safety might include SAR limits for different countries [e.g., American limits vs. European limits for accepted SAR-values] and their different ways to measure it)

9. equipment type approval processes/requirements

10. how to identify and locate appropriate industry technical standards, applicable codes, and other pertinent requirements

11. cost calculation models

B.7   Fundamental Knowledge

Basic knowledge that a wireless communications engineer would use in order to perform tasks across all domains. Apply basic concepts related to electrical engineering, communications systems, and general engineering management.

Knowledge related to electrical engineering

1. fundamental AC/DC circuit analysis

2. mathematics including linear algebra, probability, statistics, and Boolean arithmetic

3. operation of complex test instruments, including oscilloscopes, spectrum analyzers, network analyzers, TDRs, and signal generators

4. frequency spectrum and Fourier transforms

5. basic printed circuit board design considerations

6. transmission theory and lines, antennas, and basic electromagnetic wave theory and applications

7. power calculations (examples might include dB, dBm, and dBx)

8. basic concepts of queuing theory and traffic analysis

9. basic signal processing (examples: analog and digital processing; quantization; linear filtering theory, concepts, and design)

10. basic concepts related to optical communications

11. basic electronic system-level block diagrams

12. basic power supply design

Knowledge related to communication systems

13. basic communications and information theory (analog and digital)

14. basic telephony (including signaling, switching, and transmission)

15. noise impairments

16. basic EMI, EMC, and interference

17. frequency allocations and reuse

18. modulation techniques for analog (examples might include AM, FM, and PM)

19. modulation techniques for digital (examples might include FSK, PSK, and QAM)

20. wireless multiple-access schemes (examples might include FDMA, TDMA, CDMA, and variants)

21. basic satellite communications

22. digital data transmission formats (examples might include E1/T1 and OC-n/SDH)

23. basic components of RF circuitry

24. basic RF circuit design, including filter design

25. basic RF coupling, radiation, and antenna theory concepts

26. measurements for RF circuits and sub systems, such as output power, receiver sensitivity, noise figure, linearity performance, and spectral performance

Knowledge of general engineering management:

27. project management methods and processes

28. fundamental engineering economics

29. design and configuration for ease of maintenance

30. documentation and configuration control schemes

31. IEEE Code of Ethics

Appendix C

Creating WEBOK 2.0

C.1 Development of Knowledge Areas for the Wireless Industry

Global communication is a defining political and economic force in the world today. It requires new ways of thinking and responding. For engineering professionals, recognizing and understanding this phenomenon is fast becoming a job requirement.

To address the worldwide wireless industry's growing and ever-evolving need for qualified communications professionals who can demonstrate practical problem-solving skills in real-world situations, the IEEE Communications Society (IEEE ComSoc) developed the IEEE Wireless Communication Engineering Technologies (WCET) certification program. Individuals receiving this certification are recognized as having the knowledge, skill, and ability to meet challenges in various wireless industry, business, corporate, and organizational settings.

The first task in the creation of any certification program is to identify the areas of expertise to be tested. To this end, ComSoc convened a meeting of 16 industry experts in December 2006. This group, called the Practice Analysis Task Force (PATF), was directed to draft a list of all the tasks and knowledge statements that a practicing wireless engineering professional with at least three years of experience should know. These statements were then grouped into seven technical areas, creating a document known as the Delineation.

The Delineation was presented to 11 focus groups in seven cities in six countries. Nearly 90 volunteers examined each statement in the Delineation. The deliberations of each focus group were recorded and each comment noted. In addition, a half dozen independent reviewers examined the Delineation line by line. A second PATF meeting was held in May 2007 to review the consolidated feedback. Suggested changes to the Delineation were carefully debated, resulting in a carefully refined document.

IEEE ComSoc then collected feedback about the Delineation from practicing wireless engineers via an Internet survey delivered to over 5,000 wireless engineers. Seeking validation of the Delineation, the survey asked each respondent to evaluate the task and knowledge statements in terms of:

- How often they encountered them in their career;
- How important they were to the industry;
- Who in their organization performed the tasks;
- When in a practitioner's career the knowledge would be acquired;
- How important was each task or item of knowledge to their career; and
- What fraction of their time did they personally spend on each task.

Participants in the survey also provided extensive demographical information to enable IEEE ComSoc to assess the responses and the applicability of the Delineation.

As the Delineation was being developed, it was recognized that there was significant value in using this description as the basis for an overview book on wireless communications, *The Wireless Engineering Body of Knowledge*. The First Edition of the WEBOK, with Gustavo Giannattasio as the Editor-in-Chief, was published in 2009. It was well-received by the industry and has become a popular reference text.

C.2 Development of the Second Edition

Wireless communications is a rapidly changing field, and it was understood from the beginning that the Delineation could not be a static document, but rather would need to be updated periodically to reflect changes in wireless technologies, capabilities, services, standards, regulations, and customer expectations. In 2010, a group of wireless communications experts was convened and tasked with reviewing the Delineation and updating it to reflect developments in the industry. Acting as a new PATF, this group also had the benefit of inputs from numerous other industry experts whose comments were solicited through the WCET Industry Advisory Board.

The changes to the Delineation were not extensive, but they did involve adding new tasks and knowledge statements. Some topics (for example, LTE) were made more explicit, and some areas were modestly expanded to incorporate additional examples, helping to clarify the scope of particular tasks and/or knowledge requirements.

Although the WEBOK continued to sell well, it was quickly recognized that it should be updated to reflect the same industry developments as were included in the revised Delineation. A team of editors was assembled, headed by Andrzej Jajszczyk. The team included a mix of new and returning chapter editors, and they in turn recruited both new and returning authors to contribute to their chapters. Their specific instruction, as experts in various aspects of wireless communications, was to update each chapter as appropriate to reflect developments in that technical area. They were given the updated Delineation to serve as a guide to what other industry experts had jointly determined were among the most important recent developments.

The authors and editors took their charge seriously and thoroughly reviewed the original WEBOK. Readers familiar with the 2009 Edition will recognize that much of the information – as well as its organization and format – has been retained. However, each chapter of WEBOK 2.0 contains new information; new topics have been added, existing topics have been expanded, new or additional examples have been included to enhance the clarity of some discussions, etc. WEBOK 2.0 is not a "new" book, but it is a significantly updated book, and one with a substantial amount of new content.

C.3 Independence of the WEBOK and WCET Certification

Although the WEBOK and the WCET Certification program are both based on the same overall description of tasks and knowledge areas, they were created in completely separate efforts. None of the team involved in creating the WEBOK, either the 2009 Edition or the current version, has any involvement with the certification examination. The WEBOK is a separate product and its content should not be taken as an indication of any questions that may be asked on the exam. In particular, the WEBOK is not intended as a "study guide" to be followed as preparation for the certification exam.

Appendix D

About the IEEE Communications Society

The IEEE Communications Society is a diverse group of industry professionals with a common interest in advancing all communication technologies. Individuals within this unique community interact across international and technological borders to produce publications, organize conferences, foster educational programs, promote local activities, and work on technical committees. Website: www.comsoc.org

D.1 Conferences

Every year, the IEEE Communications Society sponsors major conferences that attract hundreds of the best quality paper/presentation submissions and attendees. Held at convenient locations around the world, these meetings attract thousands of participants who have much to share beyond their strong desire to learn. Communications Society conferences and workshops provide ideal opportunities to be a part of the latest technological developments and to network with the leaders who are changing the world of communications.

- IEEE/OSA Conference on Optical Fiber Communications/National Fiber Optic Engineers Conference (OFC/NFOEC)
- IEEE Wireless Communications and Networking Conference (WCNC)
- IEEE/IFIP Network Operations & Management Symposium (NOMS)
- IFIP/IEEE International Symposium on Integrated Network Management (IM)
- IEEE International Conference on Communications (ICC)
- IEEE International Enterprise Networking & Services Conference (ENTNET)
- IEEE Conference on Computer Communications (INFOCOM)
- IEEE/AFCEA Military Communications Conference (MILCOM)
- IEEE Global Communications Conference (GLOBECOM)
- IEEE Consumer Communications & Networking Conference (CCNC)
- IEEE International Symposium on Personal, Indoor and Mobile Radio Communications (PIMRC)
- IEEE Conference on Sensor and Ad Hoc Communications and Networks (SECON)
- IEEE International Symposium on Dynamic Spectrum Access Networks (DySPAN)
- IEEE International Symposium on Power Line Communications and Its Applications (ISPLC)

D.2 Publications

- *IEEE Communications Magazine*
- *IEEE Network: The Magazine of Global Internetworking*
- *IEEE Wireless Communications*
- *IEEE Communications Letters*

- *IEEE Transactions on Communications*
- *IEEE Journal on Selected Areas in Communications*
- *IEEE/ACM Transactions on Networking*
- *IEEE Transactions on Wireless Communications*
- *IEEE Transactions on Network and Service Management (TNSM)*
- *IEEE Communications Surveys & Tutorials*
- *IEEE Transactions on Mobile Computing*
- *IEEE/OSA Journal of Lightwave Technology*
- *IEEE/OSA Journal of Optical Communications and Networking*
- *IEEE Wireless Communications Professional e-newsletter*
- *ComSoc e-News*
- *ComSoc Digital Library*

CPSIA information can be obtained
at www.ICGtesting.com
Printed in the USA
BVOW04s0158240917

495668BV00006B/19/P